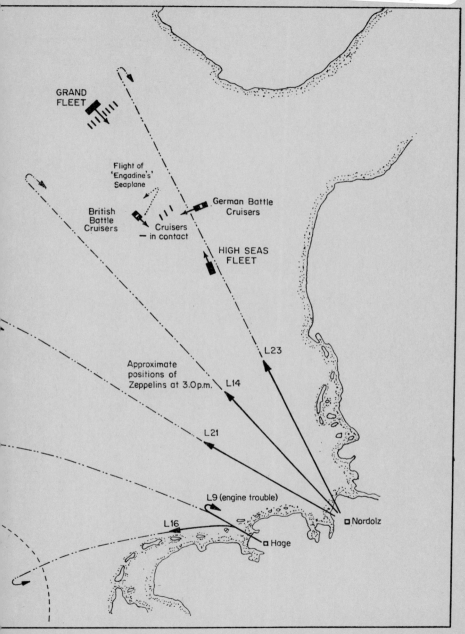

GRAND
FLEET

Flight of
'Engadine's'
Seaplane

British
Battle
Cruisers

German Battle
Cruisers

Cruisers
— in contact

HIGH SEAS
FLEET

L23

L14

Approximate
positions of
Zeppelins at 3.0 p.m.

L21

L9 (engine trouble)

L16

□ Nordolz

□ Hage

Aircraft and Sea Power

AIRCRAFT AND SEA POWER

by VICE ADMIRAL SIR ARTHUR HEZLET
K.B.E., C.B., D.S.O., D.S.C.

STEIN AND DAY/*Publishers*/New York

First published 1970
© 1970 by Vice Admiral Sir Arthur Hezlet
    K.B.E., C.B. D.S.O., D.S.C.
Library of Congress Catalog Card No. 75-108319
Printed in Great Britain
Stein and Day/*Publishers*/7 East 48 Street, New York, N.Y. 10017
SBN 8128-1308-1

Contents

v

Illustrations

PHOTOGRAPHS

MAPS AND DIAGRAMS

ENDPAPERS

Preface

No doubt the first reaction of many on seeing this book will be to ask what a submariner knows about aircraft and sea power. Although I once held a private pilot's licence and flew a light aeroplane, I have never been a professional aviator nor even served in an aircraft-carrier. In some ways this may well be an advantage. Professional aviators are likely to be biassed in favour of either the dark or the light blue uniform. My contacts during my service career were as much with Coastal Command as with the Fleet Air Arm and so I hope I can better hold the balance. In writing about aircraft and sea power, of course, one does not only need to know about aircraft: a knowledge of sea power is no less important and here I am on firmer ground.

Probably the greatest technical influences on naval warfare in the twentieth century have been submarines, aircraft, electronics and now the nuclear weapon and the missile. I dealt with the submarine in my first book* and it was while studying the subject, which involved a great deal about aircraft as a counter measure, that I became interested in aircraft and sea power. The reader will not find many new facts in this book, which is based almost entirely on information from published sources, but I hope in the analysis and comment he or she will find interest. The subject is treated internationally and considers the influence of all kinds of aircraft on sea power. It is not only concerned with Coastal Command and the Fleet Air Arm but with all the activities of air forces which are connected with sea power. Inevitably in a book of this size it cannot be a complete history. Much has had to be left out and the choice of what to put in has been governed by what is relevant to the study.

* *The Submarine and Sea Power* (Peter Davies, 1967)

The word aircraft is used throughout as the generic term for everything that flies and includes lighter as well as heavier than air machines. Aeroplane or landplane is used for machines with wheels; seaplane or floatplane or, early on, hydro-aeroplane is used for machines with floats. Other terms such as flying-boat, helicopter and rigid airship are, I think, self-explanatory.

I would like to express my thanks to the Publishers, who commissioned this book, and whose deadlines and restraints have kept me to my task. My thanks are also due to Captain A. H. Barton, R.N.(Retd.), my erstwhile colleague at the Royal Naval Staff College, and now my neighbour, who has read, checked and corrected the drafts and whose comments have been most valuable. As usual the staff of the Admiralty Library and the Library of the Royal United Service Institution have been courteous and interested, as have the staff of the photographic section of the Imperial War Museum. Finally my thanks are due to Miss Hilary Bacon and Miss Barbara Anderson of the administrative staff of the New University of Ulster, who have typed the fair copies of this book for me in their spare time.

Bovagh House, Aghadowey,
Co. Londonderry, Northern Ireland

30th April 1969

I

The Birth of Naval Aviation

BY THE BEGINNING of the twentieth century man had already been ascending in balloons for well over a hundred years. The first manned flight took place in France as early as 1783 and eleven years later balloons were used for the first time in war at the Battle of Fleurus. They were employed again in the Peninsula in 1808–14 and in the American Civil War of 1861–5 and by the end of the century many of the armies of the world had regular balloon detachments. They do not seem, however, to have been considered for use in war at sea and in 1900 the navies of the world were still uninterested. This is not altogether surprising as free balloons, being at the mercy of the wind, could never return to report their sightings and the use of captive balloons to increase the horizon of a ship at sea would have been of marginal value in the days of coal-fired ships with their huge columns of smoke which could be seen at great distances. No doubt interest would have been greater if the attempts to propel and steer free balloons had had more success. A steam 'dirigible', as this type was called, flew as early as 1852, but its speed, which was only 6 m.p.h., meant that it could not compete with the lightest wind. In 1884, the electrically-propelled airship *La France* was faster and was able to make a circular fight of five miles, returning under control to its starting-point, but by then its battery was exhausted. There was nothing, therefore, in the performance of aircraft to suggest to the sailors of the nineteenth century that they would help them in sea warfare. Would-be inventors of flying-machines, unlike the submarine designers, made few claims for them in the maritime role. Although some said, among many other things, that aircraft might be able to attack ships, in no case did their protagonists assert that flying-machines would be able to command the sea or even to help ships to carry out their time-honoured functions. The general attitude, as late as 1898, is typified by the refusal of the U.S. Navy's Board of Construction to provide any money for experiments with aviation on the grounds that, if anyone's, it was the army's business.

The key to the further development of aircraft both heavier- and lighter-than-air depended upon the provision of a sufficiently powerful power plant which was light enough to allow the machine to become airborne. Ample power had been available in steam engines for many years but these were far too heavy. It was not, therefore, until the internal combustion engine appeared at the end of the century that real progress could be made. In France Santos Dumont and later the Lebaudy brothers soon began to build small airships propelled by these engines which beat all former records. It was in Germany, however, that the greatest contribution to aviation was made. In 1898 the Zeppelin company was formed and in July 1900 their first airship, *LZ1*, flew at Lake Constance. All earlier airships had relied on the pressure of gas inside them to maintain their shape and the passenger car with the engines was simply slung below the envelope on wires. Count Zeppelin saw that the speed and weight-carrying capacity of such airships was bound to be limited and *LZ1* was of a completely new type known as a 'rigid'. Seventeen gas bags supported a long aluminium girder structure or keel and were enclosed by a streamlined aluminium frame covered tightly with fabric. *LZ1* was over three times the size of any earlier airship and was not unsuccessful. She was, however, difficult to control and the hull was found to be not strong enough. The two 15 h.p. Daimler engines only drove her at 16 m.p.h. and were very heavy. Count Zeppelin, after a few hours' flying, wisely decided to dismantle her and build a better one.

Nevertheless, by 1901 lighter-than-air craft were flying and well in the lead, although none of them had developed to the stage when they could be of any use. Experiments with heavier-than-air craft were made by many inventors in Europe and America without much success except that a few gliders succeeded in gliding. The most successful gliders were the Wright brothers in America who also produced a 12-h.p. engine with the remarkably low weight of 179 lbs. At the end of 1903 they made the four historic powered aeroplane flights which are generally accepted as the true beginning of the age of aviation. The first flight lasted only twelve seconds but the last was for nearly a minute and covered a distance approaching half a mile.

Europe was far behind the Wright Brothers: it was not until 1906, three years later, that Santos Dumont in France flew an aeroplane of his own design for a distance of a third of a mile and not until the following year that Henri Farman, also in France, managed to keep his machine in the air for a minute and make a circle. Nevertheless it was in France, from 1904 onwards, that considerable progress was made with non-rigid airships: the Astra, Clement-Bayard and Zodiac companies as well as the Lebaudy brothers built them in appreciable numbers and then the British Army balloon establishment at Farnborough obtained permission to build one too.

In 1905 Count Zeppelin completed his second rigid airship *LZ2*. She was similar to *LZ1* but stronger and had two much more powerful engines totalling 170 h.p. On her second flight in June 1906, her engines failed and she was carried off in a high wind and was wrecked. She was not, therefore, properly tested but had succeeded in climbing to a height of 1,800 feet and had made 26 m.p.h. Meanwhile the Wright brothers, with their third machine, produced the first really practicable aeroplane which was able to manœuvre freely and remain airborne for half an hour, covering some 25 miles, but after this they made no further progress for some years and their invention was refused by both the U.S. Army and Navy. By 1906, therefore, three types of aircraft had flown: the rigid airship, the non-rigid airship, and the aeroplane. Great steps though these were, the energy of the inventors and aviators was expended in solving the problems of flying and the business of putting their machines to a practical task could not yet be considered. General naval opinion was by no means convinced that aircraft had a role in naval warfare and many still doubted whether they would prove more than expensive toys. The policy of most navies, therefore, was simply to watch developments and not to help in any way.

It was in 1906 that the German Navy first began to take an interest in the Zeppelin for naval reconnaissance. Count Zeppelin completed his third airship *LZ3* in October 1906 and a German naval officer was present at her trials. Count Zeppelin had hoped to sell her to the Army but they refused her in spite of the fact that she remained airborne for eight hours and made a flight of over 200 miles. Admiral von Tirpitz considered that the Zeppelin had to have a greater endurance and be considerably faster if it was to make useful scouting trips and compete with the winds normally encountered at sea. In any case, he adhered to the policy he had applied to torpedoes, submarines and all other new weapons, that they should be proved and fully efficient before being adopted.

Nevertheless, in April 1908 definite proposals were made within the German Admiralty for Zeppelins for scouting, but stipulated that a range of 1000 miles was necessary and a speed of between 35 and 45 m.p.h. It was also suggested that they could be used to drop bombs. Admiral Tirpitz seems to have accepted these proposals but stuck to his policy that the airships should be tested and found thoroughly reliable before the Navy would buy them. Count Zeppelin produced his fourth airship *LZ4* in June: she had been designed to meet the specifications of the German Army and was slightly larger and more powerful than her predecessors. She flew over the Alps to Lucerne and back and kept in the air for twelve hours, but was still too slow. In attempting to meet the German Army's twenty-four-hour endurance requirement in August her engines failed and she was caught in a thunderstorm and burnt, fortunately without loss of life. In 1909 Count Zeppelin completed *LZ5* to Army specifications and in May she flew a distance of 820 miles; as she stayed in the air for the stipulated 24 hours, she was accepted. In August the Zeppelin company completed *LZ6* hoping the Army would buy her too, but when they refused Count Zeppelin formed a subsidiary company called Delag to carry passengers in Zeppelins.

In March 1907 the Wright brothers, unable to sell their invention in their own country, offered it to the British Admiralty. The Admiralty refused it and this act has often been quoted as an example of their reactionary and unprogressive attitude. It would, however, be unfair to suggest that the administration, which was already introducing the Dreadnought battleship and transforming naval warfare, was narrow-minded. The Wright brothers had made their trials in great secrecy and their claims were thought by many to be exaggerated and this seemed to be confirmed by their rejection by the U.S. Army and Navy. Even if their claims were substantiated, a powered box-kite with a 25-mile range, no payload and a single pilot whose whole attention was taken up keeping the machine in the air, did not seem to contribute much to naval warfare. Clearly many years of expensive development would be required to make it of the smallest value. The British Admiralty still preferred to remain spectators in aircraft development and not to hasten it. If they had any interest in aircraft, it was in any case centred in the rigid airship which seemed at this time far more likely to be of use than the aeroplane.

In July 1908, Captain Bacon, the Director of Naval Ordnance, forwarded a paper proposing that a rigid airship should be built for naval reconnaissance. Of the three types of aircraft that had actually flown, the rigid airship seemed by far the most promising for scouting at sea. It offered a greater endurance, speed and load-carrying capacity than the non-rigid airship, and could carry a wireless transmitting set and a crew in sufficient comfort to navigate. The airship had the added advantage over the aeroplane that if its engines failed it would not crash into the sea but could hover. Aeroplanes of the day, with their endurance of half an hour or so, were scarcely capable of flying farther than the horizon and back again; furthermore no purely British aeroplane had yet even flown. Captain Bacon believed that geography favoured the use of airships for reconnaissance over the North Sea, because whatever the direction of the wind, sorties could be made from bases in Scotland or in southern England. His proposals had the full support of Mr McKenna, the First Lord of the Admiralty and Sir John Fisher, the First Sea Lord. Mr Asquith, the Prime Minister, worried by the way Great Britain was lagging behind other countries in aviation, readily agreed, and the proposal was approved. Vickers were invited to tender and a joint design team with the Admiralty began work. The requirements stated were similar to those of the German Navy: the airship was to be able to do 1000 miles at 40 knots, to carry wireless and a crew of twenty and to alight on water. This was a great step from the ungainly sausage-shaped non-rigid airship *Nulli Secundus* which, after many delays for lack of funds, had just been completed by the Army Balloon Factory at Farnborough and which flew at 16 m.p.h. over London in October.

In January 1909, a sub-committee of the British Committee of Imperial Defence sat to consider aerial navigation. They agreed with the decision to order a rigid airship, not only because it would be useful for naval reconnaissance, but because, unless Great Britain possessed one, it would be impossible to learn its true capabilities and so assess the menace and devise a defence against it. As a scout, they saw that the airship would have the great advantage that it would enable a blockading fleet to see into enemy harbours and should be able to spot enemy torpedo attacks approaching. The members of the sub-committee rightly believed that bombing would not be very effective, although attacks on ships at sea and on dockyards were seen to be a possibility. They therefore recommended that the decision to build a rigid airship be confirmed and the contract was placed with Vickers at Barrow on 7th May. *Naval Airship No. 1* or *Mayfly*, as she was nicknamed by the wags, was to be roughly the size of a contemporary Zeppelin and it was somewhat optimistically planned to complete her by the autumn. At the same time Captain Sueter,[1] an officer with much experience of submarines, mines and torpedoes and of an inventive turn of mind, was appointed as Inspecting Captain of Airships. Considerable attention was given to the problem of air defence: although the sub-committee considered that aeroplanes were still experimental and that it was too early to assess their value in war, they foresaw that they might be developed to attack airships. At the same time experiments with anti-aircraft guns were made; shoots were carried out at a towed balloon and the Navy began the design of a 3-inch anti-aircraft gun for mounting in ships.

While interest in Germany and Great Britain was concentrated upon the rigid airship, in the United States Navy some of the more 'air-minded' thinkers were saying that aeroplanes would be of use to a navy for scouting and spotting mines and submarines, but the Navy Department took no action. In September both the U.S. Army and Navy attended a demonstration by Orville Wright but the plane crashed and his passenger was killed. Nevertheless, the Wright brothers could now keep in the air for an hour and a half and cover sixty miles or so. During 1908 Wilbur Wright visited Europe and by the end of the year had broken all records, staying in the air for two hours and twenty minutes. European aviators learnt much from him and heavier-than-air craft began to show that they were also likely to develop so as to be of value in war.

Heavier-than-air craft improved during 1909 especially in Europe, and in July Blériot flew the Channel. His aeroplane was a small monoplane of his own design with a 25-h.p. engine. This exploit made a great impact in England where it awakened the public to the fact that 'Great Britain was no longer an island'. In August an international air meeting at Rheims was attended by naval officers from Great Britain, the United States and France, who all sent in enthusiastic reports. There were two basic types of aeroplane at the meeting, high-wing tractor monoplanes of the Blériot type and pusher biplanes of the Farman type. Three records were established at Rheims: a speed record of 47 m.p.h. by Blériot, a height record of 500 feet by Farman and a distance record of 112 miles by Antoinette. The Wright brothers, who had now succeeded in selling a plane to the U.S. Army, had lost their monopoly but, in spite of French progress, were still well in the lead. Curtiss, their nearest rival in the U.S.A., had only just flown a kilometer and the first all-British machine built by Short Brothers got to the stage during the year of flying a circular mile. Aeroplanes of these and many other types were now flying successfully but they were difficult to control, extremely frail and unreliable, and of very limited endurance. They had yet to prove that they could be of use in war at sea and were no rivals to the rigid airship.

The year 1910 proved a bad one for the Zeppelins. Only the passenger-carrying *LZ7* for Delag was built and she was wrecked in June. The Army's *ZII* (ex *LZ5*) had already been wrecked in April and the Zeppelin company's *LZ6* was burnt in September. It is remarkable that no lives were lost in any of these accidents. However, the German Navy began to show more interest in the air and appointed a naval architect to study airship construction with the Zeppelin company.

During 1911 four rigid airships were built in Germany, three by the Zeppelin company and one by the new Schütte-Lanz concern, formed two years before. The Schütte-Lanz ship and one of the Zeppelins were bought by the German Army and the other two by Delag; *LZ9*, one of the army airships, reached a speed of 47·2 m.p.h. Unfortunately in May, but again miraculously without casualties, one of the Delag airships was wrecked. During the summer Admiral Tirpitz was subjected to pressure by the Zeppelin company, by the Press and public opinion and finally by the Kaiser, to equip the German Navy with airships. He allowed himself to be persuaded and set in train measures to acquire a Zeppelin. Negotiations and arguments about its size and characteristics dragged on into the following year and it was not finally ordered until April 1912.

The British rigid airship *Mayfly* had suffered many set-backs. By the beginning of 1910 she had not even been started because her shed was not yet completed. Her designers soon found that there were many problems with strength, materials, gas-bags and fabrics to be solved and all these caused delay. The enthusiasm of Captain Sueter was somewhat damped in January by the appointment as First Sea Lord of Sir Arthur Wilson who was strongly opposed to the whole project. Mr McKenna was, however, still in office. But although he was able to further the *Mayfly* project he could not persuade the Sea Lords to embark upon a regular programme of airship construction. The erection of the *Mayfly* began in June, over a year behind schedule, and in September the cruiser *Hermione* was sent to Barrow to act as depot ship for the project.

In May 1911 the *Mayfly* was nearing completion after the best part of a year under construction. It was eighteen months since she had been designed and in this period the Zeppelins had improved considerably; but in some ways she was ahead of them. Her frame was made of duralumin instead of aluminium, which doubled the strength and saved a ton in weight, and she was better streamlined. Yet when she came to be inflated it was found that she was greatly overweight. On 22nd May she was moved out of her shed and lay in the Cavendish dock at Barrow resting on the water and moored to a mast erected on a pontoon. In this position she survived winds of 45 m.p.h. but with everything on board she was found to be three tons too heavy and would not lift at all. *Mayfly* had been designed partly by what could be learnt from the Zeppelins, which was not much, and partly by naval architects who had experience of submarines. Whilst their knowledge of three-dimensional vessels was obviously useful, they were now working in a medium 800 times less dense than water. They had also to compete with requirements laid down by seamen, who stipulated equipment such as anchors and cables which, however light, could not be afforded. During June and August drastic lightening took place but when she was again inflated on 17th August she was still too heavy and further lightening had to be carried out. When finally inflated on 22nd September she just lifted with only two tons of petrol and ballast on board. On 24th September when being eased out of her shed, she was caught by a cross wind and broken in two. She was further damaged getting her back into the shed and became a total wreck; so ended the first attempt by any navy to equip itself with aircraft.

Mayfly's demise was a cause of barely concealed satisfaction to the reactionary elements of the Royal Navy. Sir Arthur Wilson, the First Sea Lord, made sure that the court of inquiry would be presided over by one of the nautical faction. Admiral Sturdee, the officer in question, is reputed on seeing the wreck, to have remarked, 'It is the work of a lunatic.' The court found that the cause of the disaster was that the frame of the airship was not strong enough and recommended that experiments with airships be abandoned. Early in 1912 the Admiralty approved this recommendation, rejected Captain Sueter's pleas that *Mayfly* be reconstructed and, to rid themselves of discordant voices, disbanded the airship section and placed Sueter on half pay.

It is easy in these days of air power to ridicule the nautical faction and to dub them short-sighted, narrow-minded bigots, but their views were held widely by men with considerable intellect and, if we disregard our knowledge of the actual development of air power, with a logical and well-thought-out case. They believed and were dedicated to the theories of sea power and above all those of Mahan on battlefleet supremacy. Any new device such as the submarine or aircraft which might upset this supremacy must in their view be discouraged. They were absorbed with the problems of the Dreadnought battleship, which was an immense innovation in itself; with the tactics and control of huge fleets and of long-range gunnery, and with a host of other problems. They believed it vital to keep a superiority of Dreadnoughts over Germany and were determined that nothing should stand in the way of this policy. They were naturally impatient over the diversion of funds to airships, especially as in the case of the *Mayfly* she cost twice as much as estimated and then wouldn't even fly. Even if airships were successful and flew, they could point to their vulnerability. In peace with no enemy trying to destroy them, of the ten Zeppelins built so far six had come to untimely ends. In war they believed that the guns of the fleet could easily dispose of airships which attacked them and that the vast majority of fleet operations would take place well outside the range of aeroplanes. In all this, as it stood in 1911, they were of course right. Where they were wrong was that they failed to see that the development of

aircraft would leap ahead in the next few years, that this was
inevitable and that a policy of discouraging aviation would not
stop progress but would merely put Great Britain behind other
countries.

The aviation faction led by Captain Sueter, before he was put
on half pay, and supported generally by Mr McKenna and by
Admiral Jellicoe, the Second Sea Lord, was certain that air
power at sea would be vital in the future. Admiral Jellicoe had
recently been up in the German passenger-carrying Zeppelin
Schwaben and was convinced of its value. Lord Fisher, from
retirement, was also a supporter: in January 1912 he went so
far as to assert that aviation had dispensed with the need for
light cruisers. Captain Sueter felt that accidents such as the
Mayfly disaster were inevitable in developing a completely new
weapon. He could point to the success through perseverence of
Count Zeppelin: in his opinion the *Mayfly* set-back should have
led to greater efforts to put matters right and not to the abandonment of the project. Captain Sueter believed that heavier-than-
air machines were required as well as airships and made the
profound statement that Great Britain must command the air
as well as the sea. Mine detection, submarine hunting, reconnaissance of enemy harbours, bombing, scouting for the enemy
at sea as well as spotting for ships' gunfire were all tasks in
which aircraft could help to command the sea. But at present
all this was based more on faith than on performance and aviation had a long way to go before it could carry them out.

There was a third faction in this controversy and this was led
by Mr Winston Churchill who had replaced Mr McKenna as
First Lord soon after the *Mayfly* disaster. The members of this
faction were firmly against the airship but very much in favour
of heavier-than-air machines. Winston Churchill was later to
say 'I rated the Zeppelin much lower as a weapon of war than
almost anyone else. I believed that this enormous bladder of
combustible and explosive gas would prove to be easily de-
structible.' He was ultimately to be proved right too, but at the
time this view was based more on intuition than anything else.
It was to be six years before aeroplanes really got the measure
of the Zeppelin and many more before they equalled it as a
vehicle for naval reconnaissance.

If airships suffered set-backs in 1910, it was the reverse with
heavier-than-air machines. In America Curtiss equalled the
European distance record and flew from Albany to New York,

which is over 100 miles. In France an advance of even greater significance for naval warfare took place when Henri Fabre fitted floats instead of wheels to his aeroplane and succeeded in taking off and landing on water. In those days of extremely unreliable engines this was a major step: it was clearly of cardinal importance that, if aeroplanes were to operate over the sea, they should be able to alight on it. It seemed natural that army aircraft should have wheels and naval aircraft floats.

In 1910 the U.S. Navy decided to appoint a Captain named Chambers to the Navy Department to keep in touch with aviation. Captain Chambers saw that the increased range of heavier-than-air craft coupled with the ability to operate from water made their use possible in naval warfare, especially if they could be carried to the scene of action in ships. In October the Navy Department seriously considered the inclusion of aircraft in the equipment of the new cruisers then being designed. Chambers and Curtiss then put their heads together and on 14th November a Curtiss biplane successfully took off from a platform on the bows of the U.S. cruiser *Birmingham*. On 18th January 1911, the same pilot landed down-wind on a wooden platform built on the stern of the battleship *Pennsylvania* and succeeded in taking off again. Both these feats were with the ships at anchor in harbour. Finally on 17th February Curtiss took off the water in a 'hydro-aeroplane' as float-planes or seaplanes were then called, landed alongside the U.S.S. *Pennsylvania* and was hoisted on board. These remarkable and successful trials were made before the U.S. Navy owned a single aircraft. They pointed the way to the operation of heavier-than-air craft at sea and it is odd how slowly they were followed up. The U.S. Navy did not order its first three planes until the following July; then two of them were land-planes, and they were all based ashore at Annapolis. An aeroplane did not land on a ship's deck again until 1917.

During 1911, heavier-than-air machines steadily improved in performance. 'Hydro-aeroplane' development continued in France and the U.S.A. where Curtiss took the lead in this type. In France experiments began with the old torpedo depot ship *Foudre* as a seaplane mother ship. In Great Britain the Admiralty, under the control of the nautical faction, reluctantly allowed four officers to accept the generous offer of Mr Francis McLean to teach them to fly on his own machines at his own expense. Private enterprise also helped to develop the 'hydro-aeroplane'

and Commander Schwann, while Captain designate of the
Mayfly, had made experiments with his own aeroplane at his
own expense. The four officers learned to fly during the summer
of 1911 and the Admiralty were persuaded to buy two machines
from Mr McLean in October and set up a flying school at
Eastchurch. By the end of the year they owned six machines and
had trained ten pilots. In collaboration with Shorts one of the
machines was fitted with floats and landed successfully on the
Medway, and thereafter Shorts took up the design of a 'hydro-
aeroplane' seriously. The air enthusiasts began from October
onwards to receive more encouragement. Press and Parlia-
mentary pressure was mounting and they gained a staunch
advocate with the arrival of Winston Churchill at the Admiralty.
Admiralty policy slowly began to change to a willingness to
experiment with aircraft so as to keep pace with other nations.
In November Mr Asquith, the Prime Minister, nervous at the
lack of progress, requested the Committee of Imperial Defence
to look into the whole problem of the development of aviation
for the services.

The period 1910–11, therefore, was one in which the principal navies of the world began to take to the air, admittedly
somewhat cautiously and often under pressure from outside.
In this first phase, they did not get much further than learning
to fly and beginning to think of the uses and problems of flying
over the sea. The Germans persevered with the rigid airships,
originally believed to be the only type that would be of any use
at sea, but the British perhaps prematurely had abandoned it.
The new type, the 'hydro-aeroplane' or seaplane, was now seen
to be practicable and Great Britain, France and the U.S.A. all
put their faith in it and had it under development.

If the year 1911 was spent by the principal naval powers in
learning to fly, 1912 was a year of experiment to find out how
aircraft could best be used at sea. On 10th January a British
Short biplane flew off a platform on the bows of H.M.S. *Africa*
in Sheerness harbour thus repeating the American feat of just
over a year before. At the same time there was intense activity
in Britain to produce a satisfactory 'hydro-aeroplane' which
was believed to be the key to operations over the sea. In March
no fewer than four types flew: Shorts and Sopwiths each

produced a professional solution to the problem; Commander Schwann's amateur project was finally successful; and the Royal Aircraft Factory at Farnborough fitted floats to one of its aeroplanes. Experiments were begun with wireless sets in seaplanes in both Great Britain and the U.S.A., and although at first they could only communicate at a few miles, they showed promise. In March very early experiments were made at Eastchurch with bombing and 100-lb dummies were dropped. Trials to spot submerged submarines took place both in the U.S.A. and Great Britain; the results varied, as was to be expected, with the clearness of the water. At the review of the fleet at Weymouth in May some aircraft were present for the first time and a Short biplane flew off a platform on the bows of H.M.S. *Hibernia* whilst she was under way at 10 knots. This was repeated in July from H.M.S. *London* steaming at 12 knots and in the U.S.A. they went further and experimented with a launching catapult. In October, one of the U.S. Navy's planes made an endurance record and stayed in the air for over six hours. Encouraged by these practical results, the U.S. Navy General Board's interest in aircraft was renewed not only for scouting from ships at sea but also for coastal patrol from shore bases. There was talk in the U.S. Navy of torpedo-carrying planes and the General Board expressed a requirement for an 'aeroplane destroyer', or what we would now call a fighter, to shoot down other aircraft.

Gradually during the year, as these technical advances were made, the British worked out what the functions of naval aircraft in war would be. They foresaw their use for distant reconnaissance for the fleet at sea and off the enemy coasts; screening the fleet against air attack at sea or in harbour; locating mines and hostile vessels in British waters; hunting submarines and defending the dockyards from air attack. Some of these functions could be exercised from shore bases but it was clear that for others seaplanes would have to be carried in ships of the fleet. The Admiralty had decided as early as July to establish a chain of seaplane bases around the coasts, this being their first attempt at a war deployment for the Royal Naval Air Service. The first plans were for five stations stretching from Scapa Flow down the east coast to the Thames estuary and for others in the Clyde and at Weymouth and Pembroke; by the end of the year the first of these coastal stations at the Isle of Grain in the Thames estuary was opened. There was, however, little progress in carrying seaplanes to sea with the fleet; the early platforms from which aircraft had flown put half the ship's main armament out of action and as they would take some time to dismantle they were unacceptable. To the air department the solution seemed to be to design a special seaplane-carrier ship. This was considered but was not approved by the Board of Admiralty. Seaplanes with folding wings suitable for stowing aboard ship were, however, designed, but this was the only progress towards carrying seaplanes to sea with the fleet.

As early as April 1912, seven months after the *Mayfly* disaster, the British Admiralty began to have misgivings about their decision to abandon airships. They were being pressed to change their mind by the Committee of Imperial Defence, who could point to the great success of the Zeppelins, of which the civil ones were now carrying passengers regularly. It was in any case clear that heavier-than-air craft were of too limited a range to meet the needs of naval scouting and they began to put out feelers to see if they could buy a Zeppelin, but the Germans would not hear of it. In May Captain Sueter had been re-installed as Director of the Air Department and in June he was sent with the Superintendent of the Royal Aircraft Factory to tour Europe and report on aviation. They flew in the Delag airship *Victoria Louise* and a Parseval non-rigid, and their report, apart from showing how far Great Britain was behind France and Germany, was wholeheartedly in favour of airships for the Royal Navy. They pointed out that all the hopes expressed in 1908 for the *Mayfly* had now been more than realized by the Germans, who thereby had a scout capable of wide reconnaissance off their coasts, which could keep its distance from ships' guns and which no aeroplane of the day had the power to destroy. In July 1912 the Admiralty gave way and asked for a supplementary estimate for airships, but this time decided to start with the smaller non-rigid type. On 25th September the naval airship section was reformed and sent to Farnborough to learn as much as possible from the Army, which had just used its airships with great success in the annual manœuvres. By the end of the year, a small commercial airship built by Willows of Cardiff had been acquired for training, an Astra-Torres non-rigid had been ordered in France and a Parseval in Germany; at the same time there had been considerable discussion about acquiring a new rigid airship and in September Vickers had again been asked to tender.

In April 1912 the British put into effect the recommendations of the Committee of Imperial Defence for what was virtually a combined air force. It was called the Royal Flying Corps and had a central flying school but separate naval and military wings. The intention was that this should be a flexible force, the whole of which, as occasion demanded, was to be available to support either the Army or Navy. It was very soon realized that training, equipment, armament and functions in the two wings were so different that this was out of the question. The military wing, consisting of light, short-range aeroplanes for tactical reconnaissance and artillery observation, would be useless over the sea and the naval wing seaplanes could not operate far inland. Administration, procurement, manning and ownership of the planes, although co-ordinated in theory, remained in effect separate and so ultimate control remained with the Admiralty and the War Office. The Admiralty were thereby able to sabotage the decision of the Committee of Imperial Defence by keeping their wing quite separate. They began unofficially to refer to the naval wing of the Royal Flying Corps as the Royal Naval Air Service. It was not until July 1914, however, that the title 'Royal Naval Air Service' became official and the naval wing was separated from the Royal Flying Corps.

In Germany, four Zeppelins were completed in 1912, one for the Army, two for Delag and one at last for the Navy. Count Zeppelin had wished to give the Navy one of the same type as he was making for Delag and the Army, and which was by now well tested. Admiral Tirpitz, however, demanded a larger ship and in the end *L1* was a compromise. On completion in October 1912 she was the largest airship in the world; she was driven by three engines totalling 495 h.p., giving her a speed of 47·4 m.p.h. Actually she was only of some 12 per cent greater volume than the *Mayfly* and the great German lead in airship design is illustrated by the fact that she had a net lift of $9\frac{1}{4}$ tons against less than 3 tons for the British ship. On 13th October she made an endurance flight in which she was airborne for 30 hours, travelled 900 miles and spent six hours at 5000 feet. In spite of this, in October, the Kaiser, believing progress to be intolerably slow, expressed his dissatisfaction with the state of naval aviation. The British on the other hand looked with apprehension at the total of seven military and civil Zeppelins available and in May the Admiralty ordered some of the new 3-inch A.A. guns for mounting in battleships and cruisers as a defence against them. On the night of 14th October their anxiety was increased when a Zeppelin was reported flying over Sheerness. This coincided with the endurance cruise of the German *L1* but the Germans strenuously denied having been anywhere near the British coast. The incident had a great effect, because it was brought home forcefully that, even if *L1* had not been over Sheerness, she had the endurance to have made such a trip and with a considerable payload at that. It stimulated thought on air defence and helped the protagonists of airships in Great Britain to secure approval for a new rigid airship.

Much, therefore, had happened during 1912 and the naval aviators of the principal navies had, with remarkable foresight, predicted the functions of aircraft over the sea. Their theory was, however, far ahead of material development and the number of machines available to them was still very small. At the end of 1912 the naval wing of the Royal Flying Corps, which was the largest naval air arm, consisted only of one small airship based at Farnborough, thirteen aeroplanes at East-church and three seaplanes at the Isle of Grain. Naval aviation was therefore still largely experimental with many problems remaining to be solved.

In January 1913 the handful of aircraft which comprised the U.S. Navy's flying arm joined the fleet in Guantanamo for the annual exercises. They kept themselves very busy practising dropping bombs and taking photographs; they spotted sub-merged submarines and scouted for surface ships, finding them without being seen themselves. Funds, however, were very scarce and by August the U.S. Navy still only had five seaplanes and three small flying-boats. In October ambitious plans were made for expansion and a Board under Captain Chambers recommended a force of 50 planes, one to be carried in each warship, and another 50 to be held in reserve as well as six for an advanced shore base. The Navy Department were, however, unwilling to spend so much on aircraft and little came of it.

In Great Britain on the other hand under the drive of Winston Churchill progress was rapid during 1913. Seaplane bases were commissioned at Calshot, Felixstowe and Great

Yarmouth and new and improved designs of seaplanes were pressed forward. It was by now accepted that to replace aeroplane wheels with floats was not enough and that seaplanes had to be specially designed as such. Captain Sueter realized that the torpedo would give naval aircraft a ship-sinking weapon and that more powerful aircraft engines were essential to lift such weights. It was under his encouragement that engines, already of 100 h.p., increased rapidly in power for naval aircraft. In May the old cruiser *Hermes* was commissioned as the headquarters of the naval wing. Originally this was simply an administrative arrangement, but before long she was rigged as a makeshift seaplane-carrier with a taking-off platform stretching out over the bows.

In July 1913 the Royal Navy's annual manœuvres were held in home waters and naval aircraft took part for the first time. The superior Blue Fleet under Admiral Callaghan represented the British and its territory comprised the whole of the British Isles except the coast from Dungeness to Yarmouth, which belonged to Red. The Red Fleet under Admiral Jellicoe represented the Germans. The *Hermes*, twelve seaplanes and an aeroplane took part, Short, Sopwith, Maurice Farman, Borel and Caudron types being represented. For the purpose of the exercise aircraft were restricted to reconnaissance, but only two of them were fitted with wireless. Blue had six of the seaplanes, three based at Cromarty and three on Loch Leven near Rosyth. They were given the task of locating any enemy ships which approached within 100 miles. Red had three seaplanes and an aeroplane based at Yarmouth, where also the *Hermes*, with two seaplanes on board and one in reserve ashore, was stationed. The *Hermes* with her seaplanes was to try and simulate a Zeppelin on reconnaissance with a range of 800 miles. She was therefore allowed to go 300 miles from Yarmouth before launching a seaplane.

The manœuvres lasted from 23rd July to 1st August and the shore-based seaplanes persevered with local patrols as weather and serviceability permitted. A Borel seaplane from the *Hermes* was wrecked trying to take off in a rough sea and the new Short Folder seaplane was damaged. She was repaired and began operations again from Yarmouth which culminated in a forced landing fifty miles out at sea and being picked up by a merchant vessel. The Borel seaplane was replaced in the *Hermes* by a Caudron which was light enough to take off the launching platform with the ship steaming at 10 knots. On a day when the weather was too rough for the seaplanes to take off, the one aeroplane which took part was able to fly up the coast from Yarmouth and sight two 'enemy' submarines. The seaplanes of both sides sighted opposing submarines on the surface but could not have been said to have exerted much influence on the operations; this was because the huge fleets, ranging far out over the open sea, never needed to come close enough to the seaplane bases to be seen. However, a great deal of valuable experience in operating aircraft was gained. It was obvious that seaplanes had to be improved: they must be able to operate in rougher water and all of them must be fitted with wireless. The aeroplane, with wheels, to everyone's surprise, showed that it could make useful coastal patrols and had a use in naval warfare.

With the success of the flying-off platform on the *Hermes*, it might be thought that the problem of operating seaplanes with the fleet had been solved. This was far from the case. The Caudron seaplane was the only type that could be used and was light with flimsy floats. Already the new seaplanes such as the Short Folder were much heavier and their development to carry torpedoes and land on rougher water was tending to make them heavier still. The solution still seemed to the Air Department to be to build a new seaplane-carrier with a much longer flying-off platform. Captain Sueter in collaboration with Lord Graham, a volunteer reserve officer, proposed a design which was worked out in detail by Messrs Beardmore. The Board were reluctant to spend so much on something which might be a failure but late in 1913 took over a merchant ship building at Blyth for conversion to a seaplane-carrier.

In January 1913 proposals for a new experimental British rigid airship were approved. The debate leading to this decision was stormy. Sir Arthur Wilson was no longer First Sea Lord but was in the background as a member of the Committee of Imperial Defence. He still held that airships could easily be shot down by guns or aeroplanes and was supported in this by Winston Churchill, who now realized, however, that British rigid airships were essential if counter measures to the Zeppelin were to be properly studied. The protagonists of airships pointed out that as a close blockade of the German Fleet was now impracticable because of mines, torpedoes and submarines, the Zeppelins would be able to help their fleet to put to sea unchallenged; British airships could help offset this advantage.

In April the new German Army Zeppelin *ZIV* force-landed in France. During her short stay the French obtained a great deal of information which they kindly passed to the British and which was used to good effect in the design of the new rigid airship. In the early part of the year, there was pressure from public opinion, Parliament and the Navy League for more airships and Winston Churchill now openly admitted that Great Britain needed them. In the late spring and early summer, the Astra-Torres airship arrived from France and the Parseval from Germany. In June, partly in reply to the large German Zeppelin programme announced early in the year, a supplementary estimate was asked for a further two rigids and six non-rigids; the two rigids were to be built by Armstrong-Whitworths and the non-rigids were to be three of the Parseval type and three of the Italian Forlanini type. In the British Army manœuvres of 1913, their small non-rigid airships were again found very useful for reconnaissance; it was soon realized, however, that if they were to fly low enough to obtain results they would prove to be too vulnerable. Towards the end of the year, therefore, the Army proposed to rely on aeroplanes and to hand over their four small non-rigid airships to the Navy.

In January 1913, the German Navy's plan for the expansion of their airship service had been approved. Ten Zeppelins were to be built to form two squadrons of four each with two in reserve. A new base was to be built for them at Nordholz near Cuxhaven. In May, *L2* a new Zeppelin was begun. She was designed by the naval constructor who had been appointed to work in conjunction with the Zeppelin company. With *L2* it was hoped to meet the full naval requirements and still get her into the existing sheds. She embodied many new features, the principal of which was an internal instead of an external keel and she made her first flight in September. In the same month *L1* participated in the High Seas Fleet manœuvres in the North Sea and although for the first two days she was unable to fly because of bad weather, she made a highly successful scouting flight on 8th September. Next day as she was returning to base she ran into a storm off Heligoland and was wrecked with the loss of 14 lives. . . This was bad enough, but a few weeks later the *L2* on her tenth flight caught fire in the air and all 28 men on board, including her designer, were killed. The Zeppelin company blamed the design and relations between them and the German Admiralty became strained. These disasters very nearly led to the disbanding of the German naval airship service. Comfort was however taken from the astonishing success of Delag's passenger-carrying Zeppelins which by the end of 1913 had made nearly 700 trips and carried over 14,000 passengers. In the end the civil Zeppelin *Sachsen* was hired from Delag to continue training until a new naval airship could be built. The large programme of the beginning of the year for ten Zeppelins was, however, little heard of again. In the spring of 1913 the Germans ceased to put all their faith in the Zeppelins and began to acquire seaplanes. A station was opened at Putzig and by the end of the year they had a dozen or so assorted machines, most of which were converted from German aeroplanes or were of foreign design.

In the spring of 1913, Germany had a total of eight large rigid airships, one naval, five military and two civil; all were capable of reaching Great Britain carrying bombs. Ever since the Zeppelin scare over Sheerness the year before, the problem of air defence had been an anxiety to the British, and in April the Prime Minister had to allay public disquiet. The problem had been placed firmly on the shoulders of the military wing of the Royal Flying Corps, but they had done very little about it, partly because they preferred to concentrate on aircraft for support of the Army in the field and partly for lack of funds. Winston Churchill was particularly concerned about the defence of the Admiralty, the Dockyards, the Naval Magazines and the Fleet in harbour. There seemed little point in spending millions to ensure a superiority of Dreadnoughts at sea if their shore support and control could not be guaranteed. The First Lord offered therefore to take on responsibility for air defence and the naval wing was given this task. In May, Winston Churchill made his famous 'swarm of hornets' speech in which he asserted that aeroplanes were the answer to the Zeppelin. Thereafter he considered aeroplanes operating from shore bases to be an essential part of the naval wing as well as seaplanes. Experiments were made with guns of various kinds in aeroplanes and also of grenades for use against Zeppelins. At the same time a requirement was stated for a fighting seaplane that could be carried in a ship at sea. In the spring of 1914 air defence trials and exercises were held at the Nore and progress was made firing Vickers and Lewis machine-guns but few of the aeroplanes had been designed for fighting and many had been bought in the open market. By the outbreak of war only two aeroplanes had machine-guns and there was a very poor stock of grenades and bombs. Out of the total of 40 aeroplanes there were no fewer than 21 different types; so the effectiveness of this heterogeneous force was very questionable especially as the average machine could only go a few miles per hour faster than a Zeppelin.

Of all the naval air arms that existed in 1914 before the outbreak of war that of Great Britain was the largest and best equipped. They had a total of 7 airships, 31 seaplanes and 40 aeroplanes. The airships were based at Farnborough and a new station at Kingsnorth on the Medway but only two of them were fit for operations, the others being too small and only of use for training. The seaplanes were based at Calshot, the Isle of Grain, Felixstowe, Great Yarmouth, Killingholme and Dundee. No standardized type had yet been evolved and there were twelve different types: Shorts, Wights, Maurice Farman, Sopwith and Henri Farman all being represented. Although only sixteen of them had wireless they were only really of use for reconnaissance. On 28th July, a torpedo had been successfully dropped from a Short seaplane with a 160-h.p. engine, but this was not yet standard armament; the torpedo was of the smallest 14-inch picket boat type and the observer and much of the fuel had to be sacrificed to get the seaplane into the air at all. The aeroplanes were based at Eastchurch, Fort Grange (Gosport), Felixstowe and Great Yarmouth. Eight of them were used for training but their principal duty was the defence of the Admiralty in London and the Dockyards at Chatham and Portsmouth against Zeppelin raids.

All these aircraft were shore based and of comparatively short range. The *Hermes* had been paid off in 1913 and the headquarters of the R.N.A.S. moved ashore where it could control the seaplane stations more effectively. The *Ark Royal*, the merchant ship converting at Blyth, was not yet finished so there were no seaplane carriers in the Royal Navy and there were no plans to carry any aircraft in warships. There were in theory twelve airships building which would have had a greater range, but of these only two were nearing completion and the second Parseval ordered in Germany seemed unlikely to be delivered. The British rigid *Naval Airship No. 9*, building at Vickers, was progressing very slowly: although ordered over a year before, the erection of the frame had not even started at the outbreak of war.

The German naval air service did not place a contract for a new Zeppelin until March 1914 but she was completed in May which was in sharp contrast to the time taken to build airships in Great Britain. For *L3* the Zeppelin company flatly refused to depart from proved designs so that she was practically a replica of *L1* but with more powerful engines of 630 h.p. She was accepted after an endurance flight lasting 35 hours and on 18th July made a scouting flight past Heligoland and Norderney along the Frisian Islands to the Dutch coast. On the outbreak of war she was the only German naval airship and there were no others even ordered. The naval seaplane section had grown by this time to a total of 36 planes, but there were only six planes ready for operations at Heligoland and three at Kiel. These seaplanes had a radius of action of 75 miles but had no wireless, and to carry bombs had to leave the observer behind and accept a reduced endurance. Little was expected of them and faith in the future of German naval aviation rested in building up a force of Zeppelins.

In January 1914 U.S. naval aviation was transferred from Annapolis to the old navy yard at Pensacola and the battleship *Mississippi* was assigned as base ship. In April in the troubles with Mexico the naval air arm first saw action. Three float-planes were sent in the U.S.S. *Birmingham* to Tampico and two float-planes in the U.S.S. *Mississippi* to Vera Cruz. They were used with success from the ships in harbour to make scouting flights inland. At the outbreak of war in Europe, the U.S. Navy had still only a dozen or so planes all fully occupied with training; nevertheless successful bomb-dropping trials and spotting for ships' gunfire had been tried out and Curtiss were building a very large flying-boat which they hoped would be able to fly the Atlantic.

Besides the three navies which have been dealt with in detail, there were seven others which had naval air arms. France had six coastal seaplane stations and three for non-rigid airships and the depot ship *Foudre* was used for experiments: Italy had three seaplane stations and twenty seaplanes, mostly of French design: Austro-Hungary had a station at Pola with five small airships, all of foreign design, and four French seaplanes: Russia had seven non-rigid airships and a force of seaplanes of French and American design with bases in the Baltic and the Black Sea, and lastly Japan, Turkey and Spain had each acquired a small airship.

Of the four distinct types of aircraft that had been evolved for use in naval warfare since 1900 (see Table 1, p. 24), by far the most important was still the rigid airship. Their endurance, reliability, wireless communications and navigation were vastly superior to the seaplanes and they were capable of providing reconnaissance over a large part of the North Sea; they were, however, expensive and needed a very large crew to handle them on the ground; furthermore, the weather was often too bad for them to leave their hangars. The non-rigid airship had some of the same advantages but to a smaller extent and it was much cheaper. Although it was extremely vulnerable and some-what frail, it could do useful coastal patrols in good weather. The seaplane and its variant, the flying-boat which in 1914 was no larger, had the advantage over the aeroplane that they could land on the sea in the event of engine failure which was still a common occurrence, but they were severely limited by rough water and could only land when it was practically calm. Sea-planes could carry a wireless set but their endurance was such that they could not go more than 70 miles or so out to sea. Although experiments had been made with ways to extend their range by taking them to sea in ships, there were no seaplane carriers yet at sea. Aeroplanes, that is land-planes, were now also seen to have a function in sea warfare. Up to 1913 this had scarcely been considered as they were certain to be lost if their engine failed, but their speed, rate of climb and ceiling were better than a seaplane's, so they were preferable for the air defence of ships in harbour and of dockyards and naval bases, and they could also be used for patrolling the line of the coast.

In general, the theory of the use of aircraft in naval warfare was well developed. Reconnaissance, spotting, bombing, tor-pedo dropping, air defence and other roles were understood and were being developed; the practical side was, however, a long way behind the theory and a great deal had to be done before aircraft performing any of these functions could have an effect on operations at sea. Although in the Review of the Royal Navy at Spithead in July 1914 five flights, totalling eighteen seaplanes, flew past in formation and two airships escorted the Royal Yacht out of harbour, the air arm was only just noticed by the huge fleet of Dreadnoughts and other types and was not taken very seriously. A few, notably Lord Fisher and Admiral Sir Percy Scott, believed that aircraft and submarines would revolutionize naval warfare, but the general opinion of the service, with a few exceptions such as Admiral Jellicoe and Captain Sueter, was that they were the First Lord's playthings. There was no doubt that sea power depended completely on the Dreadnought battleship with its attendant warships, and their operations were likely to take place well outside the range of all shore-based aircraft except the Zeppelin. The aircraft's weapons were in such an early stage that they could be practically ignored; even the Zeppelin's bomb load was only equal to a single salvo from a light cruiser. The fleets of the world therefore did not yet depend on aircraft for any purpose and had little to fear from them.

Table 1 – Types of Aircraft Used over the Sea, 1914

	Short type 74 Seaplane	B.E.2a Aeroplane	Astra-Torres Non-rigid Airship	L3 Zeppelin Airship
Speed (m.p.h.)	60	70	51	$47\frac{1}{2}$
Engines	One 100 h.p.	One 70 h.p.	Two totalling 400 h.p.	Three totalling 630 h.p.
Endurance	3 hrs 180 miles	3 hrs 200 miles	12 hrs 500 miles	30 hrs 1500 miles
Ceiling	4000 ft	10,000 ft	7000 ft	9300 ft
Armament	One m.g. Two 100-lb bombs	One m.g. Four 20-lb bombs	One m.g. Two 100-lb bombs	Three m.g. Eight 100-lb ten 25-lb bombs
Crew	2	2	6	20
Wireless	Yes	No	Yes	Yes

II

Aircraft at Sea in the Early Part of the Great War, 1914-15

ON THE OUTBREAK of war the Royal Naval Air Service was subjected to a bewildering series of orders, which perhaps with so young a service was understandable. The original intention for the seaplanes operating from their chain of bases on the east coast was to provide a complete coastal patrol system. On 29th July, however, just before war was declared, the First Lord had accepted responsibility for the air defence, not only of the Dockyards and Naval Establishments, but of the whole country. His directive to the R.N.A.S. was that air defence was to have first priority and that machines were not to be needlessly worn out by scouting. But on the second day of war, the German minelayer *Königin Louise* laid mines off the Norfolk coast, and on 8th August the Admiralty ordered a complete coastal patrol to begin from Kinnaird Head to Dungeness. Two squadrons of the Royal Flying Corps, which had not yet gone to France, were used to help. Regular patrols were made mostly by aeroplanes along the line of the coast, with instructions to report ships, aircraft or submarines. This system, however, did not last long for the R.F.C. aircraft were withdrawn and it was then decided, with the German advance into Belgium, to concentrate on the coast from the Humber to the Thames. At the same time the British Expeditionary Force was crossing to France and on 10th August a two-hourly seaplane patrol was instituted between Westgate in Kent and Ostend while the two operational airships patrolled across the Dover Straits. These patrols were kept up throughout August without incident, and when the passage of the B.E.F. was completed the seaplanes returned to the patrol at dawn and dusk along the line of the coast south of the Humber. On 26th

August yet another role was found necessary: the Eastchurch squadron of ten assorted aeroplanes was ordered to Ostend to co-operate with the Royal Marine Brigade which had been landed to try to hold Antwerp.

The German Naval Air Service, caught in a parlous state by the outbreak of war with *L3* the sole Zeppelin and only nine seaplanes available for operations, took energetic measures to remedy matters. The Parseval non-rigid airship building for the Royal Navy was requisitioned and four other airships were acquired from commerce or the army. The High Seas Fleet fully expected a heavy British attack in the Heligoland Bight on the outbreak of war and it assumed a defensive posture: the Naval Air Service under Admiral Hipper, the Commander of the Scouting Forces, co-operated. At first *L3*, based at Fuhls-büttel near Hamburg, was used to patrol the inner destroyer line in the Heligoland Bight at dawn and dusk; the seaplanes based at Heligoland had refuelling stations at Borkum and Sylt and were used to patrol to seawards out to 75 miles. It was hoped that the seaplane patrol would be continuous in daylight but this was found to be quite impossible with the number of machines available.

Gradually *L3* was used for more distant reconnaissance and in mid-August flew first as far as Terschelling and later to the Skagerrak, on both occasions to look for British forces which intelligence reported to be at sea. On 28th August, the Harwich Force advanced to attack the German patrols and this resulted in the Battle of the Heligoland Bight. No German aircraft were up to report the British approach, but *L3* took off as soon as the action began. Visibility was poor and she was fired on by her own side and returned to base, missing the chance to sight Beatty's battle cruisers which were a little farther to seawards – if she had made the sighting she might have saved three German light cruisers from being sunk. There were only two German seaplanes serviceable and they also took off, but had to fly so low that they could get no connected view of what was happening.

Mr Winston Churchill was behind a vigorous production drive to expand the R.N.A.S. Considerable numbers of aeroplanes, mostly of British and French and in one case American origin, were ordered and were of types already on the market. They were nearly all single- or two-seater general purpose aeroplanes with a single engine, the distinction between bombers and fighters being as yet hardly discernible. The sea-planes on the other hand were of three distinct types, the first of which was to meet the Admiralty requirement for one that could carry a torpedo: Shorts, Sopwiths and Wights all produced designs; at the same time, a seaplane fighter was copied from the Sopwith aircraft that had won the Schneider Trophy in April 1914 and two Curtiss twin-engined flying-boats for patrol work were purchased in America. The flying-boats were, however, no larger and of no greater endurance than the torpedo seaplanes. In order to free resources for this expansion Winston Churchill cancelled five airships, including the two new rigids from Armstrong-Whitworths.

The cancellation of the rigid airships meant that, if the fleet was to have any air support at all, something would have to be done quickly about converting ships to carry seaplanes. The *Ark Royal* was not even to be launched until September and it was clear that with her low speed of 11 knots she would never be able to keep up with the fleet. Three 22-knot cross-channel packets, the *Empress*, *Engadine*, and *Riviera*, were therefore commandeered on 11th August for conversion to seaplane-carriers. Each had a hangar built aft to house four seaplanes. They were too small to have the long flying-off platform necessary for modern seaplanes as in the *Ark Royal*: they were therefore fitted with derricks to hoist the seaplanes in and out so that they could take off and land on the sea. Although originally intended to carry aircraft for the Grand Fleet, Captain Sueter had more ambitious plans for them, seeing them as a means of carrying an offensive to the enemy coasts and the enemy fleet in harbour. On completion, therefore, they were not sent up to Scapa Flow but were kept in the south at Harwich. Although at the same time the cruiser *Hermes* was taken in hand for permanent conversion to carry three seaplanes, it was fairly clear that what was really required was a ship which combined the high speed of the cross-channel packets with the long flying-off deck of the *Ark Royal*. The liner *Campania* of 22 knots was therefore purchased in October and put in hand for conversion with the intention that she should join the Grand Fleet as soon as possible.

Although the German Navy had only one airship, there was a threat of air attack to the British Admiralty and the Dockyards from the five German army Zeppelins which were stationed behind the western front. They had, however, no intention of raiding Great Britain at this stage but were used in support of their land forces. They were operated somewhat rashly at low altitude by day and in the first month four of them were destroyed by anti-aircraft gunfire. This left only Z/X and the ex-Delag *Sachsen* in the west, other than the naval L_3. But there was no room for complacency in Britain. On the outbreak of war a crash Zeppelin construction programme had been put into effect for the building of ten of the L_3 type, half to go to the Navy and half to the Army, and delivery was expected to be better than one a month; three of the Schütte-Lanz type had also been put in hand.

At the beginning of September, Mr Winston Churchill, realizing the inadequacy of the British defences against Zeppelins, had decided that it would be better to take the offensive against them. He therefore entered into negotiations with the French to station thirty to forty aeroplanes at Dunkirk with the aim of preventing the Zeppelins being based close enough to attack the United Kingdom. The nucleus of this force was provided by the squadron already sent in support of the Royal Marines at Antwerp; its aim was indirectly to protect the Admiralty in London, the Dockyards, Magazines and oil fuel depots at Portsmouth and Chatham and the Arsenal at Woolwich: it was therefore contributing to the command of the sea which was the keystone of British strategy.

On 22nd September four aeroplanes from an advanced base near Antwerp flew to attack the Zeppelin sheds at Düsseldorf and Cologne; only one of them found its target at Düsseldorf but did no damage. On 8th October two Sopwith Tabloid aeroplanes tried again and succeeded in destroying Z/X in its shed at Düsseldorf. Immediately after this raid, Antwerp fell and the aircraft had to retire to Dunkirk which was out of range. For the rest of the year they were used to support the Army in the coastal sector and make reconnaissances along the coast for the Navy. On 21st November four new Avro machines, which had been sent in crates to Belfort in France, made an attack on the Zeppelin company's works at Friedrichshafen on Lake Constance. Armed with four 20-lb bombs each, they had to fly a 250-mile round trip. Three of the aircraft finally attacked and one was shot down; they did some minor damage but the new naval Zeppelin L_7 which was completing was uninjured.

The German naval Zeppelin sheds near Hamburg and Cuxhaven being out of range of any aircraft from Great Britain, it was decided to use seaplanes from the carriers recently converted from cross-channel steamers. An attempt was made to raid Cuxhaven on 25th October but heavy rain prevented the seaplanes leaving the water. The Zeppelin *L4* was out on reconnaissance but did not sight the British due to the poor visibility. In December 1914 the plan was renewed: nine Short seaplanes carried in the three seaplane carriers escorted by the Harwich Force were to proceed into the Heligoland Bight. The seaplanes were to try to bomb the Zeppelin sheds or, if they could not find them, any military target and on the way back they were to reconnoitre Kiel, Wilhelmshaven and the Schillig Roads. The Grand Fleet was to take up a position in the middle of the North Sea to support the operation in the hope that the High Seas Fleet would come out and that it could be brought to action. Ten submarines were to be in positions close off the enemy coast. The Germans now had four naval Zeppelins, *L3* and *L4* at Fuhlsbüttel and *L5* and *L6* at the new base at Nordholz, eight miles south of Cuxhaven, but there were only two seaplanes available at Borkum and a few at Heligoland.

On Christmas Eve the force sailed from Harwich, with the *Engadine* and *Riviera* escorted by the light cruiser *Arethusa* and four destroyers in one group and the *Empress* escorted by the light cruiser *Undaunted* and another four destroyers in a second group. They set course for the launching position 12 miles north of Heligoland which they were to reach at 6 a.m. next day. The Zeppelins *L5* and *L6* left their bases on Christmas Eve to scout to the westward but had to turn back over the inner Bight because of the weather. Had they not had to do so, they would probably have sighted the British force, for these Zeppelin reconnaissances now extended as far as 50 miles west of the Texel. The British, however, had been seen by a U-boat during the night and on reaching the launching position were sighted from Heligoland itself. Christmas morning was calm and sunny, the nine seaplanes were hoisted out and seven of them took off and made for Cuxhaven. The other two would not lift and had to be hoisted in again.

The German reaction was immediate: the Zeppelin *L6* took off from Nordholz as soon as she could be got out of her shed and *L5* prepared to follow. At Fuhlsbüttel *L3* was under repair and *L4* could not leave because of dense fog but a seaplane was sent up from Heligoland. The seaplane dropped four bombs on the British ships but missed and the attack was opposed with guns and small arms. She was hit and although she got back to

Heligoland she was a 'write off'. *L6* then arrived and after reconnoitring climbed to 5,800 feet and aimed a 110-lb bomb at the *Riviera* which missed by 100 feet. *L6* was heavily engaged by the squadron with its main armament as well as machine guns and rifles and the *Undaunted* burst a six-inch shell close to her at extreme elevation at a range of 11,000 yards. *L6* decided, wisely, that the fire was too hot to attack again and returned to the Schillig Roads to report as her wireless was out of action. When she landed at Nordholz she was found to have nine rifle bullet holes in her. Two seaplanes from Borkum also attacked the British Force with 50-Kg bombs; their aim was good but they too missed, probably because of the intense gunfire.

L5 left Nordholz an hour after *L6* and sighted three of the attacking British seaplanes on an opposite course over the River Weser. As the British machines crossed the coast, however, they ran into dense fog and only one of them found the Zeppelin base at Nordholz where she dropped her bombs on the hydrogen plant but missed. Two of the others reconnoitred Schillig Roads and sighted units of the High Seas Fleet which they reported. Two of the seaplanes returned soon after 10 a.m. and were hoisted in: a third landed alongside the destroyer *Lurcher* and three more, short of fuel, had to force-land in the sea off Norderney. The submarines *E11* and *D6* were fortunately at hand; the former surfaced but whilst rescuing the pilots was forced to dive by the Zeppelin *L5* which bombed both of them. The seventh seaplane had engine trouble and was picked up by a Dutch trawler after coming down in the sea. On the expiration of the endurance of the four missing seaplanes, the British Force, shadowed by *L5*, searched along the Frisian Islands and finally returned to base in the afternoon. *L5* then doubled back and machine-gunned the three abandoned British seaplanes which were finally sunk by the British submarine *D6*.

So ended the first carrier strike in history. From the British point of view it was a failure: they had done no damage for the loss of four seaplanes and the High Seas Fleet had not been lured to sea to be destroyed by the Grand Fleet. All that could be claimed was that the raid engendered uneasiness in the German Fleet, part of which was moved into the Baltic. The Germans had not much to be pleased about either. They had left the counterattack entirely in the hands of their Naval Air Service which had lost one of its few seaplanes and had failed to dislodge the British who remained in the Heligoland Bight for eight hours.

They could not claim to have repulsed the raid which failed more because of the weather than the opposition.

Nevertheless, historically the Christmas Day Action is of great interest. Not only was it the first ship-borne air strike ever to take place, but was also the first time that surface ships had seriously joined action with aircraft. The R.N.A.S. were not discouraged and were sure that they would have succeeded but for the weather. The surface forces were confident that given sea room they could compete with Zeppelin or seaplane attack. The British were therefore determined to try again as soon as possible. To the great fleets on which sea power depended there seemed, however, little in this unsuccessful sideshow to indicate that such operations would one day become the dominating factors in war at sea.

While this air offensive against the Zeppelin sheds was in progress, British seaplane patrols from their bases on the east coast continued spasmodically but were hampered by a shortage of replacement aircraft, by the weather and by maintenance problems. Engine failure and forced-landings at sea were not uncommon. The patrols were in a strict sense coastal, generally along the shipping channels and did not go very far to seawards; they sighted very little but helped in small ways to spot mines and to inconvenience the few U-boats which operated on the east coast. In October the two British non-rigid airships had also been used to make short scouting trips in the southern North Sea for submarines and mines.

On 3rd November 1914 the German battle-cruisers had bombarded Yarmouth and Gorleston. $L5$ had made a preliminary scouting flight to within sixty miles of Yarmouth two weeks earlier but on the day the weather prevented the planned reconnaissance by Zeppelins. The German ships were also unopposed from the air: there was not a single serviceable R.N.A.S. aircraft at Yarmouth at the time. On 16th December the German battle-cruisers had raided Hartlepool, Whitby and Scarborough and again the Zeppelins were unable to support them because of the weather. The raid was not detected by the R.N.A.S. either, for the same reasons as before; their pattern of coastal patrols at dawn and dusk would not have given any warning in any case. When the normal dusk patrol was up the German ships would still have been at a distance of over twice the radius of action of a seaplane out to sea, and by the time the dawn patrol was airborne the bombardment would have begun.

On the other side of the North Sea, the German seaplanes were still few in numbers but persevered with their patrols over the Heligoland Bight and went farther to seawards than the British. In September 1914 the British submarine *E3* turned the tables on them when she surfaced and captured the pilot of a seaplane which had force-landed in the sea. In October and November they had many more contacts with British submarines and sighted the British forces in the abortive operation against the Zeppelin sheds on 24th October, although they did not get their report through in time to be of any use. On 24th November, a seaplane from Sylt dropped five bombs near H.M.S. *Liverpool* of the 2nd Cruiser Squadron, which had penetrated into the Bight.

When the Germans had occupied Belgium in October, they established U-boat bases at Zeebrugge, Ostend and Bruges and the aeroplanes of the Dunkirk squadron added attack on these bases to its anti-Zeppelin and other functions. On 31st October a British seaplane base was also established at Dunkirk and it was whilst returning from here after ferrying some aircraft from Portsmouth that the *Hermes*, just recommissioned as a seaplane-carrier, was torpedoed and sunk by a U-boat. On 21st November yet another seaplane base was established at Dover after *U.12* had sunk the gunboat *Niger* in that area. On 6th December, the Germans sent two of their few seaplanes to Zeebrugge; these took on many duties including reconnaissance and spotting for the coastal batteries, and on 21st December one of them raided Dover.

By the end of 1914, the German airship service had eight airships, six of which were Zeppelins of the *L3* type. The other two were non-rigids stationed in the Baltic. The building of sheds had failed to keep pace so that *L7* and *L8* could not yet be deployed to the North Sea coast and had to remain inland at Leipzig and Düsseldorf. Airships were able to operate on only one day in four, however, because of fog, storms or cross-winds which prevented them leaving their hangars, yet in these five months the Zeppelins had made forty-eight scouting trips in the North Sea and the non-rigids had made ten flights in the Baltic without loss. There were still fewer than a dozen German seaplanes spread thinly between Borkum, Norderney, Zeebrugge, List and Heligoland. The Royal Naval Air Service on the other hand had grown to a total of 70 seaplanes and 109 aeroplanes and production had been greater than this expansion suggests, with the result that the force was now composed generally of more modern types. There were but seven British airships, all non-rigids, only one having been completed since the outbreak of war.

The Germans had been discussing since the beginning of the war whether to use their Zeppelins to bomb Great Britain but it was not until 1915 that they had enough of them to attempt it. The High Seas Fleet was willing to support the use of airships for bombing attacks but was insistent that naval reconnaissance should remain their primary duty. On the night of 19–20th January the first Zeppelin raid was made on King's Lynn, Yarmouth and Sheringham by *L3* and *L4*, *L6* having had to turn back with engine trouble. The raid was indiscriminate, did little damage and did not seem to show an intention to strike at the infrastructure of British sea power as had been feared. All the subsequent Zeppelin raids were of the same character and therefore are not strictly relevant to our subject. However, the facts that naval aircraft were charged with defence against them and that many of the counter measures took place over the sea bring many of the raids within any discussion of sea warfare in the North Sea. In any case, counter measures to Zeppelins when they were scouting were difficult to distinguish from those directed against Zeppelins on raids.

Anti-Zeppelin measures became one of the principal duties of the R.N.A.S. in home waters for the first part of 1915. No further raids took place until March, however, and the counter-measures did not really get under way until then. In the interval the German submarine campaign against commerce began with the declaration of a war zone on 18th February, thus providing a second and important task. At the same time the attack on the Dardanelles led to the diversion of aircraft to that theatre and there was a need for aircraft to help destroy the German light cruiser *Königsberg* in East Africa. All these urgent needs had to be met at a time when naval aviation was achieving very little in the North Sea and could scarcely afford to be weakened, an issue which was one of the many disagreements between Lord Fisher, the First Sea Lord, and Mr Winston Churchill.

The German naval seaplanes on the other hand were more successful during the same period, mainly because British forces frequently penetrated into the Bight within range of their bases. Not only did they continue to surprise British submarines; they gave warning of the approach of the Harwich Force when it made a reconnaissance into the Bight on 19th January. The Zeppelin *L5* at once took off and followed up this report, but returned to Nordholz when the Force headed for home. On 23rd January the Battle of the Dogger Bank took place: *L5* was flying a routine patrol at the time and did not realize that an action was in progress until several hours after it had begun. She closed but was engaged and driven off by British cruisers. Later she saw the *Blücher* sink and covered the retreat of the German battle-cruisers. A seaplane from Borkum also made contact and bombed a British destroyer which was in fact rescuing the *Blücher*'s survivors. German aircraft did not therefore influence or help much in the battle but this was the first time that naval aircraft had ever participated in a fleet action. No British aircraft were there at all. In the following month, however, the German airships suffered a serious reverse. *L3* and *L4* were sent to scout ahead of the supply ship *Rubens* which was breaking out of the North Sea on her way to German East Africa. The Zeppelins went as far as the Skagerrak and an unexpected southerly gale coupled with the failure of some of their engines led to the loss of both airships after force-landing in Danish territory.

On 13th January the newly-commissioned seaplane-carrier *Ark Royal* was ordered to the Dardanelles with six assorted seaplanes and two aeroplanes, and arrived just before the first naval bombardments began. The seaplanes which included Short, Sopwith and Wight types proved a great disappointment. They could not take off in choppy seas or in a flat calm, and were unable to fly high enough. They could not fly close enough to spot for the ship's gunfire during the first bombardments on 19th February because they were fired at from the ground. They were able, however, to provide some useful information about the forts and their armament and the damage inflicted by the gunfire. Admiral Carden considered aerial spotting essential if the Dardanelles were to be forced and therefore sent an urgent request home for aeroplanes which he hoped would be able to fly higher. His request was very quickly met and on

24th March, No. 3 Squadron of the R.N.A.S., composed of eighteen aircraft of six different types, began to arrive. They operated from an airfield which had been prepared for them by the *Ark Royal* on the island of Tenedos. In the interval, while they were on their way, *Ark Royal*'s seaplanes began to obtain some results with spotting and were used in the attack on the Narrows on 18th March; they were also used to try and locate the minefields and to reconnoitre the forts before and after the attacks. On 18th April two seaplanes from the *Ark Royal* took the offensive and flew over the Bulair Isthmus and bombed the Turkish battleship *Turgud Reis* lying off Gallipoli.

Early in the planning for the landings, General Birdwood suggested that kite-balloons would be very useful for observation and spotting. Again the Admiralty met the request promptly and took up and converted a tramp steamer, the *Manica*, to carry a kite-balloon. She arrived at Mudros on 9th April and on 19th spotted for the armoured cruiser *Bacchante*. The kite-balloon was found to be much more efficient for spotting than aeroplanes or seaplanes, being able to stay up longer, observe continuously and maintain better communications. Balloons were not a recent invention, as we have seen, but their use in a specially fitted ship was novel. *Manica* and other kite-balloon ships were employed throughout the campaign for observation, and without their spotting the gunnery of battleships, cruisers and monitors would have been much less accurate, indeed many targets could not have been engaged at all.

On 25th April during the landings, the aeroplanes supported the Helles sector while *Ark Royal* and her seaplanes supported the Anzac sector which was out of range of the aeroplanes from Tenedos. During the landings an aeroplane sighted the *Turgud Reis* firing across the peninsula; she was engaged by H.M.S. *Triumph* using *Manica*'s balloon for spotting, and was chased out of range. This naval engagement between battleships firing across the land was made possible by aircraft, and a number of similar encounters took place during the campaign. After the landings the aeroplanes took over all the duties necessary for the support of the army and their activities are no longer really part of this story. The seaplanes were then in general used where aircraft were required in places out of range of the aeroplanes, such as at Smyrna and Bulair. On 25th May, the German submarine *U.21* arrived at the Dardanelles and thereafter it was thought too dangerous to risk the *Ark Royal* at sea so she became practically a harbour seaplane station and only put to sea to shift base. But on 12th June the converted cross-channel steamer *Ben-my-Chree* arrived, and being smaller and faster she was able to continue to operate as a sea-going seaplane-carrier. The arrival of *U.21* added anti-submarine patrols to the duties of the sea-planes which were often carried out in the form of a search for U-boat bases on the Bulgarian and Asiatic coasts and as far south as Milo.

Operations at the Dardanelles continued up to the end of 1915. In August, the seaplanes joined the British and French submarines in attacking the communications of the Turkish Army in Gallipoli and during 1915 some 70 bombing attacks were made. On 12th August, a Short 184 seaplane from the *Ben-my-Chree* launched the first aerial torpedo in war. The Short 184, although an improvement on the Short Folder, was still unable to lift more than the small 14-inch torpedo and then only if it reduced its petrol to three-quarters of an hour's supply. The *Ben-my-Chree* was, however, able to hoist out the seaplane in the Gulf of Xeros where it had only to fly across the isthmus to find a target. The seaplane flying at 15 feet dropped her torpedo and hit a 5000-ton supply ship at a range of 300 yards. The ship was, in fact, aground having already been torpedoed by the submarine *E.14* and so did not sink. Nevertheless, this was a great milestone in naval aviation. On 17th August the attack was repeated by two seaplanes damaging a steamer and a tug. The 'bag' in all the attacks on communications was two steamers and a tug damaged and a lighter and six dhows wrecked. This was only a fraction of the damage done by the submarines in the Sea of Marmora but was more than could be achieved by ships firing across the peninsula.

In September 1914, the German light cruiser *Königsberg* had taken refuge in the delta of the Rufiji River in East Africa. She was moored out of range of the guns of ships lying off the entrance and as the Germans had defended the river mouth with shore batteries, entrenchments and mines, entry would have been very hazardous. Although the British had sunk a collier in the channel to prevent the *Königsberg*'s escape, a blockading squadron was still considered necessary. It was important to complete the destruction of the *Königsberg* not only

The Rigid Airship
(Zeppelin *L.1*)

The Seaplane
(Short No. 42)

The Non-rigid Airship
(Willows No. 2)

The Aeroplane
(B.E.2c)

PLATE I — TYPES OF AIRCRAFT USED AT SEA, 1914
(Imperial War Museum Photographs)

H.M.S. *Empress*
Cross-channel Packet
Conversion
4 Seaplanes (1914)

H.M.S. *Ark Royal*
Used mainly as a
harbour seaplane
base (1914)
10 seaplanes

H.M.S. *Campania*
Fleet Seaplane Carrier
(Second conversion
1916)
10 seaplanes

H.M.S. *Furious*
Aircraft-carrier
(Second conversion
1918)
Separate take-off and
landing on decks
16 aeroplanes

H.M.S. *Argus*
Aircraft-carrier (1918)
Clear flight deck
20 aeroplanes

PLATE II — BRITISH SEAPLANE AND AIRCRAFT-CARRIERS, 1914–1
(Imperial War Museum Photographs

to remove the threat to the trade routes but to free the blockading ships. Here there seemed to be a chance for aircraft and two civil Curtiss flying-boats were purchased in South Africa and their pilot given a commission in the R.N.A.S. The machines arrived off the Rufiji in November 1914 in the armed merchant cruiser *Kinfauns Castle*. One of the flying-boats made five flights and located the *Königsberg* twelve miles up the river. On the fifth flight her engine failed, the machine was wrecked and the pilot taken prisoner. The second machine could not be used as it had been 'cannibalized' to keep the first one flying.

The local naval Commander-in-Chief then asked for seaplanes to be sent from home to bomb the *Königsberg*. Two Sopwith seaplanes arrived in February 1915 but were found to be practically useless in the hot climate. They could only take off by leaving the observer behind and reducing their petrol until it gave them under an hour's endurance. In April they were replaced by three Short seaplanes and although they were able to locate the *Königsberg* again, they could not get above 600 feet. One indeed was shot down by rifle-fire which obviously put bombing out of the question. Seaplanes were clearly no use and a request was made for aeroplanes to be sent while an aerodrome was prepared on Mafia Island near by.

In June, the monitors *Mersey* and *Severn* arrived and also four Henri Farman and Caudron aeroplanes. In June one of the Caudrons got up to 6000 feet and bombed the *Königsberg*, but missed her. In July the aeroplanes were used to spot for the fire of the monitors and an action developed between them and the *Königsberg* which was using a shore observation post. The *Königsberg* was the loser and was finally destroyed by gunfire but not before she had shot down one of the aeroplanes and damaged one of the monitors.

The destruction of the *Königsberg* was a substantial success for naval aircraft. Without their help she could not have been even located, let alone sunk. Admittedly their bombing was not effective and the Admiralty would not allow what they considered would be a suicidal torpedo attack by the seaplanes, but the aeroplanes' co-operation with ship's gunfire was decisive. Naval aircraft had now helped ships to do something which they could not do by themselves and had made a significant contribution to the safety of the trade routes.

.   .   .   .   .   .

The German Zeppelin raids on the United Kingdom began again in March 1915 and ten attacks involving seventeen individual sorties had taken place by the end of June. The R.N.A.S. defending aeroplanes proved totally ineffective and did not make contact with a single Zeppelin. The aeroplanes could not see the Zeppelins because the raids were all at night and could not hear them because of the noise of their own engines. It was therefore decided that the attack on the Zeppelin's bases must be resumed and it was planned to repeat the Christmas Day raid in March. Between 20th March and 11th May, the seaplane carriers made four attempts to attack but all were failures because of fog or choppy seas. The failures, most discouraging to the R.N.A.S. itself, convinced many that the naval air arm was a waste of time and money, with the result that the seaplane carrier force at Harwich was dispersed.

Early in the year the Sopwith Schneider single-seater fighter seaplane became available. It had a far better performance than the heavier machines which had been designed for reconnaissance and torpedo dropping. The Schneiders had a speed of 85 m.p.h. and could climb to 10,000 feet in 35 minutes and so had a chance to catch a Zeppelin in flight. At the same time the newly converted cross-channel steamer *Ben-my-Chree* had a portable launching ramp forward which it was hoped would be long enough to be used by the Sopwith Schneider seaplanes. With this combination it was hoped to use seaplane carriers in a new way against Zeppelins. Instead of attempting to bomb them in their sheds, they could be attacked whenever they were sighted at sea by launching a fighter to chase them. On 11th May, an unsuccessful attempt was made to fly-off a Sopwith Schneider seaplane from the *Ben-my-Chree* to pursue the Zeppelin *L9* which had been sighted: but the ramp proved to be too short and its use had to be discontinued. It was still hoped that if these seaplanes could be carried in ships to the eastwards they might be able to take off from the sea and catch the Zeppelins as they approached in daylight. Seaplane carriers, the Harwich light cruisers and later specially fitted trawlers were therefore used from May onwards to take Schneider seaplanes to sea when Zeppelin raids were expected. The plan was for the ships to hoist out the seaplanes to patrol fifty miles from the coast an hour before dark and again at dawn in the hope of catching returning Zeppelins. The weather seldom permitted the seaplanes to be used, however, and only one Zeppelin was sighted and she escaped. Fortunately these failures were offset by a considerable victory for the R.N.A.S. in Belgium, when they destroyed the army Zeppelins *LZ37* and *LZ38* and damaged *LZ39*.

On 3rd July it was decided to try a new plan with the seaplane carriers *Engadine* and *Riviera*. This time the aim was to make a reconnaissance of the Ems and Borkum and hope that Zeppelins would be enticed to sea so that they could be shot down by the three Sopwith Schneider seaplanes carried in the *Engadine*. On the way to the Bight the force was reported by a U-boat and no less than six Zeppelins were ordered to investigate as soon as it was light. At dawn the British Force stopped north of Ameland and hoisted out the Short reconnaissance seaplanes, only two of which got away without engine trouble. The Zeppelin *L9* from a new base at Hage near Emden soon had contact but was engaged by gunfire and kept her distance. *L6*, also from Hage, then arrived and attempted to attack from 5,290 feet but she too was driven off by gunfire. *L10*, *L11* and *SL3* from Nordholz appeared next and *Engadine* with high hopes and four airships in sight, hoisted out her three Sopwith Schneider seaplanes. Alas two broke up their floats trying to take off in the choppy sea and were lost, and the third was badly damaged. Of the two Short seaplanes which had flown off earlier, one pilot made a successful reconnaissance of Borkum and the Frisian Islands and returned but the other got lost and was picked up by a Dutch trawler and interned.

The Germans were very pleased with these results; only the Zeppelin *L7* which went up from the other new base at Tondern in Schleswig had seen nothing. They believed that they had beaten off the attack. Little did they know that they had swallowed the bait and had acted exactly as the British had wanted them to. It was the weakness of the launching arrangements of the British fighter seaplanes which had saved them from loss. The British now certainly lost faith in seaplanes operated from carriers and no further offensive operations were attempted for some six months.

The strength of the R.N.A.S. was doubled during the first half of 1915: on 1st June it stood at 11 airships, 118 seaplanes and 236 aeroplanes. In spite of this, it could not be claimed that the effect of the R.N.A.S. at sea and in the air above it was other than marginal. Its aircraft had been principally employed without success for anti-Zeppelin operations: the use of aeroplanes in defence had so far proved abortive as had the attempts to use seaplanes offensively to raid the Zeppelin sheds. The only results had been obtained by naval aeroplanes ashore in Belgium but these were really pure air operations and little to do with the sea. The British seaplanes had proved of little more use for reconnaissance or patrols from shore bases: they had sighted practically nothing and their part in operations with the fleet and against the U-boats, as will be described in the next two chapters, had hardly begun. In the many subsidiary operations, especially at the Dardanelles and in East Africa, aeroplanes, seaplanes and kite-balloons had had rather more success. In spotting for ship's gunfire they had achieved notable results and had, so to speak, extended sea power inland a little way.

Many technical advances had indeed been made: machine guns were now mounted in most aircraft, wireless sets were more efficient and heavier bombs were available. In addition to the Sopwith Schneider fighter, new torpedo and reconnaissance seaplanes were coming forward, notably the Short 184. Nevertheless they were still seaplanes and it was the poor performance of the seaplane itself, especially the difficulty of operating it from a seaplane carrier in the North Sea, which was the main cause of the lack of results.

III

Aircraft in the North Sea and with the Fleets
1914 – 1918

AT THE OUTBREAK of war the Grand Fleet had no air support at all and relied entirely on its cruisers for reconnaissance and a very few anti-aircraft guns for defence. On 11th August 1914 two seaplanes followed by others were sent up from Yarmouth to a base at Scapa Flow and on 27th August two Henri Farman seaplanes escorted the Grand Fleet to sea. This is often taken as a landmark in air–sea co-operation, but they only had the endurance to accompany the fleet for a short distance after which, as usual, the fleet was left on its own.

The German High Seas Fleet, on the other hand, had an increasing number of Zeppelins to co-operate with it. Although the Zeppelin raids were given great publicity and were undoubtedly the most important employment of them in the eyes of Captain Strasser, their commander, they in fact spent much more of their time scouting and patrolling in the North Sea. Admiral von Pohl arrived to command the High Seas Fleet in the early spring of 1915 and made a series of short cruises into the North Sea, mainly for training (see Fig. 1, p. 42). In the sortie of 29–30th March Zeppelins were used for close tactical reconnaissance for the first time: *L6* remained in close company with the fleet, *L7* patrolled to the north-westward and *L9* to the south-westward. In another cruise of 17–18th April similar dispositions were made. The Grand Fleet had intelligence of this movement and was also at sea but the Zeppelins did not sight it before the High Seas Fleet turned for home. The British believed that the Germans had retreated because they had discovered that the Grand Fleet was at sea

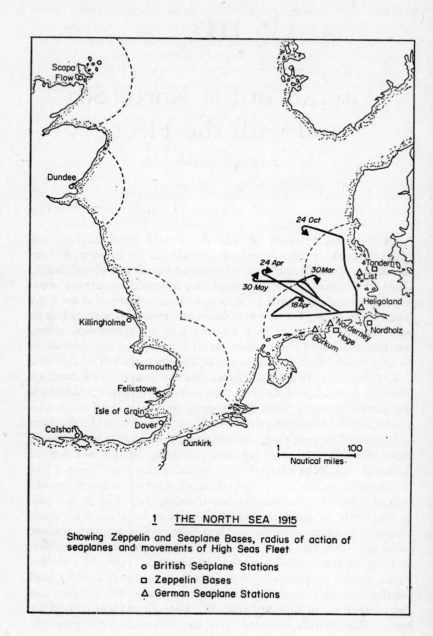

1   THE NORTH SEA 1915

Showing Zeppelin and Seaplane Bases, radius of action of
seaplanes and movements of High Seas Fleet

o   British Seaplane Stations

□   Zeppelin Bases

△   German Seaplane Stations

and that they must have obtained this information by radio intelligence. Investigation, however, revealed that this was impossible as no signals had been made; they therefore concluded that the High Seas Fleet must have been warned by Zeppelin reconnaissance.

Just before this incident, on 12th March, Winston Churchill had cancelled *Naval Airship No. 9*. She was the only British rigid airship under construction and was discontinued on the grounds that she would not be completed before the end of the war, which was then expected to be by the end of 1915. Lord Fisher does not appear to have been consulted until too late and was adamant that rigid airships were essential to give the Grand Fleet 'eyes'. Admirals Jellicoe and Beatty were also of this opinion. They pointed out that, although on paper they had a substantial superiority in Dreadnoughts, the margin in practice was quite small. The Germans could always arrange to put to sea when their full strength was available but this was not so with the Grand Fleet and it was their average strength, allowing for docking and repairs, which had to be compared. Anything which gave the High Seas Fleet an advantage might wipe out the Grand Fleet's margin altogether and an action in which the High Seas Fleet had eyes and the Grand Fleet was blind might well lead to disaster. Lord Fisher feared that the Zeppelins would make it impossible ever to surprise the High Seas Fleet and bring it to action. He pointed to von Pohl's cruise of 22–23rd April in which Zeppelins had again been in company with the High Seas Fleet and had formed a complete screen, (see Fig. 2, p. 45). All these Admirals believed that rigid airships were essential but Winston Churchill never reversed his decision and he pursued his vendetta against them until he resigned at the end of May.

It was not, of course, that Churchill failed to see the value of air reconnaissance, but that he thought that the future lay in heavier-than-air craft for this purpose. On 17th April, the *Campania* had joined the Grand Fleet. She had a speed of 22 knots, carried ten seaplanes and had a flying-off deck 120 feet long. It was, however, very soon found that 120 feet was not long enough and that she would have to stop to hoist out or recover seaplanes as with the earlier seaplane-carriers. On arrival she was given a station five miles astern of the centre of the battlefleet and so 20 miles behind the leading light cruisers. It was at once apparent that with the constant stopping to operate seaplanes she could not keep up with the fleet and she was left far astern. This was bad enough but it was also found that the sea was seldom smooth enough to operate seaplanes at all. During June and July, *Campania* took part in a series of exercises with the fleet. When the seaplanes could be launched they were found to be very useful, especially for tactical reconnaissance, but the chances of launching them in a fleet action were clearly remote.

The orders issued as a result of these exercises confirmed her station, but said that if the advanced light cruisers made contact with the enemy she was to proceed at full speed to the front and prepare to hoist out her seaplanes. The first duty laid down for the seaplanes was to scout ahead of the fleet for U-boats and minelayers and to report them and attack with bombs. If nothing was seen, they could then scout towards the enemy fleet and report its position and composition. If Zeppelins were sighted and *Campania* had any fighter seaplanes on board they were to be launched to the attack. Finally, if weather was too bad for seaplanes, she was to attach herself to a cruiser squadron and act as a surface ship.

By these orders any idea of air reconnaissance before contact was gained was given up. Even tactical reconnaissance after action was joined was to take second place to a search for U-boats and minelayers. The *Campania* was normally, therefore, to keep her seaplanes on board at sea, ready for use if an action developed and the weather permitted. Spotting for ship's gunfire and the use of torpedoes by seaplanes were not mentioned in the orders at all. So much for the aspirations of the 'air minded' officers of the R.N.A.S., but the means for a more ambitious use of aircraft were not to hand; the majority did not believe for one moment that the seaplanes were even likely to get into the air in a battle, and they were probably right. Shortly after the *Campania* arrived, the *Engadine* joined the Battle Cruiser Fleet and the same orders applied to her except that Admiral Beatty stationed her in the van with the light cruiser screen.

The two seaplane carriers with the Grand Fleet could not, therefore, make up for the lack of rigid airships. Mr Balfour, soon after his arrival at the Admiralty as First Lord, called a conference on airships. He decided to reverse his predecessor's decision and to recommence *No. 9* and to build more rigid airships. The design of an improved type similar to *No. 9* was begun and three were ordered in October. Mr Balfour also decided to cut down an order for fifty of the small S.S. anti-submarine airships[2] and to substitute thirty of a new and larger coastal type. One of these had already flown, having evolved in much the same way as the S.S. type but using an Astra-Torres envelope with a car consisting of two Avro fuselages joined together. The purpose of these coastal airships was not originally to hunt submarines but to provide a more constant and reliable patrol for the east coast.

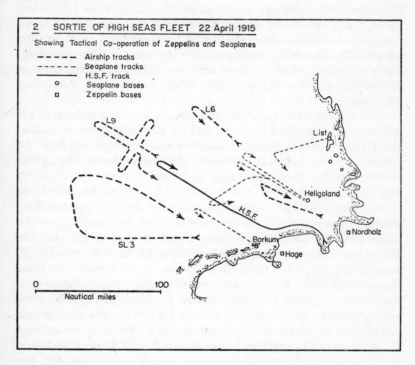

2   SORTIE OF HIGH SEAS FLEET   22 April 1915

Showing Tactical Co-operation of Zeppelins and Seaplanes

– – – –   Airship tracks
– – – –   Seaplane tracks
————   H.S.F. track
    o   Seaplane bases
    □   Zeppelin bases

L6

L9

List

Heligoland

H.S.F.

□ Nordholz

Borkum

SL 3

□ Hage

0   100
    Nautical miles

The German Navy had by midsummer 1915 eleven airships and all except two, which were in the Baltic, were stationed at Nordholz, Hage, and Tondern on the North Sea Coast. Three of the Zeppelins were of the new *L10* class and were greatly improved. They were nearly half as big again as the *L3* class and could reach the west coast of England with two tons of bombs. They had four engines giving a speed of 58 m.p.h. and could climb to 12,800 feet. Admiral von Pohl had no doubt of the value of Zeppelins in naval warfare and stated a requirement for eighteen of them in the North Sea. He estimated that six patrol sectors had to be covered when the High Seas Fleet put to sea, six more airships were needed as reliefs and six in reserve to cover casualties, breakdowns and repairs. This figure was approved by the German Admiralty who began a new base at Ahlhorn near Bremen for them.

In the middle of 1915, Admiral Jellicoe was seriously worried by the German use of Zeppelins for scouting. They could always evade ships and their guns, and all the seaplanes, except the Sopwith Schneiders, were too slow and could not climb fast enough to catch them. He suggested that the fleet should carry aeroplanes and that they should be brought back from France and the Dardanelles to compete with the menace. The problem, however, was to operate them with the fleet and, at the time, it is doubtful if more aeroplanes at home would have made any difference.

At the same time the Admiralty, in view of the success of air spotting against the *Königsberg* and on the Belgian coast and at the Dardanelles, wished to introduce this development into the Grand Fleet. Admiral Jellicoe, although seeing the great advantage this could give him in battle, was not particularly optimistic; he pointed out that the chances of his seaplanes being used in action were very small, and that in any case their communications were not yet good enough for spotting; at the same time he asserted that the Zeppelins could spot for the High Seas Fleet against him and that he had practically no means of preventing them from doing so. Although the *Campania* could carry Sopwith Schneider seaplanes the chance that the weather would be good enough to allow them to be used was negligible. She had been sent in July as a mere mobile seaplane base to Pierowall Harbour in the Orkneys to co-operate in anti-U-boat operations in the Fair Island Channel and then to Cromarty to try to locate some enemy minefields. On her return in August, while steaming at 17 knots, she successfully launched a Sopwith Schneider seaplane from her flying-off deck using a wheeled trolley. This was a great step forward and meant that fighter seaplanes could now shoot down Zeppelins at sea in choppy weather which hitherto precluded their use. If it was rough, however, the seaplane would be lost on landing, but the pilot could probably be saved and a Schneider for a Zeppelin would be a good bargain.

This advance did not, however, solve the spotting problem. The single-seater Schneider had no wireless and could not be used for this purpose. In August Admiral Beatty suggested kite-balloons might meet the case, pointing out that they could be used in weather in which the reconnaissance seaplanes could not take off and in which even Zeppelin operations could be difficult. In September, a section of kite-balloons was sent to Rosyth and the *Engadine* towed one at a height of 3000 feet whilst steaming at 22 knots in a rough sea. Admiral Jellicoe was not keen to have special fast kite-balloon ships to accompany the fleet but suggested that the *Campania* might carry one. She was sent back to Cammell Lairds in November to be given a longer flying-off deck so that her reconnaissance seaplanes would also be able to take off; she was to be fitted at the same time to carry kite-balloons.

In August 1915 the Zeppelin campaign against London began. Five airships of the latest *L10* type were to be used for the attack on the city whilst the older *L9* and two Schütte-Lanz ships were to raid northern England. Only *L6*, *L7* and the Parseval blimp *PL25* were originally to have been kept back for scouting, but in the event the Zeppelins were used much more for reconnaissance than for raiding. During the whole of 1915, 297 scouting trips were made in the North Sea against 51 raiding sorties. Only two Zeppelins were lost in action on raids but the weather was a potent adversary and four other airships were lost by accidents of some sort. Therefore, although nine new airships were built in the second half of 1915, overall strength only rose by two.

On the British side, a new seaplane-carrier, the *Vindex*, arrived at Harwich in September. Like most of her predecessors she had been converted from a cross-channel steamer and had the usual hangar and cranes aft, but this time for five seaplanes. She had also been given a flying-off deck forward, 64 feet long with a small hangar for Sopwith Schneider seaplanes. It was realized even before she was finished that it would be too short for them and on 3rd November, in a successful trial, a Bristol Scout aeroplane was flown off; from then, when on anti-Zeppelin patrol, *Vindex* carried two of these aeroplanes which had, however, to be dismantled to stow them in the hangar. Bristol Scouts were single seaters and had a speed of 93 m.p.h., a ceiling of 15,000 feet and were armed with 48 Ranken anti-Zeppelin darts. It was possible to launch these aircraft in seas in which seaplane operations were out of the question and, with their high performance, they stood a very good chance of bringing down any Zeppelin they sighted. They could not, of course, be recovered but could often fly back to a land base or, if this was not possible, could 'ditch' alongside. They were fitted with flotation bags and the pilot could be picked up and the aeroplane or parts of it salvaged. This system had hardly got into its stride before the end of 1915 but obviously was very promising.

There were few naval operations in the North Sea during the second half of 1915, the only one of interest to our subject being the sortie of the German minelayer *Meteor*. On 6th August she sailed to mine the Grand Fleet base in the Cromarty Firth and on her voyage northwards, the airship *SL3* scouted ahead during daylight. The *Meteor* laid her mines but her presence was detected and four groups of light cruisers set off to intercept her. The Harwich Force, steaming at full speed to cut her off, was sighted and reported by a seaplane from Borkum, and the airships *L7* and *PL25* which went up from Tondern were able to shadow it and give the *Meteor* an accurate picture of the situation. The *Meteor* being a slow ship was, however, unable to avoid the light cruisers closing in from four directions and she was sunk. This does not detract from the excellence of the German air reconnaissance which, if the Germans had sailed surface ships in support, might have saved the minelayer.

In the second half of 1915, the strength of the Royal Naval Air Service again more than doubled: at the end of the year it had 44 airships, 603 aeroplanes and 319 seaplanes and flying-boats. There were now nine seaplane-carriers, to which were allocated 64 seaplanes and 7 aeroplanes. Five of the new Coastal airships were running trials and thirty-three of the smaller S.S. blimps had been completed. The rigid airship *No. 9* was being erected at Barrow and three others, *Numbers 23, 24* and *25*, were on order. In December the Admiralty sought covering approval to build a total of sixteen rigid airships for fleet reconnaissance and set in train bases for them on the east coast. A new fighter seaplane, the Sopwith Baby, was under development to replace the Schneider; the first of these had already been delivered and was capable of 100 m.p.h. with a ceiling of 10,000 feet. New torpedo seaplanes were slower coming forward but were to be able to carry an 18-inch torpedo and have a much greater endurance. The hope was that they would be able to attack the High Seas Fleet by night in its bases. Many naval officers were, however, losing faith in the seaplane

altogether and were turning to the flying-boat. Much larger
flying-boats were already on order from Curtiss in America and
it was hoped that they would be able to land and take off in
rougher seas.

In the first half of 1916, the movements of the main fleets in the
North Sea were influenced considerably by air operations. The
British, depressed by their failure to intercept Zeppelins, decided that they must take the offensive again. On 18th January,
the *Vindex*, escorted by the Harwich Force, set out to raid the
Zeppelin base at Hage but the attempt failed because of fog. A
second attempt on 29th January was called off because, as
Vindex was hoisting out her seaplanes, the light cruiser *Arethusa*
was narrowly missed by a torpedo from a U-boat.

On 18th January Admiral Scheer became Commander-in-
Chief of the High Seas Fleet. On the return from a Zeppelin
raid on the Midlands on 31st January *L19* was lost. A fog had
prevented any British counter measures at sea and it would
seem that her loss was a combination of engine failure and a
strong southerly wind. Admiral Scheer believed that she could
have been saved if German surface ships had been at sea and
he resolved to support Zeppelin raids with warships in future.
He realized that British light warships, seaplane-carriers and
trawlers often put to sea to intercept Zeppelin raids and he
hoped that he could destroy them at the same time. But he had,
in fact, a more profound intention. Always an advocate of a
more offensive use of the fleet, he believed that, with close
co-operation with Zeppelins and U-Boats, it was worth trying
to cut off and destroy a portion of the Grand Fleet itself. If the
strength of the Grand Fleet could be decreased in this manner
a general action on equal terms might become possible with
the hope that command of the sea could be regained.
Zeppelin reconnaissance formed an important part of this plan
and their scouting flights were now more often made in co-
operation with the fleet. Only two of the nine Zeppelin raids
made between March and May were, however, supported by
sorties of the High Seas Fleet and no contacts were made.

In February 1916 the Royal Flying Corps took over responsibility for the air defence of the United Kingdom but the
Royal Naval Air Service's role continued to include attack on

Zeppelins over the sea. In mid-March the British decided to try another seaplane attack on the Zeppelin sheds, but this time at Tondern. The *Vindex* sailed on 24th March, escorted by the Harwich Force and with the Battle Cruiser Fleet in support. At 05.30, 25th March, three Short and two Sopwith Baby seaplanes were hoisted out inside the Vyl light vessel and took off successfully in very cold weather. Surprise was achieved as the German seaplanes at List considered the weather too bad to take off. Only one of the five seaplanes found the sheds at Tondern, but her bombs were iced up and failed to release; three had engine trouble and came down in enemy territory and only two of the machines returned. German seaplanes from List thereupon took off and bombed the ships, but missed. There was subsequently a brush between British and German light surface forces in a gale. Although both the Grand Fleet and the High Seas Fleet actually left harbour, there were no further contacts. The raid, therefore, as in every operation of this kind to date, failed to destroy any of the Zeppelins or their sheds.

On 25th April the German battle-cruisers bombarded Lowestoft; the operation was supported by the High Seas Fleet and co-ordinated with a heavy Zeppelin raid. The Zeppelin *L7* scouted ahead of the battle-cruisers leaving the Heligoland Bight, but when the *Seydlitz* struck a mine she was detached to escort her back to harbour; *L5* covered the northern flank of the High Seas Fleet in the Dogger Bank area and *L9* was stationed between the battle cruisers and the High Seas Fleet itself. Some reconnaissance value was also obtained from the seven Zeppelins which set out to raid the United Kingdom. The British as usual were warned by radio intelligence that some German attack was imminent and ordered aircraft from the east coast bases to scout at dawn, and all available machines to attack the enemy as soon as located. *Vindex*'s seaplanes were also ordered to fly up the coast to help, and when the bombardment began, three seaplanes and some aeroplanes took off from Yarmouth and Felixstowe. They bombed the German ships but missed and one was damaged by anti-aircraft fire; they also attacked the German submarine *UB.12* and incidentally a British submarine too. The Zeppelin *L9* reported the approach of the Harwich Force from the south but was attacked by two BE2c aeroplanes from Yarmouth. They caught her at low altitude but their bombs missed, their Ranken darts had no effect and she escaped. Aeroplanes from Yarmouth also chased

L13 and *L16* for sixty-five miles out to sea but failed to bring them down. *L6*, for some reason, failed to sight Beatty's battle-cruisers which were approaching from the north but nevertheless screened the retirement of the High Seas Fleet successfully.

The Zeppelins were undoubtedly of great value to Admiral Scheer in this sortie and gave him a feeling of security against surprise by superior forces. They had, however, several narrow escapes when they tangled with aeroplanes from the British shore bases in daylight. The R.N.A.S. aircraft for the first time were able to oppose a surface raid but the small bombs they carried were useless against armoured ships. Torpedoes would have been more effective but were not used, although seaplanes of the same type that had launched the attacks in the Dardanelles from the *Ben-my-Chree* were stationed at both Yarmouth and Felixstowe. It is of interest that the aircraft were alerted by radio intelligence to stand by and there was still no question of air reconnaissance providing the first indication of such a raid. Even the dawn patrol was far too late as the bombardment had already begun.

The 'Hoyer' raid, as the *Vindex* operation against Tondern on 25th March is usually known, showed that, even if no Zeppelins were destroyed, it was a way to entice the High Seas Fleet to sea. Making this the primary aim, a second raid was planned, using both the *Vindex* and the *Engadine* escorted by the First Light Cruiser Squadron and sixteen destroyers. The Grand Fleet was to be at sea and minefields laid and submarines positioned to intercept the High Seas Fleet if it came out. Eleven of the new Sopwith Baby seaplanes were embarked which, in addition to their high performance as fighters, could carry two 65-lb bombs each. When the force arrived off Sylt on 4th May, conditions seemed excellent, but eight of the eleven seaplanes bent their propellers in the sea trying to take off and had to be hoisted in again. Of the three that got into the air, one hit the mast of the destroyer *Goshawk* and crashed while another had to return with engine trouble; the single surviving seaplane flew on and dropped its bombs on the Tondern shed but missed. One of the Tondern Zeppelins, *L7*, then took off to scout towards Horns Reef and *L9* ascended from Hage. *L7* sighted the British Force and was chased for half an hour by the light cruisers *Galatea* and *Phaeton*; a parting shot from the *Galatea* hit her at long range and she caught fire and came down in the sea. The British submarine *E31*, which was submerged near by, surfaced and completed her destruction and then picked up the survivors. Thus the Tondern raid was saved from being a complete fiasco by good gunnery. The High Seas Fleet made no move and *L22* under repair in her shed at Tondern escaped damage. *L7* was the first Zeppelin to be shot down by ship's gunfire and it is of interest that this was done, not by an anti-aircraft gun, but by an ordinary 6-inch gun on a low-angle mounting. It was the last of the second series of seaplane carrier raids on the Zeppelin bases, which like the others, were all complete failures.

On the eve of Jutland there were still no British airships operational in the North Sea. In January 1916, a fifth rigid airship *No. 26* had been ordered from Vickers and six more were authorized. Sheds for them were started on the east coast at the three bases building for the coastal blimps, at East Fortune in Scotland, Howden in Yorkshire and Pulham in Norfolk. All these new airships were to be of the *23* class and it was hoped to complete them by the autumn. The *23* class were similar to *No. 9* but were lengthened and had more powerful engines; they were therefore basically of 1912 vintage and the design compared very unfavourable with the latest Zeppelins. Their disposable lift was only 8 tons against the 15 tons of the Zeppelins of the *L10* type. It is understandable that Lord Fisher, of the Board of Inventions and Research, was very annoyed at the failure to recover the wreck of a Zeppelin in January. On 8th May, however, H.M.S. *Agamemnon* at Salonika shot down the German Army Zeppelin *LZ85* and the wreckage was shipped home in the kite-balloon ship *Canning*. At the same time a Swiss national who had been employed by the Schütte-Lanz company had arrived in England. He claimed to be able to reproduce the plans of this type of rigid airship and, in collaboration with the Admiralty, he did so. As a result two new airships, *Numbers 31* and *32* were laid down to this design with Messrs Shorts and in March, the Director of Naval Construction, who was responsible for airship design, recommended a modified type *23* in which by removing the keel and making other changes, the disposable lift could be increased by $3\frac{1}{2}$ tons. The later airships building to the *23* design were therefore altered to this *23X* type and were to be completed in October and November 1916.

Although there were five coastal airships completed by the beginning of the year and although their bases at Pulham,

Sopwith 'Baby'
Seaplane
(Single-seat fighter)

S.1
(Submarine Scout
Blimp)

Torpedo Seaplane
(Short 184)

aval Airship No. 9
(Later *R.9*)

Curtiss *H.12 Large
America* Flying-boat

PLATE III — R.N.A.S. AIRCRAFT IN THE GREAT WAR, 1914–18
(Imperial War Museum Photographs)

Short *184* Seaplane
The R.N.A.S. Maid
of all work

H.M.S. *Yarmouth*
Showing Sopwith
Pup Fighter on
launching platform

Sopwith '1½ Strutt
taking off a turret
platform
H.M.A.S. *Australia*

H.M.S. *Eagle*
The first 'island'
type aircraft-carrier
1922

PLATE IV

(Imperial War Museum Photographs)

Howden and Longside were completed in February and March none of them was yet operational. Their endurance was not enough to scout like the Zeppelins across the North Sea and so in January it was decided to design a larger non-rigid airship, which would be able to accompany the fleet. Six of this *NS* or North Sea type were ordered but as an interim measure it was hoped that the Coastals could be of some assistance to the fleet if their range could be extended by towing. In May, in a trial at Harwich, *C1* was towed successfully at 20 knots by the light cruiser *Carysfort*. No further progress had, however, been made before the main fleets met at Jutland.

The *Campania* rejoined the Grand Fleet in April on completion of her refit. She had been greatly improved. The flying-off deck had been extended from 120 to 200 feet so that she could launch the two-seater reconnaissance seaplanes on trolleys in the same way as the single-seater fighters; it should now be possible to use all her seaplanes in an action whatever the weather although, if the sea was rough, all of them would be lost on landing. At the same time she had been given a rig which enabled her to recover seaplanes from the sea at a speed of six knots so that it was no longer necessary to stop. When the sea was calm enough to permit seaplanes to land, therefore, which would admittedly only be occasionally, she would now be able to operate them when the fleet was cruising and still have a chance to keep up. Finally she had been fitted with a kite-balloon and so could also be used to spot gunfire and to report a 'bird's eye' view of an action to the Commander-in-Chief. On arrival in the fleet, *Campania* showed great promise. Seaplanes were hoisted in and out without stopping and at the end of May Sopwith Babies were flown off the deck at 20 knots; successful spotting both by her seaplanes and the balloon were carried out during practice firings but the heavier reconnaissance seaplanes had not been launched from the deck before the Battle of Jutland. Insufficient experience was obtained to rewrite the Battle Orders, so the functions of her aircraft were not altered, but her station was changed to a position close to the flagship *Iron Duke* in the hope that her kite-balloon would be of use to the Commander-in-Chief.

Admiral Scheer's original plan for the sortie of the High Seas Fleet which led to the Battle of Jutland was to bombard Sunderland. The operation, in accordance with his new theories, was to be in close co-operation with U-boats and Zeppelins,

E

both of which had a vital part in the plan. Unsuitable weather kept the Zeppelins grounded for some days and rather than abandon the operation altogether Admiral Scheer altered the plan to an advance to the Skagerrak and the Norwegian coast, making air reconnaissance less important and enabling the U-boats still to play their part. The High Seas Fleet therefore put to sea on 31st May without a preliminary Zeppelin reconnaissance but ten of them were kept ready in case the weather improved (see front endpaper). The weather did, in fact, get better after the fleet had sailed and the first flight, consisting of *L16*, *L9*, *L21*, *L14* and *L23* took off to reconnoitre in a wide arc from Holland to Norway. This type of Zeppelin reconnaissance was in fact novel. Instead of being used for close tactical scouting in the same way as light cruisers, they attempted for the first time in co-operation with the fleet to make a distant strategic reconnaissance. They had not even caught up with the fleet, however, before the surface screens made contact in the afternoon.

The Grand Fleet had the usual warning of the German sortie by radio intelligence and, in fact, sailed from Scapa before the High Seas Fleet left the Jade. *Campania* was anchored in a remote part of the Flow and missed the signal to sail. She put to sea two and a quarter hours late and steamed at full speed to catch up. As she had no escort and, in the Commander-in-Chief's opinion, could not join before an action, she was ordered back to harbour and so was not present at the Battle of Jutland. The battle was fought far outside the range of the hundred or so seaplanes and flying-boats of the R.N.A.S. based on the east coast, and the air component of the Grand Fleet was therefore reduced to the four seaplanes carried in the *Engadine* with the Battle Cruiser Fleet. *Engadine,* unaltered since her conversion at the beginning of the war, had no flying-off deck and had to stop to hoist out her seaplanes. She was stationed in the centre of the light cruiser screen ahead of the battle-cruisers and there was no thought of using her seaplanes before the enemy appeared; if she had tried to operate them she would have been left miles astern. Her main hope was that a Zeppelin would be sighted and that she could use one of her two Sopwith Baby seaplanes to shoot it down.

At 2.20 p.m. on 31st May the light cruiser screens made contact and twenty minutes later Admiral Beatty ordered the *Engadine* to send up a seaplane to scout to the N.N.E. Smoke had been reported in this direction and the light cruisers only covered the arc from south to east. The seaplanes were folded in the *Engadine*'s hangar and it took nearly half an hour to get one of the Short 184 seaplanes away. The machine flew off to the N.N.E. but clouds kept her below 1000 feet. At 3.30 she reported the three German light cruisers of the Second Scouting Group with five destroyers. She closed to within 3000 yards and was heavily fired upon and at 3.33 she saw the enemy turn to the south and reported it. As she was sending a further report, a petrol pipe broke and she had to force-land in the sea. At 3.47 the *Engadine* stopped alongside and hoisted her in, the operation taking over a quarter of an hour. By this time the battle-cruisers were in action eighteen miles away to the south-eastward and she hurried after them at her full speed with all four seaplanes on board. *Engadine*, with 22 knots, was the slowest ship in the Battle Cruiser Fleet and she lost another three miles in the run to the south. At 4.33 the light cruiser *Southampton* sighted the High Seas Fleet and Admiral Beatty turned to the north. *Engadine* did not turn until the *Lion* passed her on an opposite course by which time the leading German battleships were only thirteen miles away. Twenty minutes later the range was down to ten miles but *Engadine* was fortunately screened from the enemy by the Fifth Battle Squadron. By 6.15 the Grand Fleet had come in sight and as they were deploying the *Engadine* passed to their disengaged side. At 6.40 she met the disabled armoured cruiser *Warrior* and took her in tow and so played no further part in the action.

By this time, *L14* and *L23* of the first flight of Zeppelins had reached the battle area but the cloud was so thick that they saw nothing. The first flight was then recalled to base, *L14* being eleven miles north of Admiral Jellicoe's flagship as she turned. As she passed over the two fleets at the height of the action she neither saw nor heard anything. Just after ten o'clock that evening, Admiral Scheer radioed for air reconnaissance at Horns Reef at dawn. This was a job for the German seaplanes but soon after midnight the second flight of Zeppelins consisting of *L11*, *L13*, *L17*, *L22* and *L24* took off to cover the seaward flank of the High Seas Fleet as it retired. *L22* and *L24* sighted gunfire before dawn and *L24* dropped bombs on a number of ships fifty miles off the Danish coast. After dawn she reported another force including battleships and this was believed by Admiral Scheer to be the Grand Fleet. Analysis has never established the identity or even existence of this force and it remains a mystery. At dawn *L11* found the Grand Fleet forty miles north of Terschelling and reported its position. She was heavily engaged successively by the Battle Cruiser Fleet and the First Battle Squadron and soon lost contact in bad visibility. Admiral Scheer wrongly interpreted this report as British reinforcements arriving from the Channel. These reports were, however, of little significance. Admiral Scheer had decided the night before to retire from the action and was already safe in the Horns Reef swept channel.

Subsequently it was widely believed in Great Britain that the Zeppelins had played a vital part in the battle and that the escape of the High Seas Fleet was largely due to their co-operation. This was of course quite untrue although it cannot be denied that the Battle of Jutland was influenced in some ways by aircraft. The action took the form it did solely because Admiral Scheer relied upon Zeppelin reconnaissance to bombard Sunderland and when it was not available he changed his plan. It is probably true that had the visibility been good Zeppelin reconnaissance would have given warning of the approach of the Grand Fleet and there would have been no battle at all.

It was afterwards calculated that the *Campania* could have caught up with the Grand Fleet before the action and that there were in fact no U-boats on her track. There is no doubt that if she had been with the fleet, she could have got all her seaplanes away as the sea was calm and even the two-seaters could have been hoisted out. With the very poor visibility, low cloud and smoke, however, together with primitive communications and little practice, it seems unlikely that they would have achieved very much. Certainly no great change in the outcome of the battle could have been expected from their co-operation. This view is supported by *Engadine*'s performance. She was hampered not only by the bad visibility but by an unwritten law that seaplanes were only to be used when ordered and above all by the pressing need to keep up with a fast moving surface action. With hindsight, however, one cannot resist pointing out that if, instead of launching a single seaplane at 3.48 to scout to the N.N.E., all four had been sent off to search to their maximum range between east and south, aircraft might well have altered the outcome of the battle. Nevertheless, *Engadine*'s seaplane was the first ship-borne aircraft ever to take part in a fleet action.

She reported the enemy accurately including an important change of course. That her reports, although received in the *Engadine*, never reached the flagship and were duplicated by the light cruisers, does not alter the fact that this was a milestone in naval aviation.

By the middle of 1916, the great fleets of Dreadnought battleships were still the ultimate force on which sea power depended; they had, however, accepted assistance from aircraft for scouting, spotting and attacking enemy aircraft. For scouting the Zeppelin rigid airship was by far the most efficient. It had the endurance to make reconnaissances of the whole North Sea and had been present at the Battles of Heligoland, the Dogger Bank and Jutland. Although it could not claim any spectacular successes which were solely due to its co-operation, it had become essential to the German High Seas Fleet if it was not to be surprised and brought to action by the superior Grand Fleet. Of the fifteen German naval airships lost since the beginning of the war, only one had been destroyed by ships at sea: four had been brought down during raids by anti-aircraft fire but the rest of the casualties were all due to weather, engine failure or an accident of some sort. One German Army Zeppelin had been destroyed in the air by a naval aeroplane over the land and two more in their sheds by bombing, but on the whole they had proved difficult to counter. Furthermore faster types of Zeppelin of increased endurance and a higher ceiling were just coming into service.

Seaplanes, on which the British had concentrated, had proved decidedly inferior. When operated from coastal stations they had not the endurance to accompany the fleet or even reach likely battle areas, moreover they had failed to detect any of the German surface raids on the British coast. When taken to sea in seaplane-carriers to extend their range, they had proved practically useless because of the weather. In short the High Seas Fleet had eyes and the Grand Fleet was nearly blind; the fact that weather only permitted the Zeppelins to operate on one day in four did not really matter as the Germans could always choose suitable conditions for their sorties. Efforts to produce British rigid airships were very slow and none of them had yet flown, a situation which would have had serious consequences for the British had it not been for their extremely efficient radio intelligence service.

Admittedly there had been some significant successes with

seaplanes such as the operation of torpedo planes from the *Ben-my-Chree* at the Dardanelles and the flight of the *Engadine*'s seaplane at Jutland. Measures were indeed in hand to improve both seaplanes and their carriers but this would not alter the fact that seaplanes, as flying machines, were bound to be inferior to aeroplanes. The principal justification for continuing with seaplanes was the unreliability of aero engines which deterred aeroplanes from operating over the sea. Nevertheless it was becoming clear that the future lay, not in improving the seaplane, but indeed in finding a way to operate aeroplanes over the sea. With the sole exception of the Zeppelin, therefore, the influence of aircraft on sea power was still marginal, but the rate of development of all types of aircraft was rapid and better performance could obviously alter this conclusion radically in the near future.

At the end of May 1916, *L30*, the first of the 'super-Zeppelins', had been completed and she became available for operations in July. She was very nearly twice the size of the *L10* type, with a ceiling of 17,700 feet and a useful lift of 30 tons. Her six engines gave her a speed of 62 m.p.h. Although scouting had never ceased in theory to be the primary role of the naval airships, this type was produced entirely with raiding in mind and could carry 3½ tons of bombs for 1200 miles. Nevertheless, it was very formidable in naval warfare too. The *L30* class could reach Scapa Flow, which raised the possibility that the Grand Fleet might be bombed in harbour by day, a threat against which it had no counter measure. The new Zeppelins could fly too high to be hit by anti-aircraft gunfire and even the latest fighter aeroplanes would take an hour to climb up to them and at that altitude had little margin of speed. Even if it was not attacked, the Grand Fleet could now be shadowed whenever it left harbour and its best seaplanes, the Sopwith Babies, would be unable to climb high enough to shoot the shadower down. Nine more of these formidable airships were completed during the second half of 1916, so that the threat was considerable.

The Zeppelin reconnaissances, which now extended as far as the Norwegian coast and the Moray Firth, were interrupted at the end of July and in early August for four raids on the United

Kingdom. These raids were opposed by the usual sea-going patrols. Aircraft from the R.N.A.S. at Yarmouth and the sea-plane carrier *Vindex* fired at Zeppelins, and the light cruiser *Conquest* and two trawlers engaged others with their guns, but they all failed to bring one down. Nevertheless it was an improvement that they were now at least making contact.

In the middle of August Admiral Scheer took steps to renew his plan to bombard Sunderland. As before, the High Seas Fleet was to work in close co-operation with U-boats and Zeppelins. Admiral Scheer was determined that he would not be surprised by superior forces again and it was the primary task of the Zeppelins to ensure that he was not. He deployed eight Zeppelins to hold the ring and warn him of any hostile forces which entered the area of operations (see Fig. 3, p. 60). Their patrols were co-ordinated with five lines of U-boats and with minefields. He stationed *L30*, *L32*, *L24* and *L22* on a patrol line from Peterhead to Norway to give him warning if the Grand Fleet left Scapa Flow for the southwards. He placed *L31* off the Firth of Forth which was the base of the Battle Cruiser Fleet and stationed *L13* in the Flanders Bight to report any forces approaching from the south. Finally he put *L21* between the Humber and the Wash and *L11* in the bombardment area off Sunderland. *L14*, *L16*, *L17* and *L23* were held in their bases as reliefs.

At this time the air component of the Grand Fleet was practically the same as it had been two months earlier at Jutland. Two days after the battle, however, a two-seater scouting seaplane had flown off the deck of the *Campania* and in July exercises with the Grand Fleet with seaplanes used in this way had been encouraging. Two Coastal airships, *C4* and *C11*, had arrived at Howden and the seaplanes on the east coast, some now fitted with wireless with a range of sixty miles, began to make patrols farther to seaward.

After waiting for two days for suitable weather, Admiral Scheer put his plan into effect and the High Seas Fleet sailed at 21.00 on 18th August. The eight airships took off soon after midnight, the four northerly ones arriving on station by the afternoon of 19th August without sighting anything. In fact, the Grand Fleet, as usual warned of the impending movement by radio intelligence, had sailed before the High Seas Fleet during the afternoon of the 18th August. The Grand Fleet therefore passed through the intended Zeppelin patrol line at

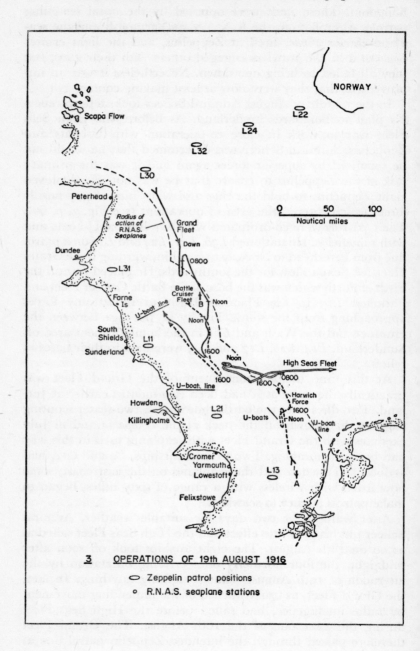

NORWAY

○ L22

○ L24

○ L32

○ L30

Scapa Flow

Peterhead

*Radius of
action of
R.N.A.S.
Seaplanes*

Grand
Fleet

Dawn

0800

0   50   100
Nautical miles

Dundee

○ L31

Farne
Is.

U-boat line

South
Shields

○ L11

Sunderland

Battle
Cruiser
Fleet

B   Noon

Noon

1600

Noon

1600   0800

High Seas Fleet

1600

U-boat line

Howden

Killingholme

○ L21

U-boats

Harwich
Force

1600

Noon

U-boat
line

Cromer
Yarmouth
Lowestoft

○ L13

Felixstowe

A

3   <u>OPERATIONS OF 19th AUGUST 1916</u>

     ○   Zeppelin patrol positions

     o   R.N.A.S. seaplane stations

night well before the Zeppelins arrived in their stations and so
was not sighted by them. At o6.oo, however, the southernmost
Zeppelin, *L13* saw the Harwich Force seventy miles east of
Lowestoft and reported its position giving the course as S.W.
which was accurate but temporary. (Posn. A on Fig. 3, p. 60.)
She was then forced up into the clouds by anti-aircraft fire. *L31*
also sighted part of the Battle Cruiser Fleet ninety miles east of
Farne Island but was engaged by the *Galatea* and also forced up
into low cloud. She reported the contact, giving the course as
N.E., again true but temporary. (Posn. B on Fig. 3, p. 60.)
Admiral Scheer knew from radio intelligence that the Grand
Fleet was at sea, but these two reports gave him the impression
that only small isolated forces were in the vicinity both unaware
of his presence and steering away from his intended track. As no
reports had been received from his northern Zeppelin patrol
line he assumed that the Grand Fleet was to the north of it.
L11 was covering the advance of Admiral Hipper in the van,
and so Admiral Scheer decided to keep to his plan and continue towards Sunderland.

In fact the High Seas Fleet was in considerable peril: the
Grand Fleet was rapidly approaching from the north and
would soon be able to force an action on it. The Grand Fleet
had, however, one serious disadvantage: it was practically
blind and did not know the position of its enemy. The *Campania*
had had to be left behind at Scapa with engine defects and only
the faithful *Engadine* with her two scouting and two fighting
seaplanes was with the fleet. The R.N.A.S. air stations on the
east coast had been alerted and at dawn two seaplanes from
Yarmouth flew across to the Dutch coast but saw nothing. The
eleven seaplanes at Killingholme and two at South Shields as
well as the Coastal airships at Howden made no searches to
seawards and so missed a chance to sight the High Seas Fleet
which came within 75 miles of them in daylight. It was
too rough to use the *Engadine*'s seaplanes and the only air
support in the Grand Fleet was the *Campania*'s kite-balloon
which had been transferred to the *Hercules* and was used at sea
from a battleship for the first time. The Grand Fleet was acutely
aware that the Germans knew its position because of the
constant sighting of Zeppelins, some of which, notably *L11*,
were shadowing.

L13 was soon in touch with the Harwich Force again and at
midday reported that it included battleships and was sixty miles

N.E. of Cromer. Admiral Scheer believed that this was the chance he had been waiting for and that he might be able to destroy an isolated portion of the Grand Fleet. He therefore abandoned the bombardment of Sunderland and turned S.E. to engage. This decision saved him. At the time the advanced screens of the two fleets were only 30 miles apart and the main bodies 65 miles. Although *L31* had actually reported the Grand Fleet as the 'main body' and *L11* and the U-boats had reported strong forces to the north, Admiral Scheer seems to have been completely unaware of the proximity of the Grand Fleet until about 16.00.

In the early afternoon the battleships *Hercules* and *King George V* of the Grand Fleet had *L11* and *L31* in sight and the Third Cruiser Squadron another Zeppelin, probably *L21*. Although the sea was still choppy, *Engadine* tried to get a Sopwith Baby away to attack them but it was too rough and she had to hoist it in again. Both fleets eventually returned to harbour without any surface contact being made. *L11* and *L31* screened the High Seas Fleet as it retired and at 17.30 all the Zeppelins were recalled to base.

The operations of 19th August are of more interest from the air point of view than the Battle of Jutland and illustrate the influence of aviation on fleets at that time. Admiral Scheer was disappointed with the Zeppelins and complained that they had not been entirely reliable. He said that they had proved negative and only showed where the enemy was not. The Germans subsequently considered that they had expected too much of them and that the hundred-mile-long patrol lines had been too great. Admiral Jellicoe, on the other hand, felt that his views, of many years' standing, on the need for airships for fleet reconnaissance had received striking confirmation. Every movement of every part of his fleet had been watched and Zeppelins were in contact off and on from six in the morning until six at night. The fact that their reports had been confusing and had not given a true picture of the situation to their Commander-in-Chief was, of course, unknown to him. The Zeppelins had indeed done very well. They had sighted all the main British forces involved and had relocated them frequently during the day, all without loss to themselves. The only really incorrect report was by *L13* when she thought she saw battleships with the Harwich Force. The German failure to obtain a clear picture of the situation was due more to a lack of experience of planning air reconnaissance and of working with aircraft: the interpretation of reports from aircraft was still in its infancy and the German command was probably more to blame than the Zeppelins.

In the six weeks following the Sunderland operation, the Zeppelins made six raids and they made another at the end of November. Against these raids the patrols at sea made contact on three occasions. The light cruiser *Conquest* hit *L13* with a 3-inch shell and she had to drop her bombs and return to base; and a Short seaplane and a Sopwith Schneider both attacked Zeppelins which they claimed to have damaged. But these skirmishes were nothing to the severe defeat which the Zeppelins suffered at the hands of the Royal Flying Corps over the United Kingdom. In the three months from October to November, five airships were shot down, four of them in flames. Except for *SL11*, an army Schütte-Lanz airship, all were 'super-Zeppelins' of the *L30* class. The development which had made this substantial victory possible was the combination of searchlights on the ground with patrolling fighters armed with machine guns firing incendiary ammunition. On 28th November the R.N.A.S. got their chance when three BE2c aeroplanes from Yarmouth, which had been trying to shoot down Zeppelins since the beginning of the war, at last succeeded and destroyed *L21* ten miles east of Lowestoft as she was returning from a raid. In addition to these casualties, five more German airships were lost by accidents and another in the Baltic during the autumn. No airships were brought down however while scouting at sea and this, in spite of the heavy raids, remained their principal occupation. During 1916 they had made 283 scouting flights against 202 raiding sorties. Although, therefore, the Zeppelin had received a sharp check as a strategic bomber, it was still supreme for scouting at sea.

The Grand Fleet had expected that British rigid airships for scouting would be available by the end of 1916. Of the eleven building, however, only *No. 9* was nearing completion. She was intended to be used for training and the Admiralty had required her to be specially strengthened for rough handling; this and other modifications reduced her theoretical disposable lift to 3·1 tons. When she first flew on 27th November, however, she could only lift just over two tons and the Admiralty refused to accept her. Subsequently they agreed to take her if Vickers could make her fit for the training role, but obviously there was a

considerable loss of confidence in British airship design and it
was clear that airships of the 23 class were unlikely to be of use
for operations. On 24th September, the German Zeppelin *L33*
had been shot down by anti-aircraft fire over the United Kingdom; she did not catch fire and was captured practically intact.
With a disposable lift of 30 tons the German lead in design was
so great that it was decided to copy her. Two of this type, *No. 33*
and *No. 34* were therefore authorized in place of *No. 28* and
No. 30 of the 23X class. As they were over twice the size of the
23 class, new sheds had to be built before they could be started,
and so they could not hope to be completed before the end of
1917.

Efficient air reconnaissance for the Grand Fleet was therefore some way off and they had to do the best with what they
had got. By December there were fourteen Coastal blimps
based on the east coast at East Fortune, Longside, Howden and
Pulham, but they only had the endurance to co-operate with
the Grand Fleet in the western half of the North Sea. Two of the
smaller S.S. Blimps were stationed at Caldale in the Orkneys
but they were only of use as an anti-submarine escort for the
fleet entering and leaving Scapa Flow. Experiments with towing were continued and in September the light cruiser *Canter-
bury* had towed *C1* at 26 knots and later successfully refuelled
her by pipe-line and changed the crew by ladder. On 30th
September *C20* from East Fortune made a patrol with the
Battle Cruiser Fleet and in October, after two light cruisers had
met coastal blimps from Longside to practice working together,
Fleet exercises confirmed their value. Plans for the airships to
work with the fleet began to be formulated and the experience
was expected to be valuable when the rigid airships came into
service.

In the southern North Sea, seaplanes and flying-boats began
to make longer patrols regularly. On 19th September, when
German destroyers came out as far as the Dogger Bank, seaplanes from Yarmouth were sent out to search for them, two of
them carrying torpedoes. On 23rd September a seaplane from
Yarmouth flew across to the Texel and back via Felixstowe. In
September, the *Campania* was out again with her seaplanes
exercising with the Grand Fleet and in October the section of
the battle orders dealing with her was rewritten. The most
significant change was that her station was altered from the
vicinity of the fleet flagship to a position twenty miles or so

ahead of the battlefleet in the centre of its light-cruiser screen. *Engadine* was moved even farther to the van and was now to lead the Battle Cruiser Fleet five miles ahead even of its light cruisers. The seaplanes were now to be used, whenever weather permitted, to scout up to fifty miles ahead for the enemy fleet; permission to use the seaplanes had to be obtained, however, from either the Commander-in-Chief or the Vice-Admiral Commanding the Battle Cruiser Fleet. This substantial advance in thinking was of course an attempt to offset the great advantage which the Zeppelins gave the High Seas Fleet. Unfortunately the means to implement it were no better than before. All depended on the weather, for unless it was practically flat calm, the seaplanes from the *Engadine* could still not be used at all, and although they could be launched from the *Campania* they would be lost on landing. The light cruisers therefore remained the eyes of the fleet in all but the very occasional periods of calm weather.

Suggestions to improve the efficiency of seaplane-carriers with the Grand Fleet had been coming in since 1915. In August of that year, two naval architects produced a design with a sloping stern to facilitate recovery. The Admiralty refused to build such a novel ship but eventually took over the *Manxman*, another cross-channel steamer, for conversion. She joined the Grand Fleet at the very beginning of 1917 and was simply an improved *Vindex*, fitted to take four reconnaissance seaplanes aft and four Sopwith Baby seaplanes forward. The Captain of the *Manxman* at once proposed that the Baby seaplanes be replaced with Sopwith Pup aeroplanes. He pointed out that these modern fighters, which had been in use in France for some months, had the performance to bring down even the latest Zeppelins. They had a ceiling of 17,500 feet and could reach it in just over half an hour and they would still be faster by 35 m.p.h. than a Zeppelin at that height. He also asserted that with flotation gear the chance of survival of an aeroplane on ditching in the sea was actually better, in all but a flat calm, than of a seaplane trying to land. After trials in January 1917 had shown that the Baby seaplanes could not fly off *Manxman*'s deck except in a strong wind his suggestion was adopted. Although aeroplanes had been flying off *Vindex* for over a year they had always hoped to land ashore, and this was the first use of them deliberately as 'one-shot' weapons with the fleet.

After the Battle of Jutland, the Admiralty had bought two large liners, which were building for Italy, for conversion to seaplane carriers of the *Campania* type. Subsequently only one was proceeded with and her original design was completely altered. She was to have had a flush deck to allow a very long run for trolleyed seaplanes with the possibility that they would be able to take off even carrying torpedoes. Her funnels were therefore trunked aft in horizontal ducts to give a clear deck. This ship, the *Argus*, therefore proved very suitable for conversion to a proper aircraft carrier in due course.

At the end of January 1917, Admiral Beatty, the new Commander-in-Chief, set up a committee to study air requirements for the Grand Fleet. Their first and most important recommendation was that strategic reconnaissance over the North Sea should be systematic and frequent whether the fleet was at sea or not. To date British aircraft had had too short a range to attempt this and the Grand Fleet relied mainly on radio intelligence and to a small extent on submarine patrols to tell them when the enemy was at sea. Ultimately this requirement would be met by the rigid airship programme, but the new N.S. blimps which were just starting their trials and the much larger Curtiss H12 flying-boats which were coming into service meant that a start could be made in the near future. Their second consideration was to provide air reconnaissance and air anti-submarine screening for the fleet whenever it was at sea and before contact with the enemy was made; for this they visualized the use of the Coastal blimps whose endurance could be extended by towing or refuelling them from ships and, in calm weather, the use of seaplanes from carriers. Their third consideration was the provision of aircraft during a fleet action and it was obvious that these would have to be carried in ships of the fleet. For shooting down Zeppelins, the aeroplane was unquestionably the best, while for spotting and action observation they considered that both seaplanes and kite-balloons were required. They estimated that twenty anti-Zeppelin aeroplanes and twenty reconnaissance seaplanes were needed to perform these functions.

The *Campania*, *Engadine* and *Manxman* in the Grand Fleet carried 24 aircraft between them. Of these, 4 were fighter aeroplanes, 8 were fighter seaplanes and 12 were reconnaissance seaplanes. The 4 aeroplanes in the *Manxman* could be used in almost any weather but were 'one-shot' weapons which would always be lost on landing. The 12 seaplanes in the *Campania*, half fighters and half reconnaissance, could be operated continuously in calm weather but also became 'one-shot' weapons in rough weather. The remaining 8 seaplanes in the *Manxman* and *Engadine*, which had to be hoisted out to take off, could not be used at all except in calm weather. The committee recommended that the 6 fighter seaplanes in the *Campania* should be replaced with Sopwith Pup aeroplanes as soon as possible.

The Committee pointed out that the provision of new seaplane-carriers was urgent not only to carry the balance of sixteen aircraft which they had recommended, but to replace the *Campania* which was 24 years old and becoming unreliable, and the *Manxman* which had proved too slow to work with the fleet. The Admiralty had, in fact, just had the new construction programme under consideration and, after some discussion, had approved four new seaplane-carriers, two to be large and two small. The *Argus*, the first of the two Italian liners, could not however be finished before the end of 1917 and the Committee suggested that the light battle-cruiser *Furious*, which was nearing completion, should be converted to a seaplane-carrier (see Fig. 6, p. 106). This ship was one of Lord Fisher's controversial designs, very fast, armed with heavy guns but practically unarmoured. After Jutland, this type of ship was out of favour and there were many who doubted the value of the main armament of the *Furious* which was only two 18-inch guns. Her speed of 32 knots would obviously be very useful for a seaplane-carrier, but clearly the forward 18-inch gun would have to be sacrificed for a flying-off deck. After much argument and a staunch rearguard action by the gunnery world, this was finally agreed upon and the conversion was put in hand.

The Committee also suggested that the anti-Zeppelin aircraft in the fleet could be increased by fitting flying-off platforms in some light cruisers so that they could carry a Sopwith Pup aeroplane. This was accepted and H.M.S. *Yarmouth* was ordered to be fitted with one for trials. The Committee also recommended that kite-balloons should be carried in the fighting ships themselves rather than in the *Campania* and that a base for them should be set up ashore from which they could be transferred as required. Admiral Jellicoe had asked for twelve more kite-balloons after Jutland and these were now arriving; battle-ship and battle-cruiser flagships were given winches to fly them and two were ordered to be tried out in light cruisers.

In January a section to organize blimp co-operation with the Grand Fleet was added to the battle orders (see Fig. 4, p. 82). When the Fleet was in the western half of the North Sea the blimps from East Fortune and Longside were to be given a rendezvous and were then to scout 40 to 50 miles ahead and 30 miles on each side of the light cruiser screen. In March 1917 more towing trials, to try and extend their range so that they could accompany the fleet into the eastern half of the North Sea, established that the strain of prolonged towing was too great and a policy of refuelling and changing the crews was preferable.

The British rigid airship programme, which was still intended to be the ultimate answer to scouting in the North Sea, was going very slowly. *No. 9* was accepted by the Admiralty in April for trials and training after she had been substantially lightened by replacing two of her engines with a German Maybach taken from the wreck of the Zeppelin *L33*. The construction of the four airships of the 23 class was progressing but they were not expected to fly until the autumn; they were a year late and it was already clear that they were going to be over-weight and of very limited value. The Admiralty, however, were still determined to force on with the rigid airship programme and in January ordered three more of the type copied from the Zeppelin *L33*.

These attempts to improve the air support of the Grand Fleet came at a time when the strategy in the North Sea was undergoing a considerable change. Admiral Scheer had given up any idea of cutting off a part of the Grand Fleet and defeating it in detail, and believed that the only hope for the future lay in an unrestricted U-boat campaign against commerce. He considered that the role of the High Seas Fleet was now entirely auxiliary to the U-boats and that its function was to defend the U-boat bases, safeguard their entry and exit and to raid the Dover Barrage. The aim of the German naval air service was now to co-operate in this strategy. In the Heligoland Bight whenever weather permitted three airships patrolled to seaward of the minesweepers to give warning if British ships tried to interfere with them. At the same time Zeppelins were used to try and spot mines from the air and they continued to harass British submarines in the Bight. Long-range daylight scouting flights to find merchant ships were also made into the U-boat blockade area off the east coast of the United Kingdom and as far north as the Moray Firth. The German seaplane section of their naval air service had increased in numbers substantially and had also improved in quality. Seaplanes with a 200-h.p. engine were able to carry better radio and two machine-guns as well as an observer. Their tasks in the Bight were unchanged by the new strategy of the High Seas Fleet and operations continued from Borkum, Norderney, Heligoland and List.

The Zeppelins, licking their wounds after their defeat in the raids of the previous autumn, were in the first half of 1917 used mainly for scouting. They made only four raids totalling sixteen sorties and lost another two Zeppelins in the process, both shot down in flames. They were, however, by no means satisfied with their auxiliary role and were determined to make a 'come back' as strategic bombers. Captain Strasser was convinced that this could be done by flying higher. The German Admiralty's plan for faster airships was therefore set aside and steps taken to improve their ceiling instead. By reducing the number of engines and machine-guns and various other measures to lighten her the new Zeppelin *L42* got up to 19,700 feet on her trials in March and similar alterations were made in all new airships. These measures, however, affected adversely their efficiency as scouts because they lost performance at low altitudes. Throughout the first half of 1917, Zeppelin strength remained at about 14 airships. Captain Strasser made a case for a fleet of 30 of them, 24 for the North Sea and 6 for the Baltic, a case made entirely on the grounds that they were needed for scouting although there is little doubt that he hoped to use them mainly for raiding.

In April, some H12 Curtiss 'Large America' flying-boats, which had been under development since 1916, came into service at Felixstowe and Yarmouth. These flying-boats were a great step forward and were superior to any seaplane. They were considerably larger, could operate in rougher water and, with an eight-hour endurance were able to make scouting flights half-way across the North Sea. The British radio intelligence always knew when Zeppelins were out scouting and had a very good idea of where they were. Moreover the new flying-boats could reach the position in which the southernmost Zeppelin

F

usually patrolled. In April, therefore, it was decided to send out a flying-boat whenever a Zeppelin was known to be in this area. On 14th May an H12 flying-boat from Yarmouth shot down *L22* at about 4000 feet off Terschelling while she was screening the German minesweepers. A week or two later, however, both *L40* and *L46* escaped a similar fate by climbing rapidly above the flying-boat's ceiling. Nevertheless on 5th June a flying-boat from Felixstowe caught *L43* at 1500 feet off Vleiland and destroyed her. At the end of June, Zeppelins on patrol were ordered to fly at 13,000 feet all the time, to keep a sharp lookout and, if attacked, to climb and call up fighter seaplanes from Borkum.

The advent of the flying-boat therefore first seriously challenged the Zeppelin as a scout even in its own waters of the Heligoland Bight. On the face of it, it seems strange that such a large and unwieldy aircraft should have been so successful as a fighter. The fact was, however, that any aircraft which could get within range with a machine-gun firing incendiary ammunition could shoot a Zeppelin down. The success of the flying-boats was mainly due to surprise and excellent radio intelligence but as soon as the Germans realized what was happening they took the counter-measures described above. These counter-measures, however, had their price. Scouting from high altitude was much less effective even in good weather and it was useless in bad weather: anti-submarine operations and mine spotting could not be attempted at all. In any case a rapid climb to maximum altitude meant the expenditure of nearly all a Zeppelin's ballast and could only be done once a patrol and then the airship had to return to base. The flying-boats, therefore, even if they did not succeed in shooting down any more Zeppelins, had secured a substantial success.

Captain Sueter, the Director of the Air Department in the Admiralty, had never ceased to look for ways to attack the enemy fleet from the air in its harbours. He had ordered the development of the Short 320 seaplane of greatly increased endurance and able to carry the larger 18-inch torpedo for this purpose and had stated a requirement for a torpedo-carrying aeroplane which was being produced by Sopwith. When the Short 320 came into service it was still unable to reach the bases of the High Seas Fleet from the United Kingdom and was too large to operate from existing seaplane carriers. The Austrian Fleet in the Adriatic on the other hand was a practicable target as the torpedo planes could be launched to attack Pola from an advanced base at Venice. The Admiralty decided that this was a better objective in the immediate future than the High Seas Fleet and took steps to build up a force of Short 320 torpedo seaplanes at Otranto.

Another way to attack the High Seas Fleet was conceived by the flying-boat protagonists at Felixstowe; it was to use Large America flying-boats, which could carry two 230-lb bombs each. They had not the endurance to make a return trip by themselves either and the plan was for destroyers to tow them across the North Sea on skids or lighters. These skids could be trimmed down to float-off the flying-boat which would then take off from the sea: the Large America would then have the range to attack and return to base on its own. Four experimental lighters were built at the beginning of 1917 and carried out successful trials in June. Yet another method of attack was conceived in 1917 when the new Handley Page bombers came into service: they could carry sixteen 112-lb bombs or double the load of the Short bombers which were the largest that had operated so far. They had still not got the range to reach the German Fleet bases but on 9th July, one of them which had been sent to the Aegean for the purpose, was able to attack the *Goeben* in Constantinople; although she dropped eight 112-lb bombs from 800 feet and hit her, the armour deck was not penetrated and they failed to cause any damage.

By 1917 there was considerable agitation in the Press and in Parliament in Britain for the R.N.A.S. and R.F.C. to be amalgamated into a separate air force. One of the main reasons for this was the overlapping and inefficiency of the supply of aircraft. Nevertheless relations between the two air services seem to have been good; better in fact than between the R.N.A.S. and the Admiralty on the one hand and the R.F.C. and the War Office on the other. The daylight raid on London by German Gotha aeroplanes in June 1917 brought matters to a head and made it clear that a separate air force which could give its whole attention to the problem of air defence and which would not be dominated by the specialist needs of either the Army or the Navy was urgent. The Admiralty was strongly opposed to a separate air force but Captain Sueter, who had directed the R.N.A.S. practically continuously since 1908, was an unapologetic protagonist. Early in 1917 the Admiralty decided to rid themselves of a difficult subordinate and appointed Captain Sueter to command the R.N.A.S. in the Adriatic.

In July 1917 the *Furious* joined the Grand Fleet (see Fig. 6, p. 106). She was a formidable surface warship with one 18-inch and eleven 5·5-inch guns, and as an aircraft carrier she was a considerable advance on the *Campania*. She was ten knots faster, she had anti-torpedo bulges which made it safer for her to slow down to pick up seaplanes, and she could catch up the fleet again quickly. Her flying-off deck forward was 228 feet long and 50 feet wide and any seaplane or aeroplane of the day could take off. The capacity of her hangar under the flying-off deck was ten aircraft and her first air group consisted of six Sopwith Pup aeroplanes which, being unable to land on, were one-shot, and four two-seater Short seaplanes. A month before the *Furious* joined, the light cruiser *Yarmouth* had returned from refit with a flying-off platform for a Sopwith Pup aeroplane. The platform extended over the conning tower and the forward gun and did not have to be unrigged to clear for action. Flying-off trials were successful and it was decided to fit an improved platform in four other light cruisers. In the autumn, two new cross-channel packet type carriers, the *Nairana* and the *Pegasus*, joined the Grand Fleet to relieve the *Manxman* and the *Engadine*. They were similar to the *Manxman* but were converted whilst still building and embodied many minor improvements. They carried four or five Sopwith Pup aeroplanes forward and four reconnaissance seaplanes aft. The Grand Fleet now carried a total of 22 Sopwith Pup aeroplanes in its four carriers and the *Yarmouth* and eighteen seaplanes. In all but calm weather, however, when the seaplanes could take off and land on the sea, it was entirely a 'one-shot' force.

In July 1917 the Grand Fleet carried out a number of exercises with aircraft. Five of the Coastal blimps took part and proved useful for scouting but an attempt by the light cruiser *Phaeton* to refuel *C15* was a complete failure and the airship had to be ripped.[3] Further exercises in September again showed their value, nevertheless it was also clear that they had serious limitations. It was found that they were by no means always able to join the fleet as fog and squally weather often prevented them leaving their bases at Longside and East Fortune.

Experience showed that these small non-rigid airships had not the strength or endurance to operate all the year round in the North Sea, moreover they would not be able to defend themselves against enemy aircraft in action and would need fighter protection if they were to survive. In a strong head-wind their speed proved scarcely enough to keep up with the fleet and it was found difficult to control their movements when wireless silence was in force. Unhappily the new NS type, which had been specially designed for fleet work, were having a great deal of mechanical trouble. *NS1* had already flown a distance of 1,500 miles and so had plenty of endurance, but these airships were at present only reliable enough for anti-submarine patrols.

In these same fleet exercises an acute controversy developed over the use of kite-balloons. Thirteen capital ships and four light cruisers were now fitted with winches to fly them. Although in good visibility the balloons greatly extended the vision of the light cruisers, this was a two-edged weapon as the balloons could also be seen by the enemy, thus giving away the position of the fleet. When used for action observation and spotting, for which their value was not disputed, it was also feared that they gave the enemy good ranging marks. The Commander-in-Chief emphasized that they must be used properly and must be hauled down if they were likely to help the enemy more than the Grand Fleet.

The aircraft section of the Grand Fleet battle orders was again rewritten in August 1917 and showed a considerable advance in the theory for the use of aircraft. At sea the blimps were to scout ahead as before, but if they could not join the Fleet for any reason, the seaplanes from the seaplane carriers, weather permitting, were to do this duty. It was, however, noted that as aircraft were likely to be ineffective in bad weather they must be considered as supplementary to, and not a substitute for, the light cruisers. If enemy aircraft were sighted at any time at sea, fighters were ordered to be launched at once so that they could shoot them down before they located the Grand Fleet. The fighters were also charged with the protection of their own reconnaissance seaplanes and airships and, for the first time, it was emphasized that they must establish air superiority over the battle area. In action, five seaplanes were to be airborne to observe the movements of the enemy battlefleet particularly if smoke screens were made, but spotting was to be left entirely to the kite-balloons.

The German High Seas Fleet did not put to sea during the second half of 1917 and so the improved air strength and organization in the Grand Fleet were not put to the test. British mining in the Bight against the U-boats continued and the fields moved progressively outwards drawing the German minesweepers and their escorting ships and aircraft farther to seawards. The *Furious*, escorted by light cruisers and destroyers, made frequent sweeps to the edges of the minefields and to the Danish coast. On 11th September, a Pup was launched after a Zeppelin but the airship was able to climb into clouds and escape and the Pup had to 'ditch' in the sea. Enemy aircraft were sighted on other occasions but they were generally able to get away before the fighters could catch them.

On 21st August the *Yarmouth* with her Sopwith Pup was at sea with the First Light Cruiser Squadron escorting a mine-laying expedition and success was at last achieved. The force was sighted off the Danish coast by both a German seaplane from List and the Zeppelin *L23* which was on routine patrol. The British turned to the northwards for half an hour to entice the Zeppelin away from her base and then turned into the wind and launched the fighter. The Sopwith Pup shot down *L23* in flames at a height of 7,500 feet and then 'ditched' near the destroyer *Prince*. The pilot was saved and the engine and machine-gun salvaged from the plane. This was the third Zeppelin shot down in three months whilst scouting and it was significant because, although *L23* happened to be caught at low altitude, the modern shipborne fighter could have shot her down even if she had climbed to her ceiling. This the flying-boats in the southern part of the North Sea would not have been able to do; they had had a number of contacts with Zeppelins since their first successes but in every case the airships had been able to climb above them and escape. In September the flying-boats tried working in company with a DH4 aeroplane, which had an endurance of six hours and a ceiling of 17,500 feet. The flying-boat was there to do the navigation and rescue the crew of the fighter if it came down in the sea. Zeppelins were sighted twice but on both occasions the DH4 had engine trouble and could not get high enough.

In October the limitations of the British reconnaissance in the North Sea were confirmed when the new fast German light cruisers *Brummer* and *Bremse* attacked a Scandinavian convoy. Although the British were warned by radio intelligence that something was afoot and deployed thirty cruisers and 54 destroyers on patrol, the German ships evaded them all and did a great deal of damage. The *Brummer* and *Bremse* in fact never came within range of any British aircraft operating from shore bases. The *Furious* was at sea but the weather was too rough for her to recover seaplanes, and her aeroplanes, which could of course only be used once, were kept back in case Zeppelins were sighted.

British radio intelligence generally gave warning that some enemy movement was impending but it seldom gave much indication of what was actually afoot. Better air reconnaissance right across the North Sea was therefore urgent. The Admiralty still believed that the answer lay in the rigid airship and they had the support of Admiral Beatty. The R.N.A.S., however, were not now so sure. The loss of the three Zeppelins *L22*, *L43* and *L23* while scouting showed how vulnerable they were and some British senior officers believed it was only a matter of time before airships were driven out of the North Sea altogether; even the airship protagonists were beginning to believe that their future lay in extending air anti-submarine operations into the western approaches rather than in co-operating with the fleet. The realization of either use for British rigids was, however, still a long way off. The first three ships of the *23* class were delivered during the autumn but they suffered from the usual malaise and were far too heavy. Their useful lift proved to be just over 5 tons. As a result their performance was no better than the Coastal Blimps and they could get only half-way across the North Sea. Their ceiling of under 3000 feet in any case made their operation anywhere near German fighter seaplanes plain suicide; they could therefore only be used for training, trials and a few local anti-submarine patrols.

In November 1917, of the nine rigid airships building, the last of the *23* class and two ships of the slightly improved *23X* class were to complete early in 1918 but were clearly going to be of little use. The two ships of the Schütte-Lanz type were likely to be much better but they were not expected before the middle of 1918. Hopes were really centred on the four airships of the *R33* type[4] copied from the Zeppelin *L33* which were also expected in the middle of 1918. The Admiralty persevered and in November approved a new rigid airship programme which consisted of eleven more of the *R33* class; but there is little doubt that airships were rapidly being overtaken by aeroplanes and flying-boats for use in maritime warfare.

The Germans were also having misgivings about rigid airships. Captain Strasser had greatly improved the Zeppelins' chances as raiders by insisting on a high ceiling. The new *L53* type, of the same size as *L30*, but very much lighter and with special high altitude engines, began to appear in the autumn and could reach well over 20,000 feet. There were, however, only three raids during the second half of 1917. The last of these attacks, known as the 'silent raid' was a major disaster for the Germans, five Zeppelins out of eleven being lost. It was the weather which did the damage and not the British but it was nevertheless a severe setback for the Zeppelin as a bomber. General Ludendorff had, in fact, already suggested that no more airships should be built and that the available aluminium and rubber should be used to increase aeroplane construction. Admiral Scheer opposed this policy vigorously and maintained that they were still essential for scouting for which he needed eighteen in the North Sea. The Kaiser supported him but nevertheless airship building was cut down, the Army airship service was disbanded altogether and the Baltic naval airship detachment was broken up before the end of the year. The Germans, after the losses during the summer, had also begun to lose confidence in the Zeppelins as scouts. Their twin-engined torpedo seaplanes, when fitted with long-range tanks, had a ten-hour endurance and could already make scouting flights to 100 miles west of Borkum; some of these machines stationed at Norderney began to take over patrols formerly done by Zeppelins. Oddly the Germans had no success with flying-boats and discontinued development of them in 1917. They had, however, produced very successful fighter seaplanes, the best of which were the two-seater monoplanes of the Brandenburg type; some of these machines were stationed at Norderney and Borkum and began to oppose the British flying-boats in the Terschelling area. Fortunately for the Zeppelins, except for the 'silent raid', their casualties in the second half of 1917 were light. At the end of the year there were still eleven operational German airships in the North Sea as well as fifty-odd seaplanes of various types.

Soon after the *Furious* commissioned, her pilots believed that it would be possible to land a Sopwith Pup on her forward flying-off deck. This feat, which involved sideslipping round the funnel before landing, was successfully accomplished on 2nd August. It was repeated next day but the aircraft crashed over the side on the third attempt and the pilot was drowned. Nevertheless, the principle had been demonstrated and was clearly the key to the operation of aircraft with the fleet. If aeroplanes could be recovered in this manner, they would no longer be 'one-shot' weapons with all its waste and danger. They would be able to operate continuously at sea and would be practically independent of the weather.

The landing of the Sopwith Pup on the *Furious* completely altered carrier policy and no more pure seaplane carriers were ordered. It was decided to remove the after turret in the *Furious* and fit a separate landing deck aft and, in October, she was sent back to the shipyard for this to be done. At the same time the Admiralty decided that the designs of the *Argus* and another new aircraft carrier, the *Hermes*, should be altered to permit aeroplanes to land on aft. The alterations necessary for the *Argus* were small but the idea of a sloping stern originally proposed for the *Hermes* was dropped. Neither of these carriers could be available for some time: *Argus* was only launched in December and *Hermes* hardly begun, and so the cruiser *Cavendish*[5] building in Belfast was ordered to be fitted with a flying-off deck forward and a landing deck aft. In October another albeit far less important advance took place when a Sopwith Pup was flown off a platform on a turret in H.M.S. *Repulse*. This meant that the fighter strength of the fleet could be greatly increased as all capital ships could now carry them. The operation of flying-off would be simple as only the turret need be trained into the wind and the fleet would not have to alter course.

In July 1917, the Sopwith Cuckoo torpedo aeroplane first flew. When Admiral Beatty heard of this type he asked for 200 to be produced for the Grand Fleet as soon as possible. Up to now the Grand Fleet had shown little interest in torpedo planes and it seems likely that the activity of German torpedo seaplanes in the southern North Sea was partly responsible for this change of heart. It could well be only a matter of time before the Grand Fleet was faced with this type of attack and although the counter measure was clearly to have more fighters, it would be most unwise when the enemy had them for the Grand Fleet not to have torpedo planes as well. At the time a great deal of thought was being given to the problem of bringing a reluctant High Seas Fleet to action and torpedo planes might be able to attack and slow the enemy down so that the Grand Fleet could catch up. A further reason for the change of heart was probably that the Cuckoo was an aeroplane and not a seaplane: it could therefore be launched in any weather from the flying-off decks of the larger carriers. The Admiralty concurred with Admiral Beatty's request but halved the number he had asked for and ordered 100 Sopwith Cuckoos in September.

In September the trials with lighters to carry flying-boats made progress and it was found that they could be towed by a destroyer at 32 knots. The original plan to use them for a sustained bombing attack on the High Seas Fleet bases still stood and was enthusiastically supported by the U.S. Navy which agreed to join this offensive with 40 flying-boats and 30 lighters which were to be based at Killingholme and be ready by March 1918. The Admiralty, however, were much more interested in these lighters as a way to extend air reconnaissance right across the North Sea. The new F2a flying-boats, just coming into service, were an all-round improvement on the H12, especially in seakeeping qualities, but their endurance of 10 hours at 60 knots was still not enough.

The British had made great advances in fleet aviation during 1917. Although they were still committed to the development of the rigid airship and were making strenuous efforts to catch up with the Germans, this was not without misgivings. The Germans, in fact, were rapidly losing confidence in the Zeppelin for reconnaissance and were turning to long-range seaplanes. The blimp was in use with the Grand Fleet as a temporary measure, but it had serious limitations. The seaplane and the seaplane-carrier were about to be replaced entirely by the aeroplane either carried in aircraft-carriers on which it was hoped that it would be able to land or in ships as a 'one-shot' weapon. For use from the shore, the seaplane was giving way to the longer ranged flying-boat which would clearly rival the rigid airship before long. In function the need for the aircraft of the fleet to be used continuously in daylight for reconnaissance ahead of the light-cruiser screen was recognized and in action they were to gain air superiority over the fleet so that spotting, action observation and tactical reconnaissance would be unimpeded. Finally interest was being shown in the use of torpedo aeroplanes as a means to bring the High Seas Fleet to action.

.   .   .   .   .   .

On 5th January 1918 the Zeppelins suffered another major disaster when five of them were burnt in an accident at their main base at Ahlhorn, leaving only six for operations over the North Sea. They persevered with scouting in March but in April and May, *L61* and *L56* only escaped destruction by flying-boats by climbing to well over 20,000 feet: operations became increasingly difficult for them and they practically had to abandon the Terschelling area and leave it to seaplanes. They took to flying the longer scouting trips at night and not surprisingly saw nothing. Weather prevented them co-operating effectively with the High Seas Fleet in its last sortie in April, and in May they lost *L62* when she exploded in the air over Heligoland. The German seaplanes were, in fact, already able to do much of their work; there were now 124 of them in the North Sea including a number of the Gotha twin-engined machines and a new Dornier Rs III with four engines which had an endurance of ten hours.

In January 1918 there were still only fourteen of the Large America H12 flying-boats at Felixstowe, Yarmouth and Killingholme for reconnaissance over the North Sea, but their numbers increased during the year and in February they were joined by the new F2a of greater endurance. The H12 could reach Terschelling and the Dogger Bank but the F2a could go as far as the mouth of the Ems. Scouting was still by no means regular or systematic; sorties were flown only when ordered by the Admiralty who relied mainly on radio intelligence and sent them out accordingly. The sole movement of the High Seas Fleet during this period was when it made its final sortie in April and then it did not come within the range of these aircraft. The flying-boats continued to be used a great deal for anti-submarine patrols and in 1918 they had many fights with the German fighter seaplanes from Borkum and Norderney. On 4th June for instance five flying-boats took off looking for trouble and were engaged by fourteen seaplanes off Terschelling, two aircraft being lost by both sides. The range of the flying-boats was on occasion extended by towing them on lighters, as when, for instance on 18th May, a reconnaissance was made as far as Horns Reef in this way.

The Anglo-American bombing offensive by flying-boats on lighters in the North Sea never took place. The secret was given away when some of them were seen by Zeppelins in the Heligoland Bight and in any case the flying-boats were of more use against the U-boats. The Admiralty had never been very keen and refused to spare the destroyers to tow the lighters. The scheme was therefore dropped and the lighters were used to extend reconnaissance and carry fighters for the Harwich Force.

The *Furious* rejoined the Grand Fleet with an after flying-on deck in March (see Fig. 6, p. 106). She now carried sixteen aeroplanes and her air group was varied between Sopwith Pup and Camel fighters and Sopwith '1½ strutters' which were two-seat reconnaissance machines. Aircraft, however, found it almost impossible to land on her deck through the funnel gasses and disturbed air abaft the bridge. Only three successful land-ings were made out of a large number of attempts and it was decided in the end that it was impracticable. A clear deck the whole length of the ship was obviously needed and the *Argus*, which by now had grown a bridge structure amidships, was ordered to be modified once again. The *Hermes* and the Chilean battleship *Almirante Cochrane*, which had just been taken over for conversion to an aircraft-carrier, were also to have unobstructed flight decks. In spite of this setback the *Furious* and a number of light cruisers carrying fighters were used during the summer in the hope that they would be able to shoot down Zeppelins, but they had no success. In June, therefore, it was decided to take the offensive again and attack the Zeppelin base at Tondern.

On 1st April 1918, the Royal Air Force had been formed. At first this made very little difference and co-operation went on much as before. The Admiralty still controlled operations and there was no change of policy in areas which were purely maritime. Nevertheless all operations were now joint between the Royal Navy and the Royal Air Force to which latter service all aircraft now belonged.

On 19th July the *Furious*, escorted by light cruisers and destroyers and covered by a division of the battlefleet, flew off seven Sopwith Camels from a position eighty miles to the north-west of Tondern (see Fig. 4, p. 82). They carried two 50-lb bombs each and hit the hangar, destroying the Zeppelins *L54* and *L6o* which were inside. Only two of the Camels got back to the *Furious* and they had to 'ditch' in the sea: the others either crashed in the sea or force-landed in Denmark. Nevertheless, this was a conspicuous success and achieved more than all the former attempts by seaplane carriers put together.

On 11th August the Harwich Force of four light cruisers and thirteen destroyers was in the Terschelling area to launch six

Coastal Motor Boats to make an attack on German ships in the Ems. The operation was supported by four flying-boats from Yarmouth and three more were towed on lighters by destroyers. Two more destroyers towed lighters on which were Sopwith Camel fighters, which were able to fly off when towed at high speed. The Zeppelin *L53*, which was on patrol, sighted and closed this force flying high. After drawing her to seawards for a bit, the Camel from the lighter towed by the destroyer *Redoubt* was launched; it took half an hour to climb up to the Zeppelin which was flying at 18,300 feet, and this was near the Camel's ceiling. Nevertheless, she got within range and shot *L53* down in flames. The Camel was rescued but the German seaplanes from Borkum and Norderney destroyed all six of the C.M.B.'s. In this operation, therefore, aircraft did most of the fighting.

Six days earlier, three Zeppelins had made what was to be the last raid on the United Kingdom. They approached in daylight and were waiting for darkness before crossing the coast. They were seen and thirteen aeroplanes went up at once from Yarmouth and adjoining airfields. A DH4 shot down *L70* at 17,000 feet and seriously damaged *L65* which had to return home at once. The loss of *L70* was a double disaster for the Germans. Not only was she of a new type capable of 81 m.p.h. of which much was expected, but Captain Strasser, the energetic and forceful Commander of the Naval Airships, was on board. Admiral Hipper, the new Commander-in-Chief of the High Seas Fleet, then ordered that seaplanes were to make all the routine flights in the Bight and that the Zeppelins would only be used for scouting when it would be of 'unusual value'. This was in fact the end of the Zeppelin for reconnaissance in the North Sea and the seven remaining airships did little for the rest of the war.

In the summer of 1918 there were six British rigid airships at bases on the east coast of the United Kingdom, but all were of the *23* or *23X* types and were useless for operations. *R31* of the Schütte-Lanz type was the only improved airship to fly before the Armistice and *R33* and *R34* copied from the German Zeppelin *L33* were not completed until 1919. In September *R29* sighted an oil slick left by *UB.115* and called up the destroyer *Ouse* which sank her and this was the only operational success that could be claimed by the whole of the British rigid airship programmes during the war. The Grand Fleet therefore never secured their co-operation and so had little effective long-range

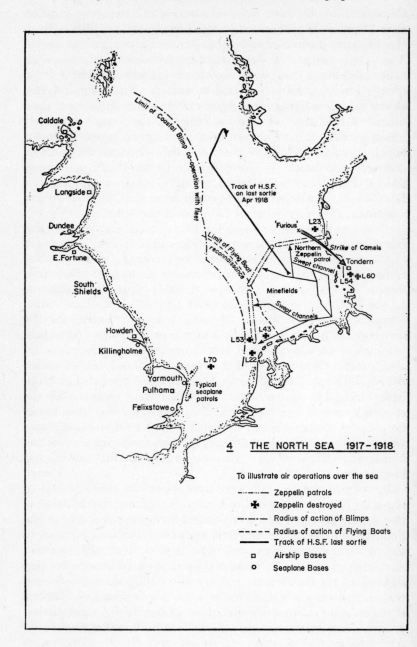

4   THE NORTH SEA   1917–1918

To illustrate air operations over the sea

- ·—·—· Zeppelin patrols
- ✳ Zeppelin destroyed
- ·—·—· Radius of action of Blimps
- — — — Radius of action of Flying Boats
- ——— Track of H.S.F. last sortie
- ▫ Airship Bases
- ○ Seaplane Bases

air reconnaissance right up to the end of the war. The Coastal blimps were no substitute; their co-operation was seldom possible and would in any case only be in the western half of the North Sea.

The failure of the *Furious* as an aircraft-carrier was a great disappointment and the result was that she could only use her aeroplanes as 'one-shot' machines. The seaplane had now practically disappeared from the fleet: the *Campania* was used for training and the *Nairana* had been sent to North Russia in July, leaving only four Short machines in the *Pegasus*. The number of aeroplanes carried in warships of the fleet was, however, very much greater than before. Twelve light cruisers carried a fighter, some on a new type of rotating platform and practically every battleship had two fighters on her turrets. Eleven battle-cruisers had a two-seater Sopwith '1½ strutter' reconnaissance machine on one turret and a fighter on another. Eighteen battleships, seven light cruisers and three battle-cruisers had kite-balloons. This meant that the Grand Fleet had some 90 fighters, mostly Sopwith Camels, and up to 20 reconnaissance aeroplanes, 4 seaplanes and 28 kite-balloons.

When the Grand Fleet put to sea in the final phase of the war it therefore carried with it a sizeable air force but it could not use it until action was joined. If the weather permitted, which was hardly ever, the four seaplanes could be used to scout ahead, otherwise reconnaissance on passage depended on a screen of light cruisers which flew their kite-balloons at their maximum height to increase their range of visibility. While the fleet was cruising, all the aeroplanes remained on board the ships ready for use and the kite-balloons in the battleships, which were for spotting and action observation, were kept hauled down. If the enemy had been sighted the orders were that the aeroplanes were to be used without hesitation and in sufficient numbers to ensure success. There is little doubt that the Grand Fleet now had sufficient fighters to deal with Zeppelins or any other enemy reconnaissance aircraft and to secure air superiority over the battle area. It could have defended itself against torpedo seaplanes if these had been used by the Germans, and ensured that its twenty-odd reconnaissance aeroplanes and the kite-balloons would be unmolested. On the other hand it could not do this for long as none of the aeroplanes could be used more than once and the few seaplanes could not be used at all except in calm weather.

In October, two weeks before the Armistice, the aircraft-carrier *Argus* joined the fleet. She had a completely unrestricted flight deck 565 feet long and 68 feet wide and could carry twenty aircraft in her hangar below it. Here at last was the means to operate aircraft with the fleet continuously and to do so, within reason, whatever the weather. She could carry Sopwith Camel fighters and '1½ strutter' reconnaissance machines and had Sopwith Cuckoo torpedo planes in her air group, but she was too late for the war and it took some years before she was fully evaluated and aircraft could operate from her as a matter of routine.

IV

Aircraft in the Defeat of the U-boats and in the Narrow Seas
1914–1918

THE SUBMARINE, like the aircraft, was a new factor in naval warfare in 1914. Initially it had greater success and on 18th February 1915 when the U-boats turned to an attack on commerce, it was obvious to the British that anti-submarine measures were urgent.

British aircraft in home waters had up to now achieved little against the U-boats. The seaplanes on the east coast fought more against their maintenance troubles and the weather than the enemy, and their patrols sighted practically nothing. The three British airships from Kingsnorth patrolled the Thames Estuary and the Channel coast as far west as Worthing but also without result. It was therefore decided to use aircraft in a concerted effort to bomb the German U-boats out of their bases in Flanders. A number of aeroplanes and seaplanes from the east coast including the two flying-boats from Felixstowe were therefore sent to Dover and Dunkirk to raid the U-boat bases at Ostend, Zeebrugge and Bruges, and also to attack Antwerp, where small U-boats were being assembled. Several raids were made in March and April but the very small bombs, mostly of 20 lbs, scarcely proved an irritant and totally failed to stop the assembly and support of U-boats in this area.

At the end of February 1915 Lord Fisher, at his wits' end for some way to counter the U-boats, had demanded the design of a small and cheap airship that could be mass-produced quickly. He wanted a large number of them and stipulated that they should be able to patrol for eight hours with a wireless set and 160 lb of bombs. He expected that they would be able to keep continually under observation those defiles through which the U-boat had to pass, such as the Dover Straits and the North Channel between Ireland and Scotland. With a speed of 40–50 m.p.h., he hoped, they would be able to surprise U-boats on the surface and sink them or at least summon anti-submarine vessels to the spot. His requirement was met in the remarkably short time of three weeks by slinging the fuselage of a B.E.2c aeroplane under the spare envelope of the No. 2 Willows airship. Fifty of these small airships or 'blimps'[6] as they were called, were ordered at once and production began in March although they did not become operational until the summer. Six had been completed by July and bases had been opened for them in the Dover area at Capel and Polegate in Sussex.

The German submarine campaign became more effective in the second half of 1915. In August merchant-ship casualties rose to 49 ships sunk which was the highest monthly total so far. The campaign was, however, concentrated in the south-western approaches where there were no British aircraft at all. On the east coast, where there were 35 seaplanes at five bases from the Thames Estuary to Dundee, anti-submarine operations generally took the form of patrols of the shipping channels from time to time; in June three seaplanes from Yarmouth made a concerted search of the Shipwash area for a U-boat which had been reported, but in general their operations were quite ineffective. There were 19 seaplanes at Calshot but there were no U-boats operating in the Channel at this time. The only aircraft on the west coast were some of the new S.S. blimps which arrived during this period at Luce Bay in Galloway, in Anglesey and at Barrow-in-Furness; their purpose was to help to prevent U-boats entering the Irish Sea through the North Channel and to harass them if they did. Here again no U-boats were found because they were passing to the west of Ireland and operating in the south-western approaches, and had temporarily abandoned the Irish Sea.

It had already been realized by the end of 1915 that the S.S. blimps were not fast enough to sink U-boats, which could always submerge with time to spare as they approached. Winston Churchill's prediction that they would only tease U-boats was to a certain extent borne out. Nevertheless, they certainly did tease them which was more than could be said for many of the other anti-submarine measures of the day. Whenever they were on patrol, the U-boats had to stay submerged and their operations were inhibited. The blimps were able to call up surface ships when they sighted a U-boat and steps were taken to improve this early air–sea co-operation: to try and compete with the U-boat sinkings in the south-western approaches, it was decided to use some of the new coastal airships and to establish a base for them at Pembroke.

In the southern North Sea and on the Belgian coast aircraft continued to be used for anti-submarine and other operations. By the autumn there were seventeen seaplanes based at Dover and Dunkirk and eleven S.S. blimps at Capel and Polegate and a new base at Marquise in France. Few U-boats were using the Straits of Dover at this time, but some of them were sighted and attacked off Zeebrugge and Ostend. Bombing attacks on the U-boat bases were made as opportunity offered, but the bomb loads were still very small and the bombers could not concentrate on the task as they were very busy attacking Zeppelin bases, enemy aerodromes and coastal batteries and co-operating with the army. In August and September, when monitors of the Dover Command were employed to bombard the U-boat bases at Zeebrugge and Ostend they were assisted by seaplanes from the *Riviera* to spot for them and the kite-balloon ship *Menelaus* was also available for this purpose. The fifteen small UB and UC-boats based in Flanders were, in fact, little inconvenienced by any of these counter measures and continued their operations without pause. Moreover, the R.N.A.S. increasingly had to attack the German seaplanes based at Zeebrugge whose strength by the end of the year had risen to fourteen machines. They were becoming very active: they had raided the Suffolk coast and had succeeded in hitting the light cruiser *Attentive* with a bomb dropped from 8,000 feet.

In September 1915, the German submarine campaign, because of political difficulties with neutral countries, entered what became known as a 'twilight period' which lasted for the best part of a year. It was not until the autumn of 1916 that, with the greatly increased total of 87 U-boats, it was renewed in all areas. Merchant-ship casualties then rose to double those of any previous period and the British realized that this form of warfare was now of first importance in the war at sea. The campaign was widespread and ships were sunk not only in the south-western approaches, the Channel and the Mediterranean, but even off North Russia and west of Gibraltar.

The bulk of the R.N.A.S. was still deployed on the North Sea coast where 111 seaplanes and flying-boats and 10 blimps were stationed between Dundee and Felixstowe; the Dover area had 48 seaplanes and 8 blimps and the Channel now 70 seaplanes and 7 blimps; the west coast still only had blimps, 10 being in the northern Irish Sea and 7 at the new bases at Pembroke in South Wales and Mullion in Cornwall. Up to now aircraft had had few contacts with U-boats and had not sunk a single one. On the east coast they continued to patrol the shipping routes but in general their operations were spasmodic and unco-ordinated. Aircraft had no way to detect totally submerged submarines in home waters where the water was murky and seaplanes generally carried only two 25-lb bombs so the U-boats were not really afraid of them. Nevertheless, the U-boats realized that if they allowed themselves to be seen, traffic would be diverted and anti-submarine vessels would be called up to hunt them. Aircraft could therefore impose restrictions on the U-boats by keeping them submerged, making them use their periscopes sparingly and forcing them to be very careful not to disturb the surface or permit oil or air leaks, but they could only enforce these restrictions in coastal waters within range of their bases and so the U-boats were able to continue to sink ships at sea with impunity in areas well outside the aircrafts' radius of action.

At the end of the year the new Anti-Submarine Department in the Admiralty had completed its analysis of the U-boat campaign and made its requirements known. It considered that the blimps were best used for area patrols and reporting any U-boats sighted, whilst seaplanes should be used to follow up these reports and to attack. The department also recommended that kite-balloons should be used against U-boats and should be towed by anti-submarine vessels to increase their range of vision so that they could sight U-boats on the surface over a wider area. The Department recommended a great increase in air patrols in the mouth of the Channel and in the south-western approaches from bases in Cornwall, South Wales and Ireland. The Admiral at Queenstown, however, who was the most influential officer in the war against the U-boats, was very lukewarm about this suggestion. He pointed out that if his patrol vessels were going to have to spend their time looking for lost seaplanes they would be more trouble than they were worth. This attitude is of interest not so much as an example of a reluctance to accept new weapons as an illustration of the ineffectiveness of aircraft against U-boats up to that time and the fact that they had frequent engine failures over the sea.

The U-boat operations against merchant shipping in the Mediterranean had become serious during the autumn of 1915 and in spite of the 'twilight period', continued to do great damage during 1916. Air anti-submarine measures in the central Mediterranean were largely in the hands of the French and Italians; there were, however, some British seaplanes at Gibraltar, some flying-boats at Malta and the *Ark Royal* was still at Mudros as a harbour seaplane tender. Admiral Mark Kerr, commanding the British naval forces in the Adriatic, asked for seaplanes and kite-balloons to patrol the new Otranto barrage which had been instituted to try and prevent U-boats leaving the Adriatic. He also asked for aircraft to bomb the U-boat bases at Pola and Cattaro and the torpedo factory at Fiume. Only a few seaplanes could be spared, however, and the R.N.A.S. anti-U-boat operations continued to be confined to local patrols in the Aegean and off Malta and Gibraltar and occasional searches for secret U-boat bases.

The German Unrestricted Submarine Campaign began in February 1917 and soon showed itself to be of paramount importance in the war at sea. In April 1917 the sinkings of merchant ships were so heavy that the British realized that if they could not check them they would have to arrange a peace by November. As called for by the Anti-submarine Department the previous autumn, the strength of the R.N.A.S. was greatly increased in the Channel and south-western approaches. In January the aircraft on the British side of the Channel were organized into the Portsmouth Group, which by April had 40 flying-boats and seaplanes based at Calshot, Portland, Bembridge and Newhaven. Their operations were co-ordinated with the French who had 37 seaplanes based at Dunkirk, Boulogne, Havre, Cherbourg and Brest. In April the South-West Group was formed with 6 blimps at Mullion and Pembroke, 5 Large America flying-boats in the Scillies and 11 seaplanes at Plymouth, Fishguard and Newlyn. Twelve Sopwith '1½ strutter' aeroplanes had to be sent to reinforce this area as there were no more flying-boats or seaplanes available. The northern Irish Sea still had its 10 blimps at Luce Bay, Anglesey and Barrow but no seaplanes or flying-boats. On the east coast there were 50 flying-boats, 37 seaplanes and a dozen blimps that could be used for anti-submarine work but they were also required for general reconnaissance.

Anti-submarine aircraft were still used mainly to patrol the inshore shipping routes and to search areas where U-boats had been reported or in which radio intelligence indicated they might be. In April the Felixstowe flying-boats established the famous 'spider-web' patrols in the southernmost part of the North Sea which were centred on the North Hinder Light Vessel and were intended to catch the Flanders U-boats on passage to their operational areas. They were designed scientifically to keep a U-boat down for ten hours and so exhaust its batteries and were used in conjunction with radio intelligence. In April 5 flying-boats made 27 patrols, sighted 8 U-boats and bombed three of them and on 20th May success was at last achieved when they sank *UC.36* off the West Hinder. Aircraft were also used extensively to patrol the lines of the barrages in the Dover Straits and the North Channel and in June there were 12 blimps and 23 seaplanes in the Dover area alone for this purpose.

In May one of the Scillies flying-boats was used to patrol the route ahead of the first ocean convoy: such patrols and the escort of convoys were to become one of the principal ways to use anti-submarine aircraft in the future. In June, in all areas, seaplanes, flying-boats and aeroplanes sighted 17 U-boats and attacked 7 of them and the blimp *C9* bombed a U-boat off the north coast of Cornwall. By mid 1917, therefore, aircraft had begun to be of value in the anti-submarine war. Nevertheless a glance at the map (see Fig. 5, p. 92) shows that the main U-boat operating areas, where most of the ships were being sunk, were still well outside their range to the westward.

In January 1917, the Admiralty had already decided that the R.N.A.S. in the Mediterranean must be reinforced to try and counter the U-boat operations there. They decided that this could best be done by strengthening the aircraft patrolling the Otranto Barrage. Six seaplanes were transferred from Dundee and by the end of April the total had risen to 24 with 6 aeroplanes, which were used to patrol in co-operation with the drifters using indicator nets and trawlers using hydrophones which constituted the barrage. In May they began to catch sight of U-boats occasionally but had no success against them in this period.

In April 1917 the German Unrestricted Submarine Campaign had brought America into the war on the side of the Allies. It will be recalled that U.S. naval aviation in August 1914 had no more than twelve planes. Since then shortage of funds coupled with a somewhat conservative attitude in the Navy Department had put them far behind the naval air services of the combatants. Plans were, however, made for coastal patrols and to carry seaplanes in battleships and cruisers, and a small blimp was ordered. The General Board in June 1916 declared that aviation was likely to be confined to a subordinate role and by the end of that year had only contracted for 60 new planes. By April 1917 the air arm still had little space in the war plans and its strength was only 6 flying-boats, 45 seaplanes which were mostly only suitable for training, 3 aeroplanes, 2 kite-balloons and a very unsatisfactory blimp. Admiral Sims, the Commander of the U.S. naval forces in Europe, recommended sending seaplanes and kite-balloons to fight the U-boats as soon as possible and although plenty of money at once became available, the U.S.N. could not do more than send a detachment of pilots and aviation personnel to Europe to help as best they could. U.S. naval aviation in April 1917 was therefore little more than a nucleus on which to base expansion, but energetic steps were taken to remedy matters and a vast programme for 4000 planes was put in hand.

In the second half of 1917, the various British anti-submarine measures first began to cope with the German U-boat campaign against commerce. Maritime aircraft started to play a noticeable part in these operations and to achieve some success. Between July and September the flying-boats, mostly from Felixstowe, sank *UC.1*, *UB.20*, *UC.72* and *UC.6* all in the southern North Sea and on 18th August a Wight seaplane sank *UB.32* off Cape Barfleur in the Channel. Finally on 11th July kite-balloons had their first success and assisted destroyers to sink *U.69* in the North Sea. In the three months July to September these five U-boats, which had been destroyed by unaided air-craft and the sixth sunk by ships assisted by kite-balloons, represented over a third of the U-boats sunk by all the British anti-submarine measures put together. Unfortunately for the British this splendid rate of sinking U-boats by aircraft was not maintained and there were no more successes.

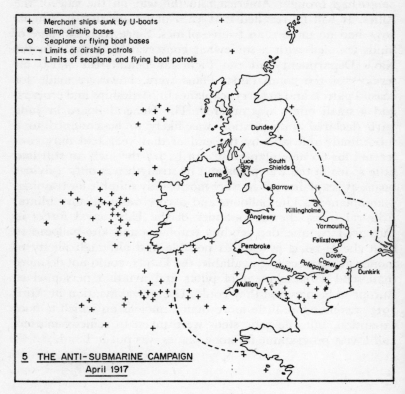

+ Merchant ships sunk by U-boats
⊗ Blimp airship bases
○ Seaplane or flying boat bases
---- Limits of airship patrols
—·—·— Limits of seaplane and flying boat patrols

Dundee
South Shields
Luce Bay
Larne
Barrow
Killingholme
Anglesey
Yarmouth
Felixstowe
Pembroke
Dover
Copel
Polegate
Calshot
Dunkirk
Mullion

5   THE ANTI-SUBMARINE CAMPAIGN
April 1917

During the whole of 1917, R.N.A.S. aircraft of all types sighted 168 U-boats and bombed 105, sinking 6, and it is of course probable that some other U-boats were damaged and delay and anxiety were imposed on many. This was a great improvement over the 1916 results but its effect on the campaign as a whole must not be over-estimated. The average number of U-boats at sea at a time during 1917 was 44 and the length of cruise approximately one month: very roughly there-fore only one in three of them was sighted by an aircraft during a sortie, and then only once. In some areas, such as the Flanders

Bight, U-boats were harassed much more than this but on the other hand many of those in the open Atlantic were not molested at all. Unfortunately there are no figures available to show how often U-boats were forced to dive by aircraft and succeeded in doing so without being seen. It is probable that this happened frequently and was, of course, nearly as important as a harassing factor and a cause of delay as a sighting.

As U-boat hunters, flying-boats were proved to be by far the best type of aircraft. They sighted 68 U-boats during 1917 and were able to attack 44 of them, sinking 5.[7] The figures for sea-planes were 66 U-boats sighted and 41 attacked but only one was sunk. There were, however, three times as many seaplanes as there were flying-boats and only half of the flying-boats were of the large H12 or F2a types which had, in fact, had all the success. Airships came third in sighting 28 U-boats and bomb-ing 16, but sinking none. Although the use of kite-balloons was greatly extended during the second half of 1917, only nine U-boats were sighted and five attacked by ships with their help, resulting in one sinking. The large flying-boats were successful because they had a crew of four, who were in reasonable comfort in cockpits with a clear view, and so could keep a far better lookout. The flying-boats were fast and so able to surprise U-boats before they had time to submerge and the 230-lb bombs which they carried made success in the actual attack more likely. The sea-planes on the other hand had a smaller crew whose view ahead was obstructed by the engine and propeller and although they were as fast as the flying-boats they did not carry such heavy bombs. Airships proved good lookout platforms but they could be seen by a U-boat a long way off and being slow gave the U-boat plenty of time to dive before they could attack. In spite of this, it was in anti-submarine work that the small non-rigid airships proved of most use. They were comparatively cheap both to build and operate, which was just as well as 37 of them were lost in 1917, mostly due to weather or engine failure or a combination of both. They were particularly useful for escorting ships and convoys as they could keep station at slow speed and could, if necessary, hover in the vicinity of a U-boat and keep it down. As a result the blimps on the east coast were drawn more into anti-submarine operations and were used less to scout for the Grand Fleet.

Sinking U-boats was not, however, the only indication of the efficiency of aircraft in the anti-submarine role. It is probable that in this period they in fact did more good by the indirect protection they gave to shipping. During the autumn the convoy system became general for both inward- and outward-bound ocean traffic which allowed aircraft to be used much more efficiently. Instead of patrolling at random they were able to search methodically the routes on which convoys were about to sail and if they sighted U-boats the convoy could be diverted and anti-submarine vessels called up to hunt them. Even if aircraft failed to sight U-boats during their patrols, their presence kept them submerged which decreased their range of vision and their mobility and so their power to intercept merchant ships. When aircraft were used to escort convoys they were often able to hold a U-boat submerged while the convoy passed in safety.

The principal problem in the use of aircraft for anti-submarine purposes was still that they could not reach the main operating areas of the U-boats, but the situation could have been much improved if more of the R.N.A.S. had been deployed to the west, especially to work from bases in Ireland (see Fig. 5, p. 92). The greatest number of maritime aircraft, including flying-boats, seaplanes, airships and aeroplanes was still on the east coast. Here they sighted and attacked many fewer U-boats than did aircraft to the west of the British Isles. The need to extend air cover westwards was by now fully recognized and the main reason for the expansion of the kite-balloon service was to give a form of air escort to convoys throughout their voyage. It was this same need which stimulated thought on using rigid airships in the campaign against the U-boats. For the present little could be done except to put in hand a large programme of anti-submarine aircraft and to urge the U.S. Navy to establish air stations in Ireland. This they readily agreed to do but U.S. naval aircraft had as yet scarcely taken part in the war and they only made their first patrol from a French base at Le Croisic in November. Nevertheless, anti-submarine operations were established by them as of first priority and over a thousand of the large H16 and smaller HS1 flying-boats were on order. There was anxiety in Britain, however, that if U-boats appeared on the American coast this would absorb the bulk of their effort and they would be unable to expand their air forces in the eastern Atlantic.

Throughout the second half of 1917, the U-boat campaign took a heavy toll in the Mediterranean as well as home waters and in August a British Commander-in-Chief was appointed to co-ordinate all anti-submarine measures. Convoys had already been instituted and the new Commander-in-Chief at once advocated the reinforcement of the R.N.A.S. so that a constant instead of an intermittent air patrol could be maintained over the Otranto Barrage, and so that intensive bombing of the U-boat bases at Pola and Cattaro could be carried out. For the present, however, the best had to be done with the aircraft available and on 2nd September, six of the new Short 320 sea-planes with 18-inch torpedoes which had been sent out to attack the Austrian Fleet, set off to attack the U-boat base at Cattaro. They did not have sufficient endurance to make the trip both ways by themselves and so had to start in tow of motor launches. After an uneventful night passage an un-expected gale sprang up, one seaplane sank and the others were damaged before they could take off and so not only did the attack fail but the number of aircraft available for patrol was reduced.

Whilst the campaign against the U-boats was being waged, aircraft from shore bases took a large part in the naval opera-tions in the narrow seas of the Flanders Bight, Thames Estuary and Belgian coast. During 1916 heavier bombers began to be available and they continued their offensive against the U-boat bases in Flanders. The British bombing squadrons in France had been re-equipped with twin-engined Caudrons and with Sopwith '1½ Strutters', both of which could carry four 65-lb bombs. The Short Bomber which came into service in Novem-ber could carry four 230-lb bombs or eight 112-lb which was three times as much as the Caudrons; there were, however, few Short Bombers as yet and they did not make many raids in 1916. As before, the bombers had many other targets besides the U-boat bases and were still unable to achieve much. They had to devote a great deal of their effort to the bases of the German aeroplanes and seaplanes in Flanders, the aircraft from which were becoming a great nuisance: in January these air-craft had bombed the British airship station at Capel near Folkestone and had raided shipping in the Thames Estuary and in February they had bombed Walmer; a flight of five twin-engined Gotha torpedo seaplanes had also arrived at

Zeebrugge and on 9th November they attacked but missed a ship off the mouth of the Thames. Altogether the German aircraft made eighteen raids on coastal towns, naval bases and aerodromes in the United Kingdom in 1916 and on 28th November one of them raided London. These operations had to be opposed not only by the R.N.A.S. bombers but also by fighters from Dunkirk which were already very busy supporting bombardments and other operations on the Belgian coast.

In early 1917 air warfare over the Flanders Bight developed considerably. Whereas the operations of the flying-boats and Zeppelins had only just begun to overlap in the Terschelling area, the British and German aircraft off the Belgian coast had been fighting each other for some time. The Germans had established a second seaplane base at Ostend and in May they had 47 seaplanes altogether in the area. They began to contest Allied air anti-submarine operations and shot down 6 French seaplanes in May so that these aircraft now had to have fighter escort. On 21st April German seaplanes shot down the British coastal airship *C17* in flames off the North Foreland and began to interfere with the Felixstowe flying-boats in the North Hinder area.

In April the German torpedo planes from the Belgian bases joined the U-boats in an attack in earnest on shipping. On 20th April they unsuccessfully raided the Downs but on 1st May they sank the S.S. *Gena* off Southwold; on 14th June they had another success and sank the S.S. *Kanakee* off the Shipwash. The torpedo aircraft were somewhat flimsy and they had a great deal of trouble with the running of their torpedoes so they were by no means always successful and at various times they missed three other steamers. The British reply to these attacks was to base fighter aeroplanes at Walmer and to use fighter seaplanes to maintain a patrol of the war channel between Yarmouth and Dover.

The bombing of the German U-boat bases in Flanders continued into 1917 and at Bruges some success against the base installations began to be apparent. There were, however, still many other calls for the bombers' services and they were unable to concentrate on this objective alone. In the autumn of 1916 German destroyers had arrived on the Belgian coast to support the U-boat campaign by raiding the Dover Barrage and $3\frac{1}{2}$ tons of bombs were dropped on them in January and February but without scoring any hits. In May 1917 German bombers based in Belgium began to raid London and the R.N.A.S. bombers had to be diverted to attack their bases, which was an added distraction.

The means to bomb the German U-boat bases in Flanders were greatly improved during the second half of 1917 and attacks continued unabated. In May the new Handley Page night bombers, which could carry sixteen 112-lb bombs, had begun to arrive, and by December there were four squadrons of them in France. In July a new day bomber arrived, the DH4 capable of carrying two 230-lb bombs. During 1917, 120 tons of bombs were dropped on Bruges, Zeebrugge and Ostend but because the aircraft were frequently diverted to other tasks this represented only 37 per cent of their effort. Considerable damage was done but no U-boats were sunk and this scale of attack was still not enough to stop them using the bases or indeed to inconvenience them seriously.

The German seaplanes in Flanders continued to be very active during this period: they bombed Calais and Dunkirk by night and vigorously opposed the British air anti-submarine operations in the southern area. On 4th July, fourteen German seaplanes bombed the R.N.A.S. base at Felixstowe destroying one flying-boat and damaging another and the raid was repeated three weeks later. Flying-boats on patrol had had fighter escorts for some time in this area but the Baby seaplanes employed were no match for the enemy and in July they had to be replaced by Sopwith Pup aeroplanes. In spite of this a Felixstowe flying-boat on anti-submarine patrol in October was shot down by five seaplanes from Zeebrugge and on 11th December German seaplanes shot down the Coastal Blimp *C27* off the Norfolk coast so that the use of these airships finally had to be discontinued in this area.

This period also saw the last of the German torpedo seaplanes: on 9th July four of them from Zeebrugge attacked a convoy in the Thames estuary but their torpedoes failed to run properly and two of the seaplanes were lost. Although on 9th September seven seaplanes sank the S.S. *Storm* with bombs and torpedoes, the Germans had lost faith not only in the torpedo seaplanes themselves which were cumbersome and very frail but also in the torpedoes, which seldom functioned correctly, and there were no more attacks.

The U.S. Navy set much store on bombing as an anti-submarine measure and believed it should even take priority over the defence of convoys. They began to build up a Northern Bombing Group at Dunkirk to attack the U-boat bases, a Group which was to have consisted of twelve squadrons. By October 1918, however, it had only 3 Caproni and 14 De Havilland aeroplanes and by the Armistice it had dropped no more than 10 tons of bombs. The R.A.F. bombers in France continued to pound the Flanders U-boat bases right up to the end of the war, 77 tons of bombs being dropped in May 1918 alone. Although no U-boats were destroyed these attacks were now able to make life difficult for them. It is of course true that better results could have been achieved if the bombers had concentrated their whole effort on the U-boat bases, but the rival uses for bombers were important to the war effort, and many of them directly to the war at sea. The U-boat bases in Germany, from which twice as many U-boats operated were, however, outside the range of even the Handley Page bombers and so bombing could never have hoped to be decisive as an anti-submarine measure on its own.

Since April when the R.A.F. was formed, bombing targets had ceased to be laid down by the Navy. Co-operation between the Royal Navy and the newly-formed Royal Air Force was first fully tested during the blocking of Zeebrugge and Ostend in April and was successful. Aircraft made a complete photographic survey before the operation, gave fighter protection, made diversionary bombing attacks during the attack and subsequently flew reconnaissances to assess the damage. The first clash of opinion between the services occurred in this southern area soon after the blocking of Zeebrugge when the Vice-Admiral at Dover did not consider he was receiving the bomber support he thought necessary to attack the resultant concentration of U-boats in Bruges. The problem however was one of priorities rather than co-operation.

In Belgium and the Flanders Bight therefore the aircraft of both sides had an important strategic task in the war at sea; the British to oppose the German U-boat campaign and the Germans to assist it. In these sea areas which were within easy range of the airfields of both sides operations inevitably led to frequent contacts and to air-fighting and the immediate aim became, as over the land, more and more one of gaining local air superiority. In such an environment the airship was soon ousted and the seaplane and even the flying-boat were outclassed. Admiral Bacon, commanding at Dover, was convinced that aeroplanes should replace seaplanes for all purposes and in September 1917 is reported to have said that he 'never wanted to see another seaplane'. The British soon adopted this policy and it was decided that even anti-submarine patrols should be made by aeroplanes as soon as they could be made available.

It is now necessary to digress for a moment from the main theme of this chapter to note three uses of aircraft at sea, which although not of very great importance at the time, are of interest to any study of aircraft and sea power. With the evacuation of Gallipoli at the end of 1915, the character of the naval campaign in the Aegean changed and so also did the function of naval aircraft. From the support of an amphibious operation it became one of blockading the *Goeben* and *Breslau* in the Dardanelles as well as one of safeguarding the supply route to Salonika from U-boats. At the beginning of 1916, aeroplanes were stationed at the island of Imbros to watch the Dardanelles and in January 1918 the *Goeben* and *Breslau* broke out: both struck mines; the *Breslau* sank and the damaged *Goeben* ran aground in the Straits as she retired. Strenuous efforts were made to complete her destruction from the air: she was heavily attacked by DH4 bombers which dropped 15 tons of bombs on her but she was only hit twice and as the largest bombs used only weighed 112 lbs, she was little damaged. The *Manxman* was rushed from Brindisi with two Short 320 torpedo seaplanes, but they did not get a chance to attack before the *Goeben* was refloated. Although unsuccessful, this attack is of interest as it was one of the few made during the war by aircraft upon an enemy capital ship.

At the same time in the Mediterranean the war against Turkey involved many minor operations stretching from Palestine and the Sinai Peninsula to the Syrian coast and even to Cyrenaica and the Yemen, and all these campaigns needed air support. This was provided by the seaplane-carriers *Ben-my-Chree*, *Empress*, *Anne* and *Raven II* which carried seaplanes to work with these minor campaigns wherever they were needed. Their seaplanes were used almost entirely for the support of the army ashore although sometimes they spotted for bombarding ships. The seaplane carrier had thereby assumed a new role by giving aircraft strategic mobility which proved very useful where the weather permitted their use especially against shore targets when the air opposition was light.

In March 1917, the R.N.A.S. took on a new function as far away as the Indian Ocean when two seaplanes in the seaplane-carrier *Raven II* were detached from the Mediterranean to search for the German raider *Wolf*. The *Wolf* had left Germany at the end of 1916 and she also carried a small seaplane which, during her long cruise lasting sixteen months, made fifty-six flights and helped capture three ships. The *Raven II* worked with the French cruiser *Pothuau* and her seaplanes made scouting flights up to 50 miles from the ship; they searched the Laccadive Islands and the Chagos and Maldive Archipelagos but without success. The *Wolf* was, in fact, in the Pacific at the time and in May the *Raven II* was recalled to the Mediterranean. Here again a sea-plane-carrier was able to bring aircraft to bear in a distant area.

By early 1918 the U-boat campaign had already begun to take a different form which had the effect of bringing operations within range of R.N.A.S. aircraft. The convoy system had made it very difficult for the U-boats to find targets in the open ocean and so they moved into coastal waters where they hoped to find unescorted ships before they joined and after they had left the convoys. In 1918 the anti-U-boat air patrols were in any case co-ordinated better and more systematically and something of the order of three times the mileage was flown on patrol. The U-boats had, however, also improved their tactics: they remained submerged all day and became adept at avoiding aircraft. As a result the sightings during 1918 did not increase in proportion to the miles flown. In all 192 U-boats were sighted and 131 were attacked. Aircraft did not sink any U-boats outright but they helped ships to destroy 6 and damage 25 of them. The effect of aircraft was as before much greater than these figures seem to show and they drove the U-boats out of some important areas altogether. Their effectiveness is confirmed by the fact that only 6 ships were sunk in convoy while air escort was present.

The average R.A.F. daily strength against the U-boats for the last six months of the war was 85 large flying-boats, 216 seaplanes, 189 aeroplanes and 75 airships. About 60 per cent of this strength was now deployed to the Channel and west of the

British Isles and the rest was still in the North Sea. The flying-boats, seaplanes and airships, as can be seen from the above figures, were reinforced substantially by aeroplanes and this was one of the advantages of a single air force. Most of these aeroplanes were DH6 training machines which were surplus to requirements and had the endurance to patrol coastal waters. A squadron of the larger Blackburn Kangaroo aeroplanes was particularly successful and sighted twelve and attacked eleven U-boats on the east coast.

During 1918 the U.S. Navy's air strength in European waters expanded rapidly. By midsummer there were five air bases in Ireland, three, with another two before the Armistice, on the French Biscay coast and one in the Azores. They also had eight stations on the Atlantic coast of the U.S.A. where U-boats had now appeared. The majority of these bases were equipped with flying-boats, the early ones of French and British types and later of the Curtiss H16 and HS1 types now being built in quantity in the U.S.A.

To the allies probably the most important role of aircraft at sea was to help defeat the U-boat and a substantial proportion of the R.N.A.S. had been engaged solely in this task for much of the war. Nevertheless, aircraft as a whole proved unimportant as actual U-boat killers. Throughout the war they sank only 6 U-boats against 83 by ships, 43 by mines and 20 by other submarines. While their contribution, as has been said before, was much greater than these figures suggest, aircraft could not claim more than to have been one of many means which defeated the U-boat.

So ended the First World War in which aircraft were first used in sea warfare. At the end of this third chapter dealing with this contest the time has come to sum up its achievements and its effect on sea power both in co-operation with the fleet and as an anti-submarine weapon.

The Great War emphasized that the main problem of operating aircraft over the sea was one of distance. The only type of aircraft which had sufficient endurance to cover the whole North Sea on its own was the rigid airship. Reconnaissance for the inferior High Seas Fleet was of the first importance to the Germans to safeguard it from surprise from the Grand Fleet.

There was therefore a considerable incentive for the Germans to develop the Zeppelin which they did with conspicuous success and which was for two years, until it was defeated by flying-boats and aeroplanes, of the greatest value. The British, on the other hand, with their excellent radio intelligence, had not the same incentive and they were so slow in producing rigid airships that they missed the war and by then they were already too vulnerable. With the Zeppelin, German air reconnaissance over the North Sea was decidedly superior to the British for most of the war. Unable to produce rigid airships, the British had to fall back on heavier-than-air craft of various types: they were less successful than the Germans in the use of seaplanes from the shore and they did not use them as systematically or to the limit of their range. By the end of the war, however, the British had developed the American flying-boat to a stage where it was far superior to the seaplane and as it was less vulnerable it also became a rival to the rigid airship.

Much British energy was also expended in devising means to carry heavier-than-air craft to sea with the fleet as a way to extend their range. The seaplane operated from a seaplane-carrier was a failure; it was very seldom that it could be used in North Sea weather and it achieved very little. This combination was of most use in the Middle East where it succeeded in carrying air power to many minor campaigns. The British therefore had to fall back on the aeroplane operated from ships or carriers as a 'one-shot' weapon. Not only was this very wasteful but it meant that aircraft could not be used until action was joined and then not for very long. It therefore did not solve the problem of tactical reconnaissance for the fleet at all; nevertheless, it provided a means by which the Grand Fleet could guarantee air superiority in action. The final stage of an aircraft-carrier on which aeroplanes could land was under development for the last year of the war and had just reached a useful point when the war ended.

As an adjunct to sea power, bombing did not prove a very great success: the fears of the pre-war thinkers that Zeppelins would become a threat to sea power by destroying the Admiralty and the Dockyards proved illusory. Bombing was seldom possible against ships in harbour and when it was it proved ineffective. The sustained attack on the U-boats based in Belgium began to tell by the end of the war but it failed to prevent the use of the bases or to interfere with U-boat operations to any extent. Moreover aircraft were of little use as anti-ship weapons at sea. Although both sides had torpedo-bombers these aircraft did not sink or seriously damage a single important warship.

The role of aircraft was, therefore, auxiliary to ships, and it was ships which were responsible for winning the war. The impact of aircraft on sea power was far less than that of the other new weapon of the twentieth century, the submarine. The battleship remained the 'unit of sea power' and although it had accepted assistance from aircraft, it did not depend upon it. Nevertheless, the rate of development of aircraft during the war had been very rapid and most people believed that its influence would increase substantially in the years to come. The future seemed to lie in the operation of aeroplanes from aircraft-carriers with the fleet, and of long-range flying-boats from the shore. In 1919 Admiral Lord Fisher, the inventor of the Dreadnought battleship only fifteen years before, was prepared to say 'The whole aspect of sea war is so utterly changed by the prodigious and daily development of aircraft . . .' and also 'All you want is the present naval side of the Air Force – that's the future navy ——'. There were few, however, who believed that the performance of aircraft up to then justified such an accolade and the general opinion was that ships, assisted by aircraft of course, would continue to command the sea.

V

The Period Between the Wars
1919 – 1939

THE ARMISTICE REQUIRED all German naval aircraft, including the Zeppelins, to be concentrated and immobilized in Germany under the control of the Allies. Sixteen Zeppelins survived the war; eight of these were eventually turned over to the Allies; one was dismantled and the remaining seven were sabotaged by their crews in June 1919. After the war the Zeppelin Company were allowed to build two small commercial airships but these too were later turned over to France and Italy as reparations.

The British ended the war with well over a thousand aircraft in use for maritime purposes. One hundred and three of these were airships which were still the property and responsibility of the Admiralty although their crews belonged to the Royal Air Force. Of these, five were the survivors of the many programmes of large rigid airships and the rest were blimps. The run down of the British airship service was rapid but a few were kept in service. In June 1919 all airships were put under the naval Commander-in-Chief of the Atlantic Fleet: *R34, R29, NS7* and *NS8* were concentrated at East Fortune to work with the fleet if there should be trouble with Germany over the Peace Treaty. The Admiralty were, however, eager to economize. The main interest in airships was now for transport and the Air Ministry took them over in October 1919. They promptly informed the Admiralty that they had no money to spend on them if carrier-borne aircraft were to be developed. The airship service was therefore disbanded and at this point fades out of British maritime history. Some airships, however, continued in use in the hope that they might be developed for commerce.

The heavier-than-air craft comprising some 450 seaplanes,

250 aeroplanes and 230 flying-boats, including all those embarked in ships, now belonged to the Royal Air Force. Demobilization and economy were the order of the day and this huge force rapidly dwindled and by the end of 1919 there were fewer than fifty aircraft. Of the ships, the aircraft-carrier *Argus* was kept in commission with the seaplane-carriers *Pegasus* and *Ark Royal*. The *Furious* and *Vindictive*, which had separate forward and after flying-decks and which had been found quite useless, were paid off. Work on the new carriers *Hermes* and *Eagle* was continued slowly, waiting for further experience with the *Argus*.

The organization, which had in fact been in operation since April 1918, was that the Admiralty had full responsibility for the ships and the Royal Air Force for the aircraft. The Navy controlled the operations of the aircraft when they were embarked but the aircraft were the property of the Air Ministry. They were paid for out of the air vote and could, in theory, be removed from the carriers and used for some other purpose if the Air Ministry thought fit. In September 1919 all R.A.F. aircraft allocated for co-operation with the Navy were grouped into a new command called 'The Coastal Area' with its headquarters at Lee-on-Solent. Coastal Area on its establishment consisted of one spotter-reconnaissance squadron, a fighter flight and half a torpedo squadron together with a flight of seaplanes and a flight of flying-boats. These aircraft, which were of wartime types, were so few in number that all they could do was to carry out some experimental work. The *Pegasus* and *Ark Royal* were, however, used to take R.A.F. seaplane contingents to operate against the Bolsheviks in North Russia and later in the Black Sea.

In 1919, Admiral Madden, the Commander-in-Chief of the Atlantic Fleet, asked for seven aircraft-carriers. He visualized three distinct types: 'Air reconnaissance ships' which were to take the place of cruisers; 'Divisional aircraft-carriers' to supply the spotting aircraft for the battle squadrons; and 'Fleet aircraft-carriers' to operate both fighters for defence and torpedo-planes for attack on the enemy battlefleet. Ideas were not very clearly formed and there was a school that believed that an aircraft-carrier should also mount heavy guns and be able to join the battle line. Practical matters however were of more immediate concern and effort was concentrated upon the development of a reliable technique for landing on a carrier and producing an

1915

As originally designed as a
Light Battle Cruiser.
Guns: Two 18", eleven 5·5",
four 3" A.A.

1917

As finally completed as a
hybrid seaplane carrier with
forward flying off deck.
Guns: One 18", eleven 5·5",
four 3" A.A.
Carried: 5 Short 184 seaplanes
and 5 Sopwith Baby seaplanes
or 5 Sopwith Pup fighter
aeroplanes

1918

As first modified as an aircraft-
carrier with after landing deck.
Guns: Ten 5·5", four 4" H.A.
Carried: 16 Sopwith '1½ Strutter,'
'Pup' or 'Camel' aeroplanes

1925

Lower flying off
deck

Funnels

As later modified as a clear
deck carrier with lower flying
off deck forward as well.
Guns: Ten 5·5", six 4" H.A.
Carried: 12 Fairey Flycatcher
fighters, 6 Avro Bison spotters
6 Fairey III D recce.
12 Blackburn Dart torpedo
planes

1939

As finally modified with 'island'
and better A.A. armament,
dispensing with lower flying
off deck.
Guns: Twelve 4" H.A.,
Three multiple pom-poms
Carried: 18 Swordfish T.S.R.
12 Skua fighter-bombers

6   THE METAMORPHOSIS OF H.M.S. 'FURIOUS'

efficient arrester gear. The rival merits of a completely flush-decked carrier, such as the *Argus*, and those with an 'island' structure, such as the *Hermes* and *Eagle*, were debated. In theory the 'island' type had many advantages: it took the funnel gasses clear of the approach path of the aircraft, it permitted a larger hangar and less heat was generated. At the same time a much more satisfactory bridge could be provided. Nevertheless, it had not yet been established that aircraft could land safely. After preliminary tests in the *Argus* with a dummy 'island', the *Eagle* was commissioned for trials before she had been fully completed and it was found that aircraft could land satisfactorily.

The demobilization of the U.S. Navy was also rapid. U.S. Naval Aviation had during the war been predominantly a shore-based anti-submarine force equipped with flying-boats. A force of land bombers was however being built up at the time of the Armistice. They had no experience of working from ships except what their battle squadron, which operated with the Grand Fleet, had been able to find out from the British. In January 1919, when the U.S. Atlantic Fleet went to Guantanamo for its annual exercises, it took with it a squadron of H16 flying-boats based on the minelayer *Shawmut* as a makeshift seaplane-tender. The battleship *Texas*, which had been with the British in the Grand Fleet, had two Sopwith Camels on her turrets and there were also six kite-balloons available. Admiral Mayo, the Commander-in-Chief in the Atlantic, had already forwarded his views on the development of the Naval Air Service. Among his recommendations was that aircraft-carriers should be built and that they should carry 25 planes. He also proposed that two rigid airships of the Zeppelin type should be taken over from the Germans.

In the middle of 1919, however, the policy was for economy and this cut down defence votes a great deal. Nevertheless the conversion of the seaplane-tender *Wright* from a merchant ship hull was authorized in addition to the conversion of the first U.S. aircraft-carrier *Langley* from the collier *Jupiter*. At the same time the building of the airship *Shenandoah* (*ZR1*), based on the design of the German Zeppelin *L49*, was authorized, as well as the purchase of the British rigid airship *R38* (*ZR2*). Later the *Los Angeles* (*ZR3*) was ordered from the Zeppelin Company in Germany. The U.S. Navy showed by these measures a desire to change the character of their naval aviation from a comparatively short-ranged shore-based force into one able to range much farther afield. Seaplane-tenders were needed to give their existing flying-boats strategic mobility and aircraft-carriers were the obvious answer to the problem of carrying aircraft with the fleet. Their interest in rigid airships after the defeat of the Zeppelins seems strange. Undoubtedly they had been given an over-optimistic view of the value of the Zeppelins in the North Sea during the war but they clearly needed very long-range reconnaissance aircraft to work in the Atlantic and Pacific Oceans. By using helium, of which supplies were available in the U.S.A., instead of hydrogen, they believed that they could surmount the problem of vulnerability, the airship's greatest weakness in war.

The development of aircraft during the war had been extremely rapid. Aircraft of several types now existed with very considerable endurance and in 1919 three types flew the Atlantic. The U.S. Navy had developed a new long-range flying-boat with four engines for anti-submarine work. On 16th May a seaplane division of three of these N.C. flying-boats took off to fly the Atlantic from west to east via the Azores and Portugal. Practically the whole American Atlantic Fleet was strung out across the ocean to help with navigation and weather reports. Two of the flying-boats failed to reach the Azores but the third completed the trip landing three times *en route* and arriving at Plymouth on 31st May. Two weeks later a British Vickers Vimy bomber piloted by Alcock and Brown flew non-stop from Newfoundland to Ireland. Finally in July the British airship *R34* flew 3,600 miles from her base at East Fortune in Scotland to the U.S.A. and a few days later returned to Pulham, thus being the first to make a double crossing. There were others hoping to cross, notably a Handley Page bomber and the U.S. blimp *C3*, but they did not succeed in doing so. These three flights, only ten years after Blériot had flown the Channel, showed the immense technical advance that had been made and the great potential of aircraft operating over the sea.

At the end of 1920 the U.S.N. began a series of trials to test the effect of bombs on ships. After they had made static trials with explosive charges of up to 1800 lbs in weight on the old battleship *Indiana,* by the middle of 1921 they were ready for sea trials with bombs actually dropped by aircraft. The targets were to be some of the surrendered Germany warships. The U.S. Navy invited the U.S. Army Air Force to take part and the First Provisional Air Brigade was specially formed for this purpose. It was commanded by Brigadier General William Mitchell who was a fanatical protagonist of a separate air force. His aim in these trials was to show that the day of the surface ship was over. One cannot but look askance at some of the methods, political and otherwise, which he used to further his views, but one cannot but admire the marksmanship and efficiency of the First Provisional Air Brigade under his command. General Mitchell had, in fact, invented a new method of sinking ships. Instead of trying to penetrate their deck armour with armour-piercing bombs or to disrupt their hulls under water by exploding torpedoes against them, he relied on the shock effect of dropping very heavy high explosive bombs close alongside and set to explode just below the surface. This 'near-miss' method of attack proved very successful and even if the bombs hit by mistake they would do immense damage to light upper works and would set the ships on fire. Tactically General Mitchell was also the fore-runner of the combined attack by high-level bombers, fighters and torpedo aircraft against ships at sea.

The first target in the trials was the German submarine *U.117* and she was rapidly disposed of by three U.S.N. flying-boats which dropped three 180-lb bombs each. The second target was the German destroyer *G102.* She was subjected to a full-scale attack by fighters and light and heavy bombers of the U.S. Army Air Force, dropping bombs from 25 lbs in weight to 300 lbs. She survived the 25-lb bombs but was very quickly sunk by the 300 lb bombs. The third target was the German light cruiser *Frankfurt* which was a fairly modern ship. This trial was designed so that the U.S.N. could investigate the damage after every hit. She survived the onslaught of 100-lb and 300-lb bombs but rapidly succumbed to 600-lb bombs. The final target was the German battleship *Ostfriesland* and during this trial there were acrimonious exchanges between the U.S.N. and General Mitchell. General Mitchell wished to mount a devastating attack and sink the ship as quickly as possible but the U.S.N. again wanted to board her after every shot. On the first day the Martin heavy bombers dropped five 1100-lb bombs in rapid succession and damaged the ship badly: she was, however, kept afloat by a boarding party. On the next day seven Martin bombers carrying 1800-lb bombs sank her after dropping four of them in quick succession in spite of the protests of the naval trial party. General Mitchell was jubilant: he took this as proof that aircraft could dominate the most powerful warships. He believed that it followed that in future, navies would only be able to operate outside the range of aircraft from the shore. From this he deduced that aircraft were all that were necessary to prevent an invasion of the United States and that therefore the bulk of the U.S. Navy, especially its battle force, was redundant.

The nautical faction strongly contested these views and they argued that the tests had proved very little. They pointed out that the ships were old and in poor condition; they were not protected by anti-aircraft guns and had no damage control parties on board. Furthermore they were at anchor and the bombing was made from a height of 200 feet or so. If the aircraft had flown at a height at which they could expect to be immune from anti-aircraft fire, they would undoubtedly have missed. The U.S. Navy ridiculed General Mitchell's suggestion that the day of the battleship was over. Nevertheless it had been proved that it was possible to sink ships with bombs, something that had never happened during the war. The General Board of the U.S. Navy, while not admitting that the day of the battleship was over, nevertheless saw that the problem of air attack was of the first importance. If the supremacy of the battleship was to continue a great deal would have to be done, not only to improve damage control and anti-aircraft gunnery but above all to develop a naval air arm with the two-fold aim of protecting the fleet and of attacking the enemy in the same way as General Mitchell had done. In a further bombing test against battleships in 1924 the unfinished modern hull of the *Washington* stood up very well to attack and this supported the views of the nautical faction.

The British also carried out trials in 1921–2 in which aircraft of the Royal Air Force attacked the radio-controlled target-ship *Agamemnon*. Dummy bombs were used against this moving target at sea and the proportion of hits was low, mainly because of lack of training. Static explosive bomb trials against the battleship *Monarch* were also carried out, and a special experimental target was built to test new methods of protection. At the same time great efforts were made to improve anti-aircraft gunnery.

Development of an eight-barrelled multiple pom-pom was begun for use against torpedo-carrying aircraft. The anti-aircraft guns of battleships and cruisers were increased from the two 3-inch or so of the war years to four 4-inch guns, and new high-angled gun control systems were installed (see Fig. 8, p. 133). In general, it can be said that the British also fully realized the dangers of air attack on battleships but believed that with better-protected ships, better gunnery and fleet fighter aircraft, they could compete with it.

Following the U.S. bombing tests the Washington Naval Conference of 1921–2 took place. Its purpose was to limit naval armaments and prevent a building race between the great powers. Attempts were made to limit all types of ship but the main discussion centred round the size and number of battleships to be allowed to each country. Agreement was eventually reached and it included clauses which limited the number and size of aircraft-carriers. The discussion of this subject seemed, however, to have been concerned more with preventing battleships being built in the guise of aircraft-carriers than the limitation of naval air power. There was no discussion of the number of aircraft that they could carry, which was what really mattered. Their size was, however, limited to 27,000 tons except for two for each country which were allowed to be 33,000 tons. This was to allow them to convert them from capital ships which were already building and which had been disallowed by the Treaty. Aircraft-carriers were defined as warships larger than 10,000 tons built for 'the specific and exclusive purpose of carrying aircraft and which must be constructed so that aircraft could be launched from and land thereon'. They were not to carry a gun heavier than 8-inch and not more than ten guns larger than 6-inch, and they were not to be replaced in under twenty years. It was at first proposed to apply the same ratio of strength to aircraft-carriers as for battleships and to limit Great Britain and the U.S.A. to 80,000 tons. Great Britain had, however, already nearly that tonnage built or earmarked for conversion. The final tonnages agreed were 135,000 tons for Great Britain and the U.S.A.; 81,000 tons for Japan and 60,000 tons for France and Italy.

The Washington Treaty therefore did not cut down the number of aircraft-carriers but set a not ungenerous limit on expansion. At the time Great Britain only had the *Argus*; the *Furious* was in the dockyard for extensive alteration and the

Hermes and *Eagle* were building and these amounted to less than half the tonnage allowed. The U.S.A. had the single converted collier *Langley* nearing completion and the Japanese were still building their first carrier, the *Hosho*. Indeed the Washington Treaty encouraged the building of aircraft-carriers as it prevented the completion of a number of battleships and battle-cruisers which were on the slips and made these hulls available for conversion.

After the Washington Conference inter-service dissension in Great Britain over the control of aircraft at sea broke out. The Royal Air Force, although it had had an independent bombing force by the end of the war, had in fact been used mainly as an auxiliary to either the Army or the Navy. The post-war R.A.F., however, was strongly an 'independent air force' and many of its senior officers held the views which were advocated by the Italian General Douhet in a book published in 1921. According to Douhet's theories, the air force in any future conflict would be more important than either the Army or the Navy. He believed that its principal task was to 'gain command of the air' by eliminating the opposing air force by bombing. Once command of the air had been achieved, he believed that it could subdue the enemy by destroying his cities and so his will to resist. He also believed that should it be necessary an air force could dominate both the opposing army and navy by bombing their bases and the infrastructure on which they depended. Douhet's theories gave no place to static fighter defence and he believed that the best use of fighters was to escort the bombers to their objectives. However, although strongly in favour of an independent air force, he admitted the necessity for what he called 'auxiliary air forces' to work with the army and navy, provided they paid for them.

The British Navy were not in the least impressed with the theories of Douhet. They had had experience with their own R.N.A.S. bombers against the German U-boat bases in Belgium. They had dropped a considerable tonnage over a period of four years and had never sunk a U-boat or driven them out of these bases. The results of bombing operations against enemy airfields were even less encouraging. They could point out that no ship of any importance had been destroyed by aircraft during the war and that the suggestion that they could challenge sea power as exercised by a superior battlefleet was quite unrealistic. Nevertheless, they saw that naval warfare would be greatly influenced by aircraft in the future. They realized that air reconnaissance by longer-ranged planes operating from aircraft-carriers could extend the vision now provided by cruisers by hundreds of miles. They believed that air spotting at the very long gun ranges now possible would make the difference between hitting and missing. In a recent exercise torpedo-bombers from the *Argus* had scored four out of six hits against a battleship proceeding at full speed and with freedom of manœuvre. Clearly fighters were necessary as a defence against such a menace and this method of attack itself would be of the greatest value to slow down an enemy fleet so that it could be brought to action. All these tasks, however, required special aircraft and above all sailors to fly them and the Admiralty began their long struggle for the control of their own air arm.

The Royal Air Force on the other hand were determined not to give up any of their independence and were most reluctant to agree to the Navy having its own private air force. They believed in the 'unity of the air' and that everything should be put into a strong offensive bomber force; and they believed this as intensely as the Royal Navy believed in battlefleet supremacy. All the Admiralty were able to achieve at this stage was to get the Air Ministry to agree that the observers in aircraft operating from carriers, who had to do the spotting for the ship's guns and identify ships on reconnaissance, should be naval officers.

After the Washington Treaty the R.A.F. aircraft working with the fleet were increased to a total of 54 machines. They were of four types: reconnaissance, spotting, torpedo-carrying and fighting. This 'first generation' of fleet aircraft consisted of wartime types such as the Sopwith Camel, Sopwith Cuckoo and Parnell Panther. They were mostly employed trying to operate from a carrier and, as an inefficient longitudinal arrester system was in use, something of the order of one in four of the planes suffered damage of some sort when landing and one in twelve proved to be beyond repair. These casualties, however, were in aircraft and not pilots and serious though the rate of loss was, it was nothing when compared with the certain loss of the 'one-shot' aeroplanes in the later days of the Grand Fleet.

In the early twenties friction between the Royal Navy and the Royal Air Force was continuous. Many of the points at issue now seem somewhat trivial but they all stemmed from the fundamental difference of outlook between the two services on the use of aircraft in war coupled with a lack of money to meet all their needs. Probably the most serious of the Admiralty's complaints was that the Royal Air Force refused to allow their pilots to specialize in carrier operations. As a result practically all the carrier's time was taken up in training them to land on a carrier and little progress could be made with more advanced exercises. By 1923 the Royal Navy's complaints were so numerous and had so much support in Press and Parliament that the Balfour Committee was set up to consider the matter. The case of the Navy was that R.A.F. control of its air arm made things as awkward as they would be if it was the Army's Royal Artillery who manned the naval guns afloat. The Royal Air Force's case was based on their theory of the 'unity of the air', by which, in its extreme form, even the carriers would be under their control.

The Balfour Committee arrived at a compromise, the most important points of which were that the aircraft allocated for operations from H.M. ships should be called the Fleet Air Arm of the Royal Air Force but should be paid for and owned by the Navy. Seventy per cent of the pilots and all the observers were in future to be naval officers. Initial training of the pilots, who would also hold R.A.F. rank, would be done by the R.A.F. The existing method by which the Admiralty stated their requirements for aircraft and the Air Ministry developed them to meet those requirements remained unchanged. In spite of the change of ownership the aircraft, when ashore, were still to come under R.A.F. command. Neither side was very enchanted with this compromise but both decided to make it work. Although the Royal Navy had hoped for a complete return to the R.N.A.S., it was nevertheless a considerable gain for them.

In fact in the five years following the First World War the record of the R.A.F. in developing naval aircraft was more than reasonable; in spite of the economy drive they had produced nine new types to meet the Admiralty's requirements and from these the 'second generation' of naval aircraft were evolved. These were the Fairey Flycatcher fighter, the Blackburn Blackburn spotter, the Fairey 111D reconnaissance plane and the Blackburn Dart torpedo aircraft, and they were a considerable advance on their predecessors. By 1925, when the Balfour Committee's findings became effective, the number of aircraft in carriers had in fact been more than doubled. One hundred and eight aircraft were embarked in the *Argus*, *Eagle* and *Hermes*, the seaplane tender *Pegasus*, the cruiser *Vindictive* and the *Furious* (see Fig. 6, p. 106) which had just completed her third alteration into a clear-deck aircraft-carrier with the funnel trunked aft. In spite of bickering and divided control, British naval aviation afloat led the world. This had been achieved however at the expense of the Coastal Area's own operational aircraft which amounted to no more than a dozen or so flying-boats and seaplanes of war-time types.

In America the agitation for a separate air force was prac-tically continuous in the years following the war. In April 1921 however President Harding ruled that aviation could not be separated from either the Army or the Navy. In March 1922 the first U.S. aircraft-carrier, the *Langley*, was commissioned and later in the year successful take-offs and landings were made, the U.S.N. being initially more successful than the British with their arrester gear. It had been hoped to include in the *Langley*'s complement some of the Martin bombers that had been used in the 1921 bombing trials but they proved to be too large. The U.S.N. policy in 1922 for carrier-borne aircraft was to develop high-speed fighters, medium-sized torpedo planes and slow speed scouting aircraft, but the first aircraft to fly from the *Langley* were existing types such as the Aeromarine 39B trainer, Vought VE7 general purpose aircraft and the Naval Aircraft Factory TS fighter; it was not until 1924 that a torpedo plane landed on her.

The American Navy began in this period to develop a new technique of dive-bombing in which they hoped to be able to secure a much greater number of hits. Plans were also made in the U.S.N. to carry aircraft in battleships and cruisers and a launching catapult was developed. The U.S. rigid airship programme however, suffered a severe setback in 1921 when *ZR2*, which was the experimental British *R38* of advanced design which they had acquired, broke up over the Humber whilst on trials and was lost. After this disaster there was at that time only one small U.S. blimp left in commission and although a number of other small airships of European design were bought for trials, few of them were found to be of much use. The *Shenandoah* (*ZR1*) built in the U.S.A., was not completed until 1923 and the *Los Angeles* (*ZR3*) was finished in Germany during the following year, crossing the Atlantic in October 1924.

Although naval aviation was still considered to be auxiliary to the battlefleet as the dominant factor in sea power, one can even at this early stage discern in the U.S. Navy the vision that aircraft might one day replace the battleship. They certainly believed that a naval action would be preceded by an air battle for air superiority and that it was essential to win this if the subsequent surface action was to have a favourable outcome.

By 1925 the U.S. Navy had over two hundred operational planes. Many of these such as the Curtiss H16 and F5L flying-boats and the Vought VE7 seaplane were wartime types or improvements on them, but the Douglas DT and Curtiss CS torpedo planes and the Naval Aircraft Factory TS fighter were new. They still had only one carrier, the *Langley*, which carried 34 planes, and about half of their total number of aircraft were shore-based. Some flying-boats were based on the seaplane tender *Wright* and so were strategically mobile, and some seaplanes were carried in battleships and cruisers. There were also the two rigid airships, with a mobile mooring mast in the tender *Patoka*, and one non-rigid airship. Although they had only one carrier, U.S. naval aviation was therefore now nearly twice the size of the British.

The British sent an aviation mission to the Japanese Navy in 1921–3 and under their guidance, the Japanese laid down their first aircraft-carrier, the *Hosho*, which was completed in 1922 and from which operations began early in 1923. In 1925 the *Hosho* carried 30 planes and there were another 130 aircraft based ashore in Japan. Many of these were British types and some were German, but all were built in Japan and the Japanese were rapidly developing their own designs.

Of the other nations of the world many of the maritime states now had shore seaplane bases. None of them however, except France, contemplated building aircraft-carriers and there were only a few small seaplane tenders amongst their fleets. France had operated the two rigid airships *Méditerranée* and *Dixmude* since the war, both of which were ex-German Zeppelins; but the *Dixmude*, the ex-German *L72*, was lost in 1923 over the Mediterranean in a thunderstorm.

The period 1925–30 was one of considerable expansion of naval aviation by all the main naval powers. Great Britain completed the conversion of her remaining two light battle-cruisers to aircraft-carriers, the *Courageous* being finished in 1928 and the *Glorious* in 1930. These ships were improvements on the *Furious*; they were of the 'island' type and were able in consequence to carry 42 instead of 36 aircraft. Their gun armament consisted of fourteen 4·7-inch H.A. guns which was the heaviest anti-aircraft armament mounted in any ship so far. Each of the three ships had two hangars, one above the other, and a lower flying-off deck forward from which fighters could fly straight off through doors in the forward end of the upper hangar.

Towards the end of this period the 'third generation' of Fleet Air Arm aircraft came into service (see Fig. 7, p. 118). These were all heavier and more powerful than their predecessors, and of improved performance. The Hawker Nimrod was an adaptation of the latest R.A.F. interceptor fighter designed for the air defence of the U.K. and was of comparable performance. The Blackburn Ripon torpedo plane was a two-seater and so better able to navigate and find its target. In two further main types we see the beginning of a policy of multiple purpose air-craft: the Fairey IIIF had the functions of reconnaissance, bombing and spotting combined and so replaced both the IIID and the Blackburn spotter; the two-seater Hawker Osprey[8] adapted from the R.A.F. Demon fighter, combined the duties of fighting and reconnaissance. The IIIF and the Osprey both fitted with floats were to be used as ship-borne catapult aircraft for the new 10,000-ton cruisers, but by the end of this period very few had been embarked. By 1930 the strength of the Fleet Air Arm had increased to 144 aircraft organized in 24 flights of six aircraft each. There were two carriers in the Atlantic and two in the Mediterranean, and a fifth one in the Far East. The sixth was generally refitting or in reserve.

Co-operation between the R.A.F. and the Navy within the terms of the Balfour ruling was reasonable during this period although the great economy drive of those days cut down sea and flying-time so that it was difficult to achieve efficiency or make very much progress in tactics. There was, however, con-siderable friction between the Admiralty and the Air Ministry over what was left of the Coastal Area; in 1926, other than the training aircraft for the Fleet Air Arm, this only comprised one flying-boat squadron at home and another, with an interim complement of float-planes, in the Mediterranean. The Air Ministry resented the Admiralty's intrusions into what they

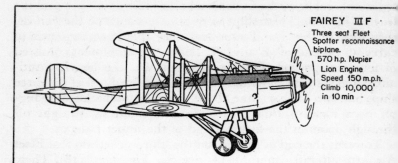

FAIREY III F

Three seat Fleet
Spotter reconnaissance
biplane.
570 h.p. Napier
Lion Engine
Speed 150 m.p.h.
Climb 10,000'
in 10 min -

HAWKER OSPREY

Two seat Fleet Fighter
reconnaissance biplane.
480 h.p. Rolls Royce
Kestrel Engine.
Speed 190 m.p.h.
Climb 20,000'
in 14 min

HAWKER NIMROD

Single seat Fleet Fighter
480 h.p. Rolls Royce
Kestrel Engine
Speed 205 m.p.h
Climb 20,000' in 10 min

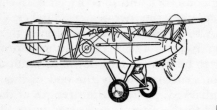

BLACKBURN RIPON

Two seat Fleet
Torpedo Bomber
biplane.
530 h.p. Napier
Lion Engine
Speed 125 m.p.h.
5,000' in 10 min

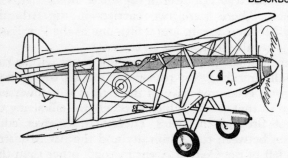

7   THE 'THIRD GENERATION' OF FLEET AIR ARM AIRCRAFT

considered was entirely their concern, but they undoubtedly gave Coastal Area a very low priority.

In 1924 the seaplane-carrier *Pegasus* had taken 6 Fairey IIID seaplanes to the Far East where they had made a survey of Malaya and visited Borneo and Hong Kong. She was however paid off in 1925 and so this eminently sensible arrangement whereby the Navy gave strategic mobility to units of Coastal Area came to an end. Coastal Area had, in 1926, received some of the new Southampton flying-boats and two of these made a tour by themselves of the Mediterranean. In 1927 four new flying-boats, each of a different type, made a trial cruise in the Baltic and, later in the same year, 4 Southamptons flew all the way to Singapore in stages and then round Australia. Successful though these flights were, the range of the Southampton of just over 500 miles was little better than the wartime F3 and was still very short for ocean reconnaissance.

By 1930 the R.A.F.'s complement of maritime aircraft had expanded to a total of 26 operational planes all except six of which were new flying-boats of the Supermarine Southampton, Blackburn Iris or Short Rangoon types. Half of these were based at home at Calshot and Mountbatten and the rest were abroad at Malta, Basrah and Singapore. Co-operation with the fleet was, however, seldom attempted and their operations seem in retrospect to have been very much an attempt to 'go it alone'. It is of interest that the Admiralty were not even consulted over the stationing of No. 203 (Flying-Boat) Squadron at Basrah in the Persian Gulf and the flying-boats at Singapore were sent there as search units for the shore-based R.A.F. torpedo-bomber squadrons, about to be sent out there, rather than to work with ships.

In 1927, the United States Navy commissioned the two giant aircraft-carriers *Lexington* and *Saratoga* of 36,000 tons, for which special provision had been made in the Washington Treaty. They had flight decks 800 feet long and 100 feet wide and they were able to carry over 70 planes each. These two carriers took part in the 1929 and 1930 exercises for the defence of the Panama Canal and in these operations the independent carrier group first began to take shape, operating on its own ahead of a supporting battlefleet. American naval aviation made great progress in these exercises.

Although agitation for a separate air force had continued in America, various boards, appointed by Congress and the Navy and Army Departments, turned the idea down. In 1925, the Morrow Board went so far as to say that the separate procurement of aircraft from industry by the Army and Navy was beneficial as it lead to competition. In 1926 a five-year programme to build a thousand aircraft for the Navy was begun and they decided against multiple-purpose types as they believed that they would be unable to carry out any of their functions properly. Emphasis was put upon high-performance fighters and dive-bombers as well as other types. There was some criticism of the great size of the *Lexington* and *Saratoga*, not only because of their cost and because they were considered to represent 'too many eggs in one basket', but also because they took up such a large proportion of the tonnage of aircraft-carriers allowed by Treaty. These ships had a heavy surface armament of eight 8-inch guns, the largest allowed by the Washington Treaty, showing that it was still thought that they might have to engage other ships.

On 3rd September 1925, the airship *Shenandoah* was wrecked in a thunderstorm over Ohio with the loss of fourteen lives. This disaster did not daunt the enthusiasm of the airship enthusiasts; indeed it seems, in spite of the earlier loss of *R38* and the French *Dixmude*, actually to have increased it. In 1928 the Bureau of Aeronautics obtained authority to build two more very large airships, each over three times the volume of the largest German Zeppelin built during the war. They were to carry in a hangar six small fighter aeroplanes which could be released and recovered in flight and which would be used to defend them against attack. It is probable that the construction in Great Britain of the large civil airships *R100* and *R101* influenced the decision to build these airships. The Admiralty did, in fact, show some slight interest in using *R100* and *R101* for reconnaissance in war provided it did not cost them anything and they had no responsibility for them. During this period the *Los Angeles*, sole survivor of the U.S. Navy's rigid airships, did a great deal of flying and experimental work and the *Akron*, the first of the new airships, was laid down in 1929.

By 1930, although the Americans had only three aircraft-carriers to the British six, they had over 200 aircraft embarked. These included the Boeing F4B fighter, the Vought O3U Corsair observation plane and the Martin T4M torpedo plane, which had similar performances to their British opposite numbers of the 'third generation'. The British however had no equivalent to the Curtiss F8C Helldiver, a dive-bomber with a range of 720 miles which could carry a 500-lb bomb. Another 200 aircraft were carried in battleships and cruisers or based on seaplane-tenders. The Americans had therefore taken the lead in naval aviation from the British who, even when the aircraft of the Coastal Area were included, had less than half this number. It must be remembered of course that the U.S.A. could afford a larger expenditure on naval aviation as they did not have to find money for an independent air force.

The Japanese had completed the conversion of the *Akagi* from a battle-cruiser hull and the *Kaga* from a battleship hull in 1927 and 1928. They now had as many aircraft-carriers as the United States. These two ships were however, slower than the *Lexington* and carried fewer planes. They were of the flush-decked type which ejected their smoke through ducts in the side and, like the *Lexington*, they carried 8-inch guns. European types of aircraft had now been superseded by Japanese designs. With the shore-based naval air squadrons based in Japan, and some float-planes carried in battleships and cruisers, aircraft strength had also risen to a greater number than the British.

In 1927 the French completed their first aircraft-carrier, the *Béarn*, converted from the hull of the battleship *Normandie*. In the following year the French merged their Navy and Army air arms into a single independent air force. Special arrangements were made for the administration and command of aircraft actually embarked in ships as well as for co-operation with maritime air units stationed ashore. The *Béarn* had an air group of forty planes which included fighters, scouting planes and bombers. Scouting seaplanes were also carried in eight cruisers, and all of these aircraft were completely under the control of the Navy. Ashore each Region Maritime had its shore-based naval air force which consisted of fighters for the defence of the naval bases, and also of bombing and reconnaissance aircraft, the majority of which were seaplanes or flying-boats. These were air force units but came under the operational control of the Admiral commanding the Region.

The Italian naval air arm had been suppressed by Mussolini in 1923 and absorbed into the Regia Aeronautica. A force called 'Aviazione per la Marina' was, however, formed and by 1930 they had established a considerable number of flying-boat and seaplane stations round their coasts. While these aircraft carried naval officers as observers and in theory were under the operational control of the Navy, there was in practice little co-operation with the fleet except with the few seaplanes carried in their new cruisers which were fully controlled by the Navy.

The Treaty of Versailles did not allow Germany to have any military aircraft at all and so the small German Navy officially had no air arm. By various subterfuges, however, they managed to carry out experiments with naval planes and even to manufacture and crate ten Heinkel flying-boats for use in war. They were also able to carry out some exercises with the fleet using civil aircraft which were allowed for towing targets for the anti-aircraft guns.

In 1927 General Douhet published a second book in which he reaffirmed his belief in the bomber as the weapon of the future and in his strategy of gaining command of the air at the outset of a war. He still maintained that air defence was futile and that everything should be concentrated on bombing. The main change in this book was that he now asserted that auxiliary air forces, by which he meant units especially intended for co-operation with the Navy and Army, were a waste of effort. He believed that they would be unable to function unless the independent air force gained command of the air, and that it was the business of the independent air force to gain air superiority over the sea as well as over the land. Having gained command of the air the independent air force could attack the enemy naval bases and prevent the opposing navy using its own auxiliary air force. From this he deduced that it was better to strengthen the independent air force rather than to divert aircraft to auxiliary purposes. But fundamentally he believed that having gained command of the air, the best way to win a war was to bomb the enemy's cities and that all other operations were irrelevant.

Between the wars we can discern that there were four schools of opinion on the future of aircraft and sea power. Between and within these four schools, of course, there were various shades of opinion. The first of these schools and the most 'air-minded' was the Douhet school, believing in an independent air force and no naval air arm at all and in its extreme form that wars of the future would be decided by air power alone and that navies were obsolete and irrelevant. In a more moderate form this school admitted the need for navies but held that the flexibility of air power would allow aircraft to operate against ships and naval bases if it proved necessary. One can discern a flaw in Douhet's arguments when he says that the only purpose of an auxiliary air force for a Navy is for use outside the range of shore-based aircraft. Nevertheless the influence of this school was very strong in all the independent air forces in Europe including the R.A.F.

The second school was that typified by the development of naval aviation in the U.S. Navy. American naval aviation agreed with Douhet that command of the air was essential in naval operations. Where they differed was that they believed that air operations were inseparable from the operations of the Army or Navy. By taking its own air force with it, the American Navy hoped to command the air wherever it went. It believed that an air battle for command of the air by the aircraft carried with the fleet would precede any naval battle. Subsequently aircraft would be provided for action observation, spotting, smoke-laying and all the other auxiliary functions necessary to co-operate in a surface battlefleet action. Official opinion in the U.S. Navy was still that the surface fleet action was what ultimately mattered, but there was a growing school which believed that the issue would already be decided by the air battle which would precede it.

The third school was that represented by the development of the British Fleet Air Arm in which all the functions of aircraft at sea were entirely auxiliary to the operations of a battlefleet and all were devoted to securing a decision by this means. This school believed that air reconnaissance, spotting, smoke-laying, the attack on the enemy battlefleet by torpedo-bombers and the defence of one's own battlefleet by fighters, were essential elements in modern naval warfare and would make the difference between victory and defeat. Nevertheless in this school's mind the battlefleet was paramount as is shown by a remark of Admiral Chatfield, the British First Sea Lord, when making his case to regain control of the Fleet Air Arm, that naval aircraft were an integral part of the Royal Navy and *second only in importance to naval gunnery*.

Finally at the nautical end of the scale were those who still believed that a Navy was capable of 'going it alone'. They considered that the power of aircraft had been enormously over-estimated, and that when the weather, the development of anti-aircraft gunfire and improved protection in ships were taken into account, ships would be able to operate and that the air would not be able to stop them. Aircraft sank no important ships during the First World War and they believed that they were unlikely to do so in the next. It is only fair to say that no responsible naval administration followed such a policy, nevertheless substantial bodies within practically every navy held these opinions and even larger bodies, although in their heart of hearts they did not believe them, certainly hoped that they were true.

A government trying to decide whether to have two separate naval and army air forces or a single independent air force, or some compromise, would have to take into account the opinions of these four schools. The opinions themselves were largely influenced by geography and it was geography which was undoubtedly the deciding factor. It was obvious that Douhet's theories were only applicable to the European theatre where ranges were short, and could not apply to the U.S.A. and Japan which were out of bomber range of any potential enemy. Thus for example Italy with shore bases throughout the Mediterranean was the only one of the Washington Treaty powers not to build any aircraft-carriers at all and to have a single air force even the maritime parts of which remained firmly under air force command.

In the U.S.A. the annual exercises of 1931 proved to be a setback for the aircraft-carriers and it was concluded by one of the participating Admirals that aircraft could not prevent an amphibious force from making a landing when covered by a battlefleet. In the 1932 exercises, however, the aircraft-carriers re-established their reputation and Admiral Yarnell, who commanded them, urged that more aircraft-carriers should be built. The cruiser float-planes in these exercises had not proved a success: although they could be launched by catapult they had to land on the sea to be recovered and this was seldom possible. There was a great deal of discussion about building a type of ship which was to be a cross between a cruiser and a small aircraft-carrier. This suggestion, combined with the 'too many eggs in one basket' argument and the high cost of operating the giant *Lexington* and *Saratoga*, led to the laying down in 1931 of a small aircraft-carrier called the *Ranger*; she was commissioned in 1934 and was of 14,500 tons but carried no fewer than 76 aircraft.

In the mid-thirties, the U.S. Navy's current generation of aircraft included the new Grumman F2F fighter, a biplane with a retractable undercarriage and an enclosed cockpit, with a top speed of 231 m.p.h. and a range of 985 miles, the Martin BM dive-bomber which was able to carry a 1000-lb bomb, and the Consolidated P2Y flying-boat with a range of over a thousand miles. The F2F was far superior to any British Fleet Air Arm fighter and the P2Y had double the endurance of Coastal Command's Southampton flying-boat. The British had no equivalent to the Martin BM which could launch its bomb with great accuracy in a terminal velocity dive, and had made excellent practice against a radio-controlled target ship at sea.

The rigid airship *Los Angeles* continued to fly successfully and although she made many scouting trips during exercises she was generally 'shot down' by carrier-borne fighters. In October 1931 the *Akron* was commissioned and she operated successfully throughout 1932, but in April 1933 she was wrecked in a storm off the Delaware coast and 73 lives were lost including that of Admiral Moffett, the Chief of the Bureau of Aeronautics. In spite of the disaster to the British *R101* the year before, enthusiasm for this type still did not seem to suffer. The *Los Angeles*, which had been paid off, was recommissioned to take the *Akron*'s place; and work on the new *Macon* was continued. *Macon* was completed in the middle of 1933 but her performance in exercises was no better and, in spite of the fighters she carried, she was usually assessed as destroyed. Early in 1935 she too was lost after a structural failure off the Californian coast. This second accident was too much: the U.S. Navy continued with blimps of various types but built no more rigid airships. The loss of the German commercial airship *Hindenburg* in America in 1937 finally put paid to any further development of rigid airships, military or commercial, in any country. In any case the Consolidated PBY Catalina flying-boat, which had been produced in 1936, had a range of over 2000 miles. The *Langley* had been converted to a seaplane tender to give these aircraft strategic mobility and two new tenders, the *Curtiss* and *Albemarle*, were laid down. This combination produced a long-range reconnaissance force superior to rigid airships and far superior to the short-range British flying-boats of Coastal Area working from fixed bases.

In 1934 the U.S. Navy laid down two medium-sized aircraft-carriers of 20,000 tons, the *Yorktown* and the *Enterprise*, and they were commissioned in 1937 and 1938. They carried as many aircraft as the *Lexington* and *Saratoga* and were only just over half the size. The Naval Expansion Act to build a two-ocean Navy was passed in 1938 and it was approved for naval aviation to be expanded to a strength of 3000 planes. The aircraft-carrier *Wasp* was also authorized under this programme and as there was not enough Treaty tonnage left to repeat the *Yorktown*, she was laid down as an improved *Ranger*. By 1938 a new generation of carrier-borne aircraft were in service or just coming into service; these included the Brewster F2A Buffalo, the Vought SB2U Vindicator and the Douglas TBD Devastator. All these aircraft were monoplanes with engines approaching 1000 h.p. U.S. naval aviation had kept to their policy of single-duty types and they were specialized as fighters, dive-bombers and torpedo-bombers, and, as a result, were greatly superior to their British counterparts and capable of competing with most shore-based aircraft.

The third generation of British Fleet Air Arm aircraft, already discussed, continued in service until the mid-thirties. The lives of some types were extended by fitting more powerful engines. In 1934 the British laid down a new aircraft-carrier, the *Ark Royal*, which brought them up to the full tonnage allowed by the Washington Treaty. The *Ark Royal* was completed in 1936 and except that she had no lower flying-off deck, was an enlarged *Glorious* carrying 55 planes. During the early thirties many fleet exercises were carried out in the Home and Mediterranean Fleets and the co-operation of carrier-borne aircraft with the fleets was much improved. Co-ordinated attacks by high-level bombers, torpedo-planes and strafing fighters were perfected but dive-bombing was only carried out by fighters using 20-lb bombs. The aim was still to assist the battlefleet and not to operate separately. Carriers did however work together on occasion, notably in 1933 when the *Furious*, *Courageous* and *Glorious* operated as a squadron under the Rear-Admiral Aircraft-Carriers. There was much discussion about the role of carriers in a fleet action and some of the more far-sighted officers suggested that their aim should be to knock out the enemy carriers rather than to attack the enemy battlefleet.

In 1934 the British Navy had 164 aircraft embarked, 138 of which were in five aircraft-carriers and the rest in five battleships and eighteen cruisers. Only one-third of the capital ships and cruisers in the Royal Navy therefore carried aircraft. In the U.S. Navy the total number of aircraft embarked was 355 which was over double the number in the Royal Navy. Of these 234 were in four aircraft-carriers and the rest in 15 battleships and twenty-four cruisers. The American carriers carried an average of twice as many aircraft as the British. Although some of this discrepancy may be accounted for by the immense size of the *Lexington* and *Saratoga,* the principal reason was that the British only carried as many as could actually be stowed in the hangars whereas the Americans not only filled the hangars but carried a large number permanently on the flight deck. Similarly the majority of the U.S. battleships and cruisers had two catapults and carried three aircraft each.

Relations between the Royal Air Force and the Fleet Air Arm at a command level were reasonably harmonious during this period, but in 1935 the Admiralty renewed its crusade for complete control. In 1937, after the First Sea Lord had threatened to resign, Sir Thomas Inskip, the Minister for the Co-ordination of Defence, decided that ship and carrier-borne aircraft should be put completely under the control of the Admiralty, but that the Coastal Area and the initial training of pilots should remain under the R.A.F. This decision has often been hailed as the emancipation of the Fleet Air Arm and a point of very great importance to its development. In fact it altered little except that the Navy now retained command of the aircraft when ashore and that maintenance was done by naval personnel instead of Royal Air Force fitters and riggers. It is true that R.A.F. personnel would no longer have to serve afloat in H.M. Ships, that the naval pilots no longer had to have R.A.F. rank and that administration was simplified. Against this the Navy had to take on the very great burden of a number of shore air stations and a huge training programme for maintenance personnel. It is significant that the Admiralty did not press for any change in the method of supplying aircraft for the Fleet Air Arm and this continued to be done by the Air Ministry.

During the negotiations for the change, the Admiralty asserted that the existing system had been proved to be a complete failure and that the Royal Air Force took only a slight interest in the Fleet Air Arm. It was certainly true that the Fleet

Air Arm had fallen behind the U.S. Navy in numbers and the performance of its aircraft. The requirements for the type of aircraft and their function, tactics and operational training were, however, already the responsibility of the Admiralty. There was no suggestion at this time that the types of aircraft provided by the R.A.F. were not what the Admiralty wanted or failed to meet their requirements. The Admiralty's desire to expand the Fleet Air Arm was hindered not so much by the R.A.F. as by money which came out of the Navy votes. The most substantial complaint seems to have been that maintenance personnel could not be found by the R.A.F. for some of the ship-borne planes which consequently had to be landed. Nevertheless the change, if nothing else, stopped a great deal of inter-service bickering and the Navy felt that it could now develop its air arm in any way it wished.

At the same time as the Admiralty took over complete control of the Fleet Air Arm, a new generation of aircraft began to come into service. The requirements for this 'fourth generation' had, of course, been stated well before 1937. The Admiralty policy was to produce a smaller number of multi-purpose types. The first step was to amalgamate the torpedo-bomber with the spotter-reconnaissance type and the result was first the Blackburn Shark, and then the famous Fairey Swordfish which went into production in 1935. The second multi-purpose type was the Blackburn Skua, or the Royal Navy's first dive-bomber, with which function it was to combine the duties of reconnaissance and those of a fighter. For ship-borne use the small Supermarine Walrus flying-boat was produced, which was a robust amphibian, albeit of low performance. For light cruisers with small catapults which could not take the Walrus, there was the Fairey Seafox, little more than a light aircraft with floats. The advantages of these types were ease of operation from ships and carriers and their flexibility: an entire carrier air group composed of Swordfish and Skuas could be used for reconnaissance or for striking the enemy. The Skuas could also be used as fighters and the Swordfish for anti-submarine patrol, smoke-laying or spotting. The Walrus could be catapulted, land on an airfield ashore or on an aircraft-carrier and on a far rougher sea than could its float-plane predecessors. The disadvantage of this fourth generation was that they were all of low performance in the air compared with shore-based aircraft. The Admiralty seems to have realized well before the war that this policy was questionable and that they were going to need better fighters. The use from carriers of high-performance types such as the Hurricane and Spitfire, now about to come into service in the R.A.F., was thought to be out of the question. The need was met therefore by the Blackburn Roc which was an adaptation of the Skua with four machine guns in a turret but it turned out to be even slower. Just before the war they acquired a naval version of the R.A.F. single-seater Gladiator biplane fighter, which, although considerably faster than the Roc, was 70 m.p.h. slower than a Hurricane.

The Admiralty's neglect of the fighter seems in retrospect very strange. It may well have been influenced by the great improvement in anti-aircraft gunnery and in the protection of the new and modernized ships of the fleet. The eight-barrelled multiple pom-pom was now in service and four were to be mounted in each capital ship. As ships were modernized their anti-aircraft batteries were doubled and the standard armament in battleships and cruisers was now eight 4-inch H.A. guns. Plans for the modernization of the battleships of the 'Queen Elizabeth' class included as many as twenty 4·5-inch H.A. guns and they had to dispense with their secondary gun armaments altogether to make room for them. The modernization of these ships included the fitting of armoured decks as well as greatly improved underwater protection.

In 1936 the Washington Treaty expired and the British decided to replace all their older aircraft-carriers. Two new ships, the *Illustrious* and *Formidable*, were authorized in 1936 and two more, the *Victorious* and *Indomitable*, the following year. These aircraft-carriers were of about the same size as the *Ark Royal* but were more heavily armoured. Their anti-aircraft armament included sixteen 4·5-inch H.A. guns and six multiple pom-poms but for these characteristics a substantial price had to be paid and the number of aircraft carried fell to 33. The Admiralty did their best to catch up the American lead in ship-borne aircraft and all the new and modernized battleships and cruisers were designed to carry three Walrus amphibians or seaplane editions of the Swordfish.

Up to 1934 the R.A.F. Coastal Area remained basically a flying-boat force and co-operated little with the Navy. The only increase of strength since 1930 was in the Far East where there were now two squadrons of torpedo bomber aeroplanes in addition to the flying-boats, all based ashore at Singapore. In

1934 the British Government, worried by the rapid growth of the new German air force, decided virtually to double the size of the Royal Air Force and the Coastal Area was included in this expansion. In 1936 Coastal Area was raised to the status of a Command and had already been increased to four flying boat squadrons at home, No. 201 Squadron at Calshot having the new London flying-boats with a range of 700 miles or so. Unfortunately two other still more modern types of flying-boat, the Stranraer and the Lerwick, turned out to be failures. The Air Ministry had decided in any case to use cheaper twin-engined aeroplanes in the general reconnaissance role at sea. Their choice for this purpose was the Avro Anson which had a slightly greater range than the London flying-boat. Two squadrons of Ansons joined Coastal Command in 1936 and at the same time the Air Ministry decided to adapt an Imperial Airways commercial flying-boat for military purposes. The Sunderland, as this aircraft was called, was a great improvement on the London and was expected to have a range of well over a thousand miles. Up to now Coastal Command at home had been entirely a reconnaissance and patrol force but it was now decided to add a squadron of shore-based torpedo-bombers to form the nucleus of a striking force.

In 1937 co-operation between Coastal Command and the Navy improved noticeably and the first experiments were made with an Area Combined Headquarters at Portsmouth. At the end of the year the Admiralty, nervous that the Ansons might be taken away in emergency to join the bomber force, obtained a Government decision that Coastal Command's role was to be confined solely to the support of naval forces. This was a decision much resented by the R.A.F. as it struck a blow at the theory of the unity of the air and the flexibility of air forces. By the end of 1937 the strength of Coastal Command had increased to six flying-boat squadrons, eight Anson squadrons and two torpedo-bomber squadrons at home, although strength abroad was unaltered. The total number of R.A.F. aircraft allocated for the support of maritime operations was now nearing 200, which was a spectacular advance from the 12 aircraft of a dozen years before.

The German Navy by 1933 when Hitler came to power had secretly developed prototype aircraft for reconnaissance, fighting, torpedo-bombing and minelaying, and were testing a dive-bomber. The decision was then made to have a separate independent air force. The Luftwaffe was formed and Göring, who was appointed its head, was determined to control all military aircraft and to oppose any form of naval air arm. In 1935, however, the Luftwaffe set up Air Command VI for maritime purposes and allowed the Navy to exercise tactical command in operations. Nevertheless Göring announced in 1938 that the Luftwaffe was responsible for the conduct of air warfare over the sea, which he considered included minelaying and attacking ships at sea as well as attacks on naval bases and ships in harbour: the operational control of aircraft by the Navy was confined to reconnaissance and planes actually working with warships. The only purely naval aircraft allowed were the few float-planes carried in their pocket battleships and cruisers. In 1935 the Anglo-German Naval Agreement had been concluded whereby Germany was allowed to build up to 35 per cent of Great Britain's tonnage of all types of ship. Shortly afterwards they laid down a medium-sized aircraft-carrier called the *Graf Zeppelin* and she was launched in 1938.

The Japanese Naval Air Force under the Fleet Replenishment Programmes began to expand after 1930. They completed the small aircraft-carrier *Ryujo* of the same size as the *Hosho* in 1933. Two larger ships, the *Hiryu* and the *Soryu* of 16,000 tons were launched in 1936–7, thereby keeping step with the United States. In the period 1936–9 they built four seaplane carriers of between 9000 and 11,000 tons, which had a speed of 28 knots, mounted six 5·1-inch guns and carried between twenty and thirty seaplanes. These ships were intended to operate their seaplanes at sea with the fleet and were not simply mobile flying-boat bases like the American seaplane tenders; except that they had catapults, they suffered from all the disadvantages of the old seaplane carriers of the First World War and it is difficult to understand why they were built except that they did not rate as aircraft-carriers by the Washington Treaty and so their tonnage did not count. The Japanese also built two hybrid cruisers, the *Tone* and *Chikuma* of 14,000 tons each of which carried six seaplanes. The Japanese at this time produced twin-engined naval aircraft to work from shore bases. The Navy 96 ('Nell')[9] is an example with a speed of 270 m.p.h. and carrying 1100 lbs of bombs for 2125 miles; in performance it compared very favourably with similar types of aircraft in other countries and had the added advantage that it was an aircraft designed specially for working over the sea. It was in 1937–9 also that the

Japanese produced the Type 97 ('Kate') and Type 99 ('Val')
carrier-borne torpedo and dive-bombers which, although
slower than their American counterparts, had a considerably
greater range.

On the expiration of the Washington Treaty in 1936 the
question of replacing the battlefleets of the world became an
urgent one. This step was bound to be a very expensive business
and it encountered considerable political and journalistic
opposition. The antagonists raised the slogan of a 'thousand
aircraft for a battleship' which had first been suggested by
Douhet in 1921 and reiterated by 'Billy' Mitchell and others.
It was based on the cost of an aircraft being £10,000 and a
battleship £10,000,000. In fact, when the short life of an aircraft was taken into account and the very large overheads in
airfields and training to keep it flying, it was calculated that
only some forty aircraft could be kept in service for the same
cost as a battleship. In this controversy the British First Sea
Lord pointed out that if Great Britain rebuilt her battlefleet
and the new ships were sunk by aircraft, money would have
been wasted, but if the battlefleet was not rebuilt and aircraft
proved unable to sink modern battleships then the other powers
who had rebuilt their battlefleets would gain command of the
sea and a future war would inevitably be lost. In the end all the
principal naval powers including the U.S.A. began to replace
their battlefleets. It does not seem that any of the naval powers
at this time seriously considered building more aircraft-carriers
instead of battleships. Even in the United States the official
naval opinion was that seapower still depended upon battlefleet
supremacy.

The influence of aircraft on the surface fleets of the world
was, however, considerable. The new British and American
battleships were designed very much with air attack in mind
and had thick armoured decks and heavy anti-aircraft armaments (see Fig. 8, p. 133). In both cases the secondary anti-
destroyer batteries were suppressed to make room for them. The
'King George V' class had a 5-inch armoured deck, sixteen
5·25-inch H.A. guns and eight multiple pom-poms. In the
crisis with Italy over Abyssinia in 1935, the British produced a
new type of warship in an attempt to compete with aircraft;
this was the anti-aircraft cruiser in which the whole armament
was composed of anti-aircraft guns. The British first converted
some of their older ships to this role and then laid down the

1914

Two 3" A.A.

H.M.S. 'IRON DUKE'
First Battleship to mount
anti-aircraft guns.
For use against Zeppelins

1918

3" A.A.

4" A.A.

H.M.S. 'BELLEROPHON'
Typical Grand Fleet Battleship
Two anti-aircraft guns.
Two Sopwith Camel fighter
aircraft on turret platforms

1929

4" H.A.

H.M.S. 'REVENGE'
As refitted. Four 4" H.A. guns
but only primitive control.
Eight twin Lewis guns.
Fighters now all in Aircraft
Carriers

1933

Multiple
pom pom

H.M.S. 'BARHAM'
As modernised. Four 4" H.A. guns
with director control for use
against high level bombers.
Two 8 barrelled multiple pom
poms for use against torpedo
planes. Fairey III F spotter
reconnaissance seaplane on
catapult

1937

Multiple
pom poms

Twin 4" H.A.

H.M.S. 'WARSPITE'
As modernised. Eight 4" H.A. guns.
Four multiple pom poms. 5"
and 3½" armoured decks.
Two Fairey Swordfish T.S.R.
seaplanes with catapult and
hangars

1940

H.M.S. 'KING GEORGE V'
New Battleship. Anti-aircraft
and secondary armament
combined. Sixteen 5·25" H.A.
guns in four separate batteries.
Eight multiple pom
poms. Air Warning
Radar.
Three Supermarine Walrus
Amphibious with catapult and
hangars.
6" and 5" armoured decks

**8   THE EFFECT OF AIRCRAFT ON
BATTLESHIPS 1914-1940**

K

'Dido' class, mounting ten 5·25-inch H.A. guns. The U.S.A. followed suit with the 'Atlanta' class mounting sixteen 5-inch anti-aircraft guns. This policy spread to the smaller classes of warship and the main armament of new destroyers in both Great Britain and the U.S.A. was dual-purpose for use against both ships and aircraft. The need to protect convoys against air attack was obvious and indeed in Great Britain aircraft seem to have been considered a greater menace to seaborne trade than the U-boat. In Great Britain a very substantial re-armament programme of existing sloops with anti-aircraft guns was put in hand. A large number of old destroyers were converted and new escort vessels were laid down, all of which were armed almost exclusively with anti-aircraft guns.

It has been seen how the fundamental policies of U.S. Naval Aviation and the British Fleet Air Arm differed and it is of interest to compare the British and American aircraft-carriers and their aircraft which were in service in 1939–40 and which were the direct result of these different policies. The most suitable ships for this comparison are H.M.S. *Illustrious* completed in 1940 and U.S.S. *Enterprise* completed two years earlier. From Table 2 (p. 136) it will be seen that the British ship is over three thousand tons larger than the American, but is slower and of shorter endurance and carries only half the number of aircraft. On the other hand she has twice as many anti-aircraft guns and is heavily armoured. The whole flight desk is protected by 3-in armour plate and the sides of the hangar and the waterline belt are $4\frac{1}{2}$ inches thick. The British policy was based on making the carrier very tough so that it could continue to operate aircraft against opposition whereas the American policy was that the carrier was more likely to survive if it could put more aircraft into the air.

Table 3 (p. 137) compares the performance of the aircraft carried. The inferiority of the British aircraft in any of their roles is at once apparent. For instance the Skua is 20 m.p.h. slower than the Vindicator as a dive-bomber and only carries half the bomb load. As a fighter it is 95 m.p.h. slower than the Buffalo and has only half the range. Finally the Devastator torpedo-bomber is 65 m.p.h. faster than the Swordfish.

Table 4 (p. 138) gives the performance of some shore-based bomber and fighter aircraft of the day. The performance of the shore-based aircraft is understandably superior to the carrier-borne types, but whereas the American Buffalo fighter can catch all the bombers tabulated, the British Fleet Air Arm fighters can hardly catch any of them. The slowest shore-based fighters, on the other hand, can overhaul all the carrier-borne strike aircraft, both British and American; the fastest fighter tabulated, the Messerschmidt 109, being two and a half times as fast as the Swordfish, produced the odd result that it was to have difficulty in flying slowly enough to shoot it down.

The suggestion by some writers that the very poor performance of the British carrier-borne aircraft was a legacy of the divided control of the Fleet Air Arm between the wars is difficult to substantiate. The Admiralty had laid down the requirements for Fleet Air Arm aircraft for the whole period between the wars, and in their case for control of the Fleet Air Arm made before 1937 did not ask for any change in the system and also appeared satisfied with the results. On the other hand, the Royal Air Force comment that it was difficult to make the Admiralty's requirements fly, clearly has some truth in it. There is little doubt that the British inferiority was more due to the Admiralty air policy than the period of dual control between the wars.

The overall development of naval aviation between the wars had therefore been very substantial. Nevertheless in the minds of all the principal navies, sea power still depended on the battleship. To some this meant the battleship supported by naval aviation; to others it meant the battleship dependent upon naval aviation; but for everyone sea power still meant the battleship.

The organization of naval air forces took different forms according to the different policies, already discussed, of the countries concerned. At one end of the scale were Germany and Italy with independent air forces and virtually no purely naval air arms at all. Although they allocated substantial forces for naval co-operation, in general they relied on the flexibility of air forces for offensive operations over the sea. In the middle were the United Kingdom and to a certain extent France with independent air forces of which substantial parts were allocated to maritime operations and which also had fleet air arms under the control of the Navy. Finally, in the U.S.A. and Japan, there were no independent air forces and as a result the naval air forces were the strongest and most highly developed of all and, of necessity, were capable of operations which in Europe would only be undertaken by independent air forces.

Table 2 – COMPARISON OF BRITISH AND U.S. AIRCRAFT CARRIERS, 1940

	H.M.S. Illustrious	*U.S.S. Enterprise*
Standard displacement	23,000 tons	19,800 tons
Full load displacement	29,100 tons	25,500 tons
Date of completion	May 1940	July 1938
Flight deck dimensions	740′ × 95′	810′ × 114′
H.P. = Speed	111,000 = 30½ kts	180,000 = 32½ kts
Fuel oil = Endurance @ 15 kts	4850 tons = 10,000 miles	5450 tons = 12,700 miles
Armament	Sixteen 4·5″ H.A. Six 2-pdr eight barrelled multiple pompoms	Eight 5″ 38 cal H.A. Sixteen 1·1″ multi-barrelled
Air group	18 Swordfish torpedo spotter reconnaissance 15 Skua fighter dive-bomber reconnaissance — 33 Total	25 Wildcat fighters 36 Dauntless scout dive-bombers 15 Devastator torpedo bombers — 76 Total
Armour	3″ Flight deck 4½″ Hangar sides 4½″ Waterline belt and machinery spaces sides 4½″ Sides ⎱magazines 2½″ Crown ⎰	4½″ Waterline belt 1½″ Deck over machinery spaces

	British				American		
Type	Blackburn Skua	Fairey Swordfish	Blackburn Roc	Gloster Sea Gladiator	Brewster Buffalo F2A	Vought SB2U Vindicator	Douglas Devastator TBD
Date in service	1938	1936	1939	1939	1940	1937	1937
Function	Fighter dive-bomber reconnaissance	Torpedo spotter reconnaissance	Fighter	Fighter	Fighter	Scout dive-bomber	Torpedo bomber
Crew Type	Two-seater monoplane	Two/three-seater biplane	Two-seater monoplane	Single-seater biplane	Single-seater monoplane	Two-seater monoplane	Three-seater monoplane
H.P.	905	690	905	840	1200	825	900
Max. speed	225	139	194	245	321	243	206
Range	760	546 miles with a torpedo	600	425 miles	965 miles	1120 miles	Approx. 500 miles with torpedo 716 miles with 1000-lb. bomb
Armament	500-lb. bomb 5 mg.	Torpedo or 1500-lb. bombs 2 mg.	4 mg. in turret	4 mg.	4 mg.	1000-lb. bomb 2 mg.	2 mg. torpedo or 1200-lb bombs
Ceiling	20,200 ft	10,700 ft	14,600 ft	32,000 ft	33,200 ft	23,600 ft	19,700 ft

Table 4 – TYPICAL SHORE-BASED AIRCRAFT

Type	British		Japanese		Italian		German	
	Hawker Hurricane	Bristol Blenheim	Army I Mk II 'Oscar'	Navy 96 'Nell'	CR42	S79	Messerschmidt 109	Heinkel III
Function	Fighter	Day bomber	Fighter	Day bomber	Fighter	Bomber	Fighter	Bomber
Crew Type	Single-seater monoplane	Three-seater twin-engined monoplane	Single-seater monoplane	Seven-seater twin-engined monoplane	Single-seater biplane	Four-seater three-engined monoplane	Single-seater monoplane	5-6-seater twin-engined monoplane
Max. speed	316	265	325	270	270	255	320	240
Range	600 miles	920 miles	950 miles	2125 miles	535 miles	1190 miles	655 miles	1510 miles
Armament	8 mg.	1000 lbs. bombs 2 mg.	2 mg.	1100 lbs. bombs 5 mg.	2 mg.	2750 lbs. bombs 5 mg.	2–20 mm 2 mg.	2200 lbs. bombs 9 mg.

VI

The Second World War – Norway to Crete

1939 – 1941

A T THE OUTBREAK of the Second World War on 3rd September 1939, the British and French Navies combined were immensely superior to the German Navy, which could not hope, as in the First World War, ever to gain complete command of the sea by defeating the Allied Navies in battle. German policy was therefore to use their ships and U-boats, which were well suited for the task, to harass British sea communications and to defend their coast line. The British, hoping that the Germans would observe international law, believed that the surface raiders were, in fact, a greater menace than the U-boats. Their main fleet was therefore stationed at Scapa Flow, not so much, as in the First World War, to be in a position to bring the German main fleet to action if it came out as to intercept individual units trying to break out into the Atlantic. The British were strong enough to divide their battle-fleet and station a Channel Force at Portland to ensure the safe arrival of their army in France.

On 3rd September 1939, the British Fleet Air Arm consisted of a total of 232 operational aircraft. Of these 176 were embarked in the six aircraft-carriers and, except for 26 which were Blackburn Skua fighter-dive-bombers and 10 which were Sea Gladiator fighters, they were Fairey Swordfish torpedo-spotter-reconnaissance machines. The remaining fifty-six were Supermarine Walrus flying-boats, Swordfish with floats, and Fairey Seafox light reconnaissance seaplanes which were embarked in battleships and cruisers. Although all these aircraft were new, in performance they were well behind modern shore-based aircraft and all except the Blackburn Skuas were biplanes. Five of the aircraft-carriers were attached to the three battle-fleets: the *Ark Royal* and *Furious* (see Fig. 6, p. 106) with the Home Fleet at Scapa Flow; the *Courageous* and *Hermes* with the Channel Force at Portland and the *Glorious* in the Mediterranean. The sixth aircraft-carrier, the *Eagle*, was stationed in the Far East. The French kept their only aircraft-carrier, the *Béarn*, at Brest with the battle-cruisers *Strasbourg* and *Dunkerque*, ready to pursue any raiders which got out into the Atlantic.

Coastal Command in Home Waters had a total of 230 aircraft at the outbreak of war. Forty of these were flying-boats, and 16 were torpedo-bombers, but the majority were general reconnaissance land-planes. Most of these aircraft were of comparatively short range, the London flying-boats and Anson land-planes having a radius of action of only 250 miles. There were, however, two squadrons of the new Sunderland flying-boats and one squadron of Lockheed Hudson land-planes recently purchased in America, which had over double this radius. The torpedo-bombers were Vickers Vildebeeste land-planes with a strike radius of only 180 miles.

The main strength of Coastal Command was deployed in the North Sea to meet the Admiralty's requirement to prevent German warships breaking out on to the trade routes (see Fig. 9, p. 141). An endless chain patrol by Ansons was maintained between Scotland and Norway but their range was insufficient to reach the Norwegian coast and so five submarines had to be disposed to fill the gap. Other patrols were flown to the north and south of this endless chain to ensure that any enemy ships which attempted to pass the patrol line by night would be located before dusk as they approached or after dawn if they had already passed. The Vildebeeste torpedo-bombers were the only striking force and were stationed in Norfolk, but with their short endurance they were of little use. Bomber Command, however, kept a force of Hampden bombers, which were of longer range, ready to attack any German ships which left harbour. The remaining aircraft of Coastal Command were disposed to escort shipping if the convoy system had to be instituted.

The German Navy depended for air support almost entirely upon the Luftwaffe, out of whose total of some 4300 aircraft, 240 were allocated for maritime purposes. Of these, 20 were embarked in the German battleships and cruisers and the rest were based ashore, half in the North Sea and half in the Baltic. For long-range reconnaissance they had the Dornier 18 and the Blomm and Voss 138 flying-boats with a range of a thousand

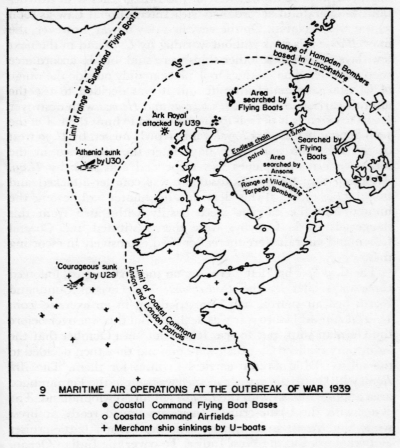

Range of Hampden Bombers based in Lincolnshire

Limit of range of Sunderland Flying Boats

Area searched by Flying Boats

'Ark Royal' attacked by U39

'Athenia' sunk by U30

Endless chain patrol

Searched by Flying Boats

Area searched by Ansons

Range of Vildebeeste Torpedo Bombers

'Courageous' sunk by U29

Limit of Coastal Command Anson and London patrols

9   MARITIME AIR OPERATIONS AT THE OUTBREAK OF WAR 1939

● Coastal Command Flying Boat Bases
○ Coastal Command Airfields
+ Merchant ship sinkings by U–boats

miles or so that enabled them to cover the whole North Sea. They also had the twin-engined Heinkel 115 seaplane, which could carry a mine or a torpedo, and the single-engined Arado 196 which was also the type of seaplane carried in the battle-ships and cruisers. Some two hundred He. 111 medium bombers were also earmarked from the main bombing force for use over the sea and they were specially trained in minelaying and in attack on ships.

The Germans had sent eighteen U-boats and the pocket battleships *Graf Spee* and *Deutschland* to sea before the outbreak of war. The U-boats sailed between the 19th and 29th August, and this movement was not detected by the British. They were deployed west and south-west of the British Isles with instructions to operate in accordance with International Law as soon as war was declared. On the very first day of war, however, the liner *Athenia* was sunk without warning by *U.30* and in the next few days sixteen or so other ships were sunk but in accordance with the rules. The sinkings took place mainly outside the range of Coastal Command aircraft and it was decided to use the aircraft-carriers *Ark Royal*, *Courageous* and *Hermes* with destroyer escort to extend air patrols to the westward to hunt them. On the 14th September the *Ark Royal* was narrowly missed by *U.39* west of the Hebrides and three days later the *Courageous* in the south-western approaches was torpedoed and sunk by *U.29*. Although the *Ark Royal*'s assailant was counter-attacked and sunk by her escort, it was clear that the hunter had become the hunted and the carriers were hastily withdrawn from this dangerous work. Convoys were then instituted and Coastal Command aircraft were thereafter used extensively in escorting them.

The *Graf Spee* had left Germany on the 21st August, followed three days later by the *Deutschland*. The Coastal Command North Sea air patrols had been tried out in an exercise from 15–21st August before war was declared but this was over before the German ships put to sea. It was not until October that the Admiralty realized that they were out and they then decided to use all available aircraft-carriers to hunt for them. The *Ark Royal* with the battle-cruiser *Renown* was sent to the Pernambuco area and the *Furious* with the *Repulse* left for Newfoundland. The *Béarn* with the battle-cruiser *Dunkerque* stayed ready at Brest while the *Hermes* in company with the French battle-cruiser *Strasbourg* went to the West Indies. To cover the Indian Ocean, the *Glorious* with the battleship *Malaya* passed through the Suez Canal from the Mediterranean and the *Eagle* with the cruisers *Cornwall* and *Dorsetshire* were withdrawn from the Far East and sent to Ceylon. The seaplane-carrier *Albatross*[10] was already at Sierra Leone and all these ships and aircraft continued to search until the *Deutschland* returned to Germany on 15th November and the *Graf Spee* scuttled herself after the Battle of the River Plate on 17th December.

On 26th September the Home Fleet had its first encounter with the Luftwaffe. They had put to sea to assist the submarine *Spearfish* which was returning damaged from patrol. The fleet was shadowed by a Do. 18 flying-boat which was shot down by one of the *Ark Royal's* Skuas. Later a single enemy bomber dive-bombed the carrier, narrowly missing her, and a number of unsuccessful high-level bombing attacks were made on the 2nd Cruiser Squadron.

At the outbreak of war the British bomber force was not allowed to bomb open cities and had little to do. On 4th September 14 Wellington and 15 Blenheim bombers were used to attack the German Fleet in its North Sea bases. The *Scheer* was hit by several bombs which failed to explode and the *Emden* was damaged by splinters, but otherwise they were unharmed and seven aircraft were lost. On 12th September the British Home Fleet, for fear of German bombing attacks, moved to Loch Ewe. It was over a month later, on 16th October, before a dozen German Ju. 88 aircraft bombed British warships in the Firth of Forth. A bomb hit the cruiser *Southampton* but also failed to explode and a destroyer was slightly damaged; two aircraft were shot down by Spitfires. Next day a small raid was made on Scapa Flow and the old battleship *Iron Duke* was hit and had to be beached. There was not another raid until 16th March, when the cruiser *Norfolk* was damaged in Scapa Flow. These early bombing attacks by both sides on ships in harbour had proved remarkably ineffective.

The Coastal Command patrols in the North Sea were continued throughout the winter but with almost a complete lack of success. On 8th October, however, they sighted the *Gneisenau*, *Köln* and nine destroyers off the southern tip of Norway and Bomber Command sent twelve Wellington bombers to attack but they failed to find the enemy ships, which had in fact returned to base. It is now known that the Germans wanted to be seen to try to draw the British Fleet within range of the Luftwaffe. On 21st November the *Scharnhorst* and *Gneisenau* sortied from the North Sea and sank the armed merchant cruiser *Rawalpindi* south of Iceland, again returning to Germany without being seen by the air patrols. The Home Fleet also failed to intercept the enemy on their way back: they had no aircraft-carrier as the *Ark Royal* and *Furious* were still hunting for raiders. In the first six months of the 'phoney war' period German ships were at sea four times and were only seen once:

Coastal Command patrols were not therefore every effective. Bad weather and long winter nights, of which the Germans took full advantage, and too short an endurance were the main reasons. Attempts to use aircraft to attack ships at sea or in harbour were also disappointing. Bomber Command had in fact dropped 61 tons of bombs on naval targets, all in daylight, but had done negligible damage.

The Coastal Command aircraft in the North Sea often sighted U-boats on passage, however, and in November they were told that attack on submarines was of equal importance to reconnaissance for ships. They were armed only with 100-lb A/S bombs which proved practically useless and they achieved no success until 30th January when a Sunderland flying-boat was able to help the *Fowey* and the *Whitshed* to sink *U.55* west of the Channel. In general the short range of Coastal Command aircraft was a great disadvantage against the U-boat too: the only success achieved by an aircraft was on 11th March when *U.31* was sunk in the Schillig Roads by a Bomber Command aircraft. Of the other 17 U-boats which had been sunk by April, 11 were disposed of by ships, 3 by mines, 1 by a British submarine and 1 was lost by accident.

In October the Luftwaffe began to attack British east coast shipping from the air and four squadrons of Blenheim twin-engined fighters were transferred to Coastal Command to deal with them. In mid-November, German aircraft joined the U-boats and surface forces in their minelaying campaign and began to drop mines from the air. Aircraft minelayers had, in fact, been used in the First World War. They have the great advantage that they can penetrate into waters where U-boats or ships cannot go and revisit areas time and again to 'refresh' them with mines. On the other hand they are apt to drop mines on land and on 23rd November the British recovered a magnetic mine complete and so found out how it worked and were able to devise counter measures.

By the end of 1939 the minelaying campaign was rivalling the U-boats and by April had sunk 129 ships of 430,000 tons but this total had of course to be shared between aircraft, U-boats and surface ships, all of which laid mines. On 9th January the German air war on shipping became unrestricted but by the end of the year only ten small ships had been sunk by direct attack by aircraft and by April the total had only reached thirty ships of 36,000 tons. The casualties caused were therefore not very great although at the time, in the atmosphere of the 'phoney war', they assumed large proportions. Fighter command gradually took over the protection of shipping on the east coast which they were able to do as the range of the shore radar stations was extended to seawards.

In the period of the 'phoney war', therefore, aircraft could not claim to have exerted a very great influence on the war at sea. The Luftwaffe had had moderate success as a weapon of attrition against merchant shipping: they had done as well as the surface raiders but had sunk no more than a quarter of the tonnage destroyed by the U-boats. The bombers of both sides had had very little success against the fleets either at sea or in harbour, although it must be admitted that they did not put much effort into the attacks and the ships took care to keep out of range as much as possible. As an auxiliary to the Navy, Coastal Command achieved little and it proved ineffective in both the reconnaissance and the anti-submarine roles. The Fleet Air Arm was dispersed for a great deal of the time hunting for raiders, a proper role but one in which it was unlucky and had no success. Consequently there were no carriers with the Home Fleet when there was a chance of a fleet action during the sortie of the *Scharnhorst* and *Gneisenau* in November.

The German decision to seize Norway in April 1940 was a considerable gamble: they had to cross the sea and to do this with a greatly inferior fleet appeared to be courting disaster (see Fig. 10, p. 148). The Germans, however, decided to risk it by using their whole surface fleet to make a surprise landing and then to withdraw to Germany, leaving the defence of the captured territory to U-boats and aircraft. Although they were inferior in ships, the Germans were able to deploy many more aircraft than the British. There were over a thousand planes available which were under the command of Fliegerkorps X: about half of these were transport machines but there were 400 medium bombers and 100 fighters. The German ships for the invasion were organized in six groups to land at Oslo, Kristiansand, Egersund, Bergen, Trondheim and Narvik. It was hoped to establish German garrisons in all these places in the early stages and then to fly in reinforcements and to bring them by sea across the Skagerrak as well. The German air plan involved a rapid move forward by Fliegerkorps X to the Aalborg airfields in North Jutland and to Stavanger, Kristiansand and Oslo and subsequently to Trondheim and Narvik. The whole plan depended heavily on the power of aircraft to dominate warships at sea.

A cardinal feature of the German plan was also, of course, surprise. In this they were not entirely successful: British intelligence detected the preparations and some bomber aircraft sighted the *Scharnhorst* and *Gneisenau* moving out into the roads three days before they sailed. At midday on 7th April, a Hudson of Coastal Command, sent specially into the Heligoland Bight to reconnoitre, sighted the first group of German warships moving north. Bomber Command aircraft were sent to attack but without result. For various reasons the Home Fleet did not get to sea until the evening of the 7th and so was already too late to intercept the German forces bound for Narvik and Trondheim. The Home Fleet steered to the northwards in pursuit but was greatly handicapped as it had no aircraft-carrier with it. The *Ark Royal* had been sent to the Mediterranean to carry out some training and the *Furious* was still in the Clyde. Admiral Forbes, however, secured the co-operation of a Coastal Command flying-boat from the Shetlands to search ahead of him on the 8th April and it sighted the cruiser *Hipper* of the Trondheim invasion force, but the fleet was not able to intercept her. In the afternoon Admiral Forbes turned south again parallel to the Norwegian coast: he did not know the enemy's intentions and so counter-marched off the Norwegian coast in some uncertainty. Early next morning there was a brush between the *Renown*, which had been supporting a British mine laying operation farther to the north, and the *Scharnhorst* and *Gneisenau*. Soon afterwards, on the morning of the 9th, the landings took place and the situation became clear, but all six German groups had arrived at their destinations virtually unopposed.

The Home Fleet was shadowed by aircraft from soon after eight o'clock in the morning of the 9th and during the afternoon an action developed between the main British Home Fleet and Fliegerkorps X. Nearly ninety He. 111 and Ju. 88 bombers attacked in a series of waves. They hit the flagship *Rodney* with a 1000-lb bomb which penetrated her armoured deck but failed to explode fully and did not do very much damage, but they sank the destroyer *Ghurka* and a number of cruisers were slightly damaged by near misses. The fleet

succeeded in shooting down four German aircraft but expended
a great deal of its anti-aircraft ammunition in doing so. The
casualties on both sides were not therefore serious but the effect
of this action was far-reaching. Admiral Forbes was persuaded
that his anti-aircraft guns alone were not enough to protect the
fleet and that he could not risk it off the southern coast of
Norway without fighter protection. At the time of this attack
the *Furious*, with the battleship *Warspite*, was hurrying up the
west coast of Scotland to join the fleet. It would, however, have
made no difference if she had been there earlier as she had left
her Skua squadron behind and had only Swordfish torpedo-
bombers on board.[11] Admiral Forbes decided, therefore, on
account of the German air superiority, to leave the southern
area to submarines and the R.A.F. Strategically, therefore,
Fliegerkorps X had secured a substantial success. They had
achieved in one action much of what they had hoped, which
was to drive the superior British fleet away and prevent it
counter-attacking the landing forces or interfering with the sea
communications to Norway. They were able, therefore, to
redress decisively their naval inferiority by the use of shore-
based air power over the sea.

All the main German objectives were in their hands by the
evening of the 9th and the British problem was now to make a
counter-attack. The first of these was made by the R.A.F. with
12 Wellington and 12 Hampden bombers on Bergen in the
early evening of the 9th. They caught the German invasion
force in harbour but did not score more than near misses and
did little damage. Early next morning, as the British destroyers
were attacking Narvik, 15 Fleet Air Arm Skuas attacked Bergen
from an airfield in the Orkneys. They dive-bombed the German
cruiser *Königsberg*, which had already been damaged by shore
batteries, obtained three hits and sank her. The *Furious* joined
the Home Fleet on 10th and early on 11th flew off eighteen
Swordfish to attack Trondheim. The *Hipper* had, however,
already left and the planes launched their torpedoes at three
German destroyers but secured no hits, probably because of the
shallow water. The same day Hudsons of Coastal Command
made a further attack on Bergen.

The Home Fleet now moved north to counter-attack at
Narvik but could not shake off the Luftwaffe, which now had
Stavanger airfield in use. The Fleet was bombed and the
destroyer *Eclipse* was seriously damaged. An attempt to relieve

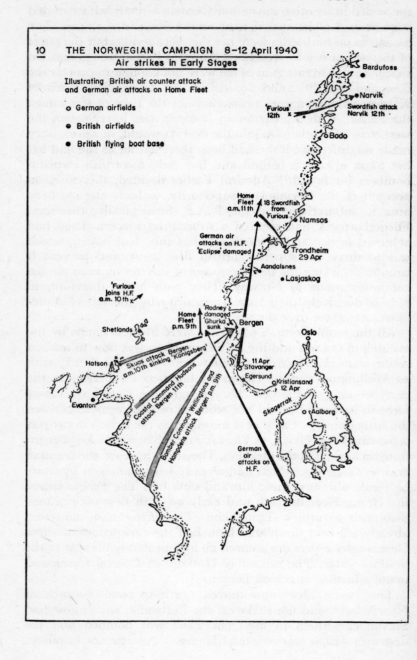

10   THE NORWEGIAN CAMPAIGN   8–12 April 1940

Air strikes in Early Stages

Illustrating British air counter attack
and German air attacks on Home Fleet

o   German airfields

●   British airfields

●   British flying boat base

Bardufoss

Narvik

'Furious'
12th   Swordfish attack
Narvik 12th

Bodo

Home
Fleet
a.m. 11th   18 Swordfish
from
'Furious'   Namsos

German air
attacks on H.F.
'Eclipse' damaged   Trondheim
29 Apr

Aandalsnes

Lasjaskog

'Furious'
joins H.F.
a.m. 10th

Rodney
damaged
Ghurka
sunk   Bergen

Home
Fleet
p.m. 9th

Shetlands

Oslo

11 Apr
Stavanger

Egersund

Kristiansand
12 Apr

Hatson

Skuas attack Bergen
a.m.10th sinking 'Königsberg'

Coastal Command Hudsons
attack Bergen 11th

Bomber Command Wellingtons and
Hampdens attack Bergen p.m. 9th

Evanton

Skagerrak

Aalborg

German
air
attacks on
H.F.

the situation by using the cruiser *Sussex* to bombard Stavanger from the sea nearly ended in disaster when she staggered back to Scapa Flow badly damaged by air attack. On 11th the *Scharnhorst* and *Gneisenau* met the *Hipper* and re-entered the North Sea. They were sighted by a reconnaissance aircraft of Coastal Command off the south of Norway and a very heavy strike of 92 Bomber and Coastal Command aircraft was sent to attack but the weather was very bad and they failed to find the German ships, which returned safely to base.

On the evening of the 12th the *Furious* was able to launch her Swordfish to attack Narvik but half the aircraft got lost in a snowstorm and had to turn back and only eight of them attacked, doing some minor damage for the loss of two of their number. Next day the *Warspite* and destroyers entered the fjord and the second Battle of Narvik took place. In this action the *Warspite*'s aircraft not only sank *U.64* but was extremely useful in reporting the positions of the German destroyers. Although ten German destroyers were sunk in this counter-attack the Luftwaffe had by now established itself at Stavanger, Oslo and the Aalborg airfields. These airfields were heavily attacked by Bomber Command but in spite of this the Germans were able to pour troops by air and sea into southern Norway under the cover of the Luftwaffe. This movement was not opposed by surface forces, but by submarines and by British aircraft using magnetic mines. They laid 160 mines in April and another 100 in May in the Elbe, off Lübeck and Kiel and in the Jade, Weser and Ems and sank 24 ships, not all of which however were involved in the Norwegian campaign.

The British now decided to land and try to retake Trondheim and Narvik. The landings took place in the middle of April and were very roughly handled by the Luftwaffe. The aircraft-carriers *Ark Royal* and *Glorious* arrived from the Mediterranean on 24th April and the *Glorious* was at once used to fly a squadron of R.A.F. Gladiator fighters to work from the frozen Lake Lasjaskog to try and protect the landing at Aandalsnes. The Germans however bombed the ice and destroyed these fighters within a few hours. Vaernes airfield near Trondheim had already been occupied by the Luftwaffe and a striking force totalling 34 aircraft from both the *Ark Royal* and *Glorious* was flown off on the 25th, doing some damage to the airfield. *Glorious* then went home for more aircraft and the *Ark Royal* attacked again with 18 aircraft on the 28th. During the

L

evacuation of Aandalsnes and Namsos, both carriers flew fighter patrols from 120 miles out to sea but they were heavily counter-attacked by the Luftwaffe and had to withdraw to seawards on 1st May after losing fifteen aircraft. Nevertheless they shot down twenty German planes but this was not enough to regain air superiority over the area. The *Furious* was still in the Narvik area but bad weather prevented her from supporting the first landings there. The *Ark Royal* joined her and the two carriers supported the operations in the Narvik area for a fortnight in ths middle of May.

On the 21st the *Furious* flew in an R.A.F. Gladiator fighter squadron to the airfield at Bardufoss north of Narvik. On the 26th they were joined by a squadron of R.A.F. Hurricanes which was flown off the *Glorious*. The Allies captured Narvik on 28th May but had to leave almost at once. Between the 4th and 7th June, the *Ark Royal* did her best to protect the evacuation not only by flying fighter patrols but by bombing the Germans as far south as Bodö. The *Glorious* was used to evacuate the R.A.F. fighter squadrons: R.A.F. pilots, who had never landed on a carrier before, performed the remarkable feat of landing ten Gladiators and ten Hurricanes on her successfully. The *Ark Royal* then accompanied the last convoy home but the *Glorious* was sent on by herself with the destroyers *Ardent* and *Acasta*. On 4th June the *Scharnhorst, Gneisenau* and *Hipper* had put to sea from Germany and were not detected by the Coastal Command patrols in the North Sea. On 8th June they sighted the *Glorious* and sank her and both her destroyers by gunfire but not before the *Scharnhorst* had been hit by a torpedo from the *Acasta*. Although she had a squadron of Skuas and some Swordfish, the *Glorious* was not flying any air patrols for her own protection. She was overcrowded with R.A.F. fighter aircraft embarked for passage and attempts to range the Swordfish to attack during the action were unsuccessful. The German ships then returned to Trondheim where twelve Hudsons attacked but failed to hit them. Two days later fifteen Skuas from the *Ark Royal* succeeded in hitting the *Scharnhorst* with a 500-lb bomb but it failed to explode and eight aircraft were lost.

So ended the Norwegian campaign which was the baptism of fire for the British Fleet Air Arm as it had been developed between the wars. It had, as part of the superior Home Fleet, failed to compete with the Luftwaffe and it had lost control of the sea and air off southern and central Norway. The fleet had tried to fight a first line shore-based air force, something which the Fleet Air Arm had not been designed to do, and of which its aircraft were not capable. The naval fighters were, as has been pointed out before, of very poor performance and slower than most of the German bombers. When the carriers did arrive, therefore, Admiral Forbes was no more able to operate off southern Norway than he had been without them. Their strike aircraft, the majority of which were Swordfish, were quite unable to compete with enemy fighters over the land. The Skua dive-bombers, however, in their sinking of the *Königsberg* achieved the first real success by aircraft in war against a major warship but this was the only substantial success in the many strike missions flown by the Fleet Air Arm, or for that matter the Royal Air Force, over Norway. The main British fighter strength based in the United Kingdom could not be brought to bear sufficiently to redress the situation. The area of operations was outside their range, except for a few Blenheims working from the Orkneys, and even these could only just reach the Norwegian coast. The carriers were therefore used in the second part of the Norwegian campaign to try to support the Army in Norway, partly by ferrying R.A.F. fighters to operate from Norwegian airfields and partly by using their own aircraft. The Luftwaffe, on the other hand, by its rapid movement forward on to Danish and Norwegian airfields, commanded the sea in the Skagerrak and off the south coast of Norway. So the stream of reinforcements to Norway continued except for a campaign of attrition by British submarines and by Bomber and Coastal Command aircraft dropping magnetic mines. These attacks had little effect on the supply of the Germans in Norway, but they did more than the whole surface strength of the Home Fleet.

The preoccupation of the Fleet Air Arm with operations over the land inevitably led to a neglect of the true functions of naval aviation. The lack of a carrier in the early stages of the campaign was sorely felt and although the weather was bad, an aircraft-carrier should have been able to tell the Commander-in-Chief more of what was happening. Much of the area of operations was out of range of Coastal Command aircraft from the Shetlands, and without aircraft he was blind; this was the main reason why the Home Fleet so often went in the wrong direction. There were no aircraft available to take part in the action between the *Renown* and the German battle-cruisers or to follow up and shadow them afterwards. Subsequently on 11th May, when the *Furious* was with the Home Fleet, the *Scharnhorst*, *Gneisenau* and *Hipper* passed them undetected and escaped back to Germany. It was not only for scouting and fighter protection that carriers were required: the Home Fleet battleships were so much slower than all the German ships that they had no chance to bring them to action unless they could be slowed down by air attack. There is no room in this book to discuss the reasons why the aircraft-carriers were not with the fleet when they should have been or why they were not used in the way they ought to have been. There is only room to point out that the Fleet Air Arm was not used in the Norwegian campaign for the purposes for which it had been designed and that this was one of the principal reasons for the failure of the Home Fleet against the German ships in the campaign.

The dominating factor over the sea as well as the land during the campaign had undoubtedly been the Luftwaffe and not the Home Fleet. Although no major British warships were lost by air attack, the only possible conclusion was that 'Billy' Mitchell's assertions had now come true. Battle-fleets, with air support of the type provided by the British Fleet Air Arm, could not compete with shore-based air power. If they were to have any hope of reasserting themselves it was obvious that they must above all have more and very much better fighters. The Admiralty already had a new fighter coming into service for the Fleet Air Arm and this was the Fairey Fulmar. It had eight machine-guns like the Hurricane but was a two-seater adapted from an earlier design and, although its range was somewhat greater, it was 50 m.p.h. slower and its ceiling was nearly 10,000 feet less. One good thing that came out of the Norwegian campaign was that the R.A.F. Hurricane fighters had not only flown off the *Glorious* but had landed on again and it could not be said in future that the Fleet Air Arm could not operate them from carriers. Unfortunately the wings of a Hurricane would not fold and they were too large for the lifts of the new aircraft-carriers of the *Illustrious* class. Nevertheless the Admiralty set in train steps to adapt the Hurricane for carrier operations. At the same time they managed to purchase some Grumman Martlet aircraft in America which were modern fighters of the type then being accepted for service in the U.S. Navy. They were 30 m.p.h. faster than the Fulmar but the early ones had not got folding wings either.

At the end of the Norwegian campaign, Coastal Command put in a considerable effort to watch the German heavy ships in Norwegian waters. Seventeen aircraft were lost in the Trondheim area on this duty in June alone. On 23rd June the *Scharnhorst* was sighted returning to Germany and was attacked by naval torpedo aircraft from the Orkneys as well as by Coastal Command but without success. The *Gneisenau* with the *Nürnberg* and four destroyers got back to Germany at the end of July undetected. In the three months August to October, Bomber Command mounted a minor offensive against all the German heavy ships in Germany and dropped 683 tons of bombs in over a thousand night sorties by Wellingtons, Hampdens and Whitleys. They did a certain amount of damage to various dockyards but only the *Lützow* and the *Prinz Eugen* were hit by a total of three bombs.

Immediately after the fall of France and the Low Countries in June 1940 the British were, in spite of their great superiority at sea, threatened with invasion. The factors which led Hitler to believe that an invasion was a practicable operation of war were firstly that the British Army, after its defeat on the continent and the loss of almost its entire equipment at Dunkirk, was in no position to oppose the German Army if it could be got ashore. Secondly it was expected that the Luftwaffe's greatly superior numbers should enable it to defeat the R.A.F. and obtain air superiority over the sea. Thirdly, the Norwegian campaign indicated that the Luftwaffe should be able to clear the Royal Navy out of the Narrow Seas and prevent it opposing the crossing. A plan was therefore drawn up involving the landing of eleven divisions in Kent and Sussex in merchant ships and barges. It was to be covered mainly by the Luftwaffe but was also to be protected by U-boats, minefields and long-range guns in the Pas de Calais.

In the past, the defence of Great Britain against invasion had been primarily the business of the Navy. It had always discharged this responsibility by stationing in the Narrow Seas a large number of small warships, or what was called a 'flotilla', to destroy the vessels actually carrying the enemy army. In the background, there was always a battle-fleet whose primary duty was to prevent the enemy battle-fleet dispersing the flotilla and escorting its army across in safety. The whole system of defence from the days of the Armada to those of the Great War of 1914–18 therefore depended ultimately on a battle-fleet even though it might not be used to defeat the invasion directly. The British naval dispositions to oppose invasion in 1940 were initially in accordance with these principles. A flotilla of some forty destroyers and over a thousand auxiliary patrol vessels was available to patrol the Channel and southern North Sea, while the Home Fleet was stationed at Scapa Flow with six cruisers farther south ready to support the flotilla.

In the summer of 1940 the new factor of air power altered this concept fundamentally. On 10th July a German Geschwader of over 300 fighters and bombers in the Pas de Calais area began attacks in the Straits of Dover with the aim of driving British warships out of the Channel. Fliegerkorps VIII in the Le Havre area attacked ships farther to the west. These attacks were opposed by R.A.F. Fighter Command and a considerable number of Ju. 87 dive-bombers were shot down. Nevertheless, even with fighter protection, it was soon found too expensive to operate ships in the Channel by day and patrols were confined to the hours of darkness. If an invasion had started, the flotilla would, of course, have been thrown into the fray by day regardless of risk, but success even then would have depended on fighter protection. These attacks by the Luftwaffe were, in fact, the beginning of the Battle of Britain and the fight soon moved inland, but the influence on an invasion of this brief contest between the two air forces over the Narrow Seas was considerable.

No one now seriously believed that if the Royal Air Force was defeated by the Luftwaffe the flotilla would be able to operate for long in the Narrow Seas. After its experiences in Norway, it was clear that the Home Fleet could not help either. It had been reluctant to face a single Fliegerkorps in that campaign and would have still less chance against the five Fliegerkorps now in France. Admiral Forbes clearly believed this to be so as he opposed the stationing of any of his battleships in the south and did his best to obtain the return of the cruisers which were supporting the anti-invasion destroyers. He even went so far as to say that whilst the R.A.F. was undefeated he believed that defence against invasion should be left to them and the Army. This could only be taken as an abdication from the Navy's traditional role and a belief that the fleet was only of use against an invasion as a last ditch suicide force if the R.A.F. was defeated. In this he was probably right: we can get some idea of the result, if the Royal Navy had been flung at the invasion forces without any air support, from what subsequently happened in the Battle of Crete. In this case the result would probably have been worse as, although the British naval forces would have been slightly greater than they were in Crete, the Luftwaffe would have been four times the size. Furthermore without air defence the replenishment or repair of ships in harbours between the Humber and Portsmouth would have been impossible and it seems probable that the flotilla would either have been destroyed or driven out of the area before the invasion fleet sailed.

The R.A.F., in addition to the engagement of Fighter Command in the Battle of Britain, was very active against the invasion preparations. Coastal Command was now primarily occupied in reconnaissance in connection with the threat and a considerable proportion of Bomber Command's strength was used to attack concentrations of barges and shipping. They sank 214 barges and 21 transports in harbour out of the 1910 barges and 168 transports that had been assembled. This was not enough to stop an invasion but it disposed of the entire reserve which the German planners had allowed. The efficacy of the R.A.F. in attacking the invasion forces directly if they had sailed is difficult to assess. The Germans would obviously have provided massive fighter cover. The performance of anti-shipping strikes in the North Sea later in the war does not suggest that the R.A.F. would have been very successful, especially as the main bomber force was trained in night operations over the land.

As the failure of the Luftwaffe to defeat Fighter Command in the Battle of Britain became obvious, Hitler began to postpone the date of the invasion until, on 12th October, he cancelled it. He was not prepared to take the risk that under an undefeated Fighter Command's cover the Royal Navy's flotilla would not be able to destroy the invasion forces before they could be got ashore. Hitler's invasion preparations were, of course, very primitive when compared with the standards achieved by the Allies in 1944 and he might well have failed anyway. What is clear is that the position of the flotilla as a defence against invasion was now dependent, not upon the support of a battle-fleet, as in the past, but upon the maintenance of air superiority over the Narrow Seas by the Royal Air Force. This would have been true whether the amphibious forces had been modern and efficient or improvised. The Home Fleet was really only of use to try and counter the operations of the German heavy warships and Hitler does not seem to have contemplated relying on them for the success of an invasion except in the somewhat minor role of creating a diversion. It is true to say that Napoleon gave up his invasion plans because his fleet could not defeat the British Navy whereas Hitler gave up his because the Luftwaffe could not defeat the Royal Air Force. Such then was the effect of aircraft upon the command of the Narrow Seas in 1940.

Nearly a month before Italy declared war on 10th June 1940, all British merchant shipping was diverted round the Cape, so adding 20,000 miles to the return journey to the Middle East. This decision was taken before the French collapse and so at a time when the Allies could bring a considerable surface superiority to bear on the Italian Fleet. The British Mediterranean Fleet then moved its base from Malta to Alexandria in Egypt and were no longer in a position to cut the Italian supply routes to their armies in North Africa or to cover the passage of Allied merchant convoys through the Mediterranean. The principal reason for these moves was the strength and proximity of the Regia Aeronautica. The British can correctly be said to have thereby surrendered the command of the sea in the central Mediterranean. They could neither use the central Mediterranean for their own traffic nor deny it to the Italians. The Italian Air Force, therefore, without a fight exerted what was probably the greatest influence of aircraft over sea power to date. With the French collapse and the departure of the Mediterranean Fleet for Alexandria, the British built up a new fleet at Gibraltar known as Force H. The British then controlled the ends of the Mediterranean and the Italians the middle.

The Italian Air Force in the Mediterranean consisted of some 2350 operational aircraft with another thousand in reserve. Just under a thousand of these were bombers, the most numerous of which were the S.79 three-engined monoplanes with a range of over a thousand miles, carrying a bomb-load of more than a ton and with a maximum speed of 255 m.p.h. Some 750 were fighters and 300 of miscellaneous types. The Aviazione per la

Regia Marina which corresponded to Coastal Command had 125 Cant 501 reconnaissance flying-boats and 85 Cant 506 three-engined bomber seaplanes. These were air force units but in theory their operations were controlled by the Navy and they had naval officers as observers. In addition there were some 64 smaller seaplanes carried in the battleships and cruisers. The air force was spread over sixty airfields and bases from Northern Italy to Cyrenaica and from Sardinia to the Dodecanese Islands, and the bombers and flying-boats could cover the whole Mediterranean except for the area immediately to the eastwards of Gibraltar.

Against this formidable air force the British had just under 200 aircraft in Egypt, 96 of which were bombers and 75 fighters, nearly all of obsolete types. They were taken up mainly with the defence of the Nile delta, including the fleet at Alexandria, and in supporting the army in the western desert. All they could do to help the fleet at sea was to give fighter protection in the vicinity of Alexandria and the coast-line of Egypt. The R.A.F. had only nine Sunderland flying-boats based at Malta available in the whole Mediterranean. Heavy bombing attacks on Malta started on 11th June and continued throughout the month. There were only three Gladiator fighters there and so the flying-boats could not be protected and had to leave for the eastern end of the Mediterranean. The Navy had therefore to rely almost entirely on carrier-borne aircraft. Force H at Gibraltar included the modern aircraft-carrier *Ark Royal* with 24 Skuas and 30 Swordfish embarked but the Mediterranean Fleet at Alexandria had only the elderly *Eagle* with 17 Swordfish and 3 Sea Gladiator fighters.

The Italian Navy had as its main aims the protection of their coast-line and of the supply routes to Africa but at the same time they desired to prevent British incursions into the central Mediterranean. Admiral Cunningham, the British Commander-in-Chief at Alexandria, was determined to seek out and destroy the Italian Fleet and to do this he had three battleships to the enemy's two but was inferior in cruisers and destroyers. Furthermore unless he could entice them out of the central Mediterranean he would have to face the whole Regia Aeronautica and a very large submarine force. During June he contented himself with a sweep along the Libyan coast, with a bombardment of Bardia and in covering convoys to and from the Aegean.

On 7th July Admiral Cunningham sailed for the central

Mediterranean to cover two small convoys carrying naval stores from Malta which he needed for his fleet at Alexandria (see Fig. 11, p. 160). Simultaneously the Italian Fleet sailed to cover a military convoy bound for Benghazi. These operations led to an action known as the Battle of Calabria which took place on 9th July and which is notable as one of the few contests between battleships in the Second World War. It is also of great interest as it was the first time that a fleet had challenged the full strength of a metropolitan air force and also as the first time in which an aircraft-carrier was involved in a fleet action.

On 8th July the submarine *Phoenix* sighted the Italian Fleet in the middle of the Ionian Sea and the Sunderland flying-boats, which were on their way back to Malta from Alexandria for this operation, were ordered to locate and shadow. During the day the British Mediterranean Fleet was heavily attacked by shore-based bombers from the Dodecanese and Cyrenaica. Some twenty high bombing attacks totalling 72 sorties were made on various components of the fleet; there were many near misses but there was only one hit, on the bridge of the cruiser *Gloucester*. Eleven of the attacking aircraft were damaged by the *Eagle*'s fighters and by anti-aircraft fire. The flying-boats sighted the Italian Fleet during the afternoon and Admiral Cunningham set a course to cut it off from Taranto. During the day the *Eagle*'s Swordfish flew anti-submarine patrols round the fleet, sighting and bombing two Italian submarines.

At dawn on 9th July the *Eagle* flew off a search of three aircraft and the flying-boats from Malta relocated the enemy fleet on a northerly course at 07.32 in the morning. At this time the Italian Fleet, which relied on the Regia Aeronautica for reconnaissance, was unaware of the British approach. The British battleships were considerably slower than the Italian and so it was important to reduce their speed by an air strike if they were to be brought to action. At 11.45, therefore, nine Swordfish were flown off the *Eagle* at a range of ninety miles but they failed to find the enemy battleships and instead attacked but missed some Italian cruisers. The battle-fleets then made contact and exchanged long-range gunfire and the Italians retired after the *Warspite* had hit the *Cavour*. During the action the *Eagle* got away nine more Swordfish but again they attacked Italian cruisers and again their torpedoes missed. One of the *Warspite*'s catapult aircraft was used with considerable success for action observation during the battle, but the other suffered a very common fate for shipborne aircraft; it was damaged by gun-blast and had to be jettisoned. The Italians withdrew at high speed for the Straits of Messina, calling on the Regia Aeronautica for support. Heavy bomber attacks developed again on the British Fleet, starting at 16.40 and continuing until dark. *Warspite* and *Eagle* were both bombed five times and other ships were attacked but again there were many near-misses and no hits. The Italian air force had recognition trouble and made a number of attacks on their own ships and failed to hit them too. These attacks, totalling 126 sorties, were made mainly by the 2nd Squadra Aereo from Sicily and also by bombers from airfields in the heel of Italy.

On the passage back to Alexandria heavy bombing attacks developed and yet again there were many near-misses but no hits and on 12th July R.A.F. Blenheim fighters came out from North Africa to give cover. These last Italian attacks involved 84 sorties from Sicily, 108 from North Africa and 56 from the Dodecanese, some of which were made on the convoys from Malta as they neared Alexandria. In the whole operation some 2000 bombs, mostly of 250 Kg and 100 Kg, were dropped with only one hit, while two aircraft were shot down by *Eagle*'s fighters and three others were damaged badly and 82 slightly by anti-aircraft fire.

The British Fleet had therefore penetrated into the central Mediterranean and put the Italian Fleet to flight and the Regia Aeronautica had been unable to stop them; furthermore the two convoys arrived safely at Alexandria. Admiral Cunningham believed that the high bombing attacks were 'more alarming than dangerous' and considered that he could penetrate into the central Mediterranean in future whenever he wished. Nevertheless he had failed to prevent the Italian convoy reaching Benghazi and as he was unable to maintain his position in the central Mediterranean indefinitely, the traffic to Libya continued unimpeded.

In this battle the *Eagle*'s aircraft were used in precisely the way for which the British Fleet Air Arm had been developed and trained. They provided air reconnaissance, action information and anti-submarine patrol adequately. Her fighters did their best but were too few to make much difference and she could not claim to have given the fleet immunity from air attack. Never was an air strike to slow the enemy down more necessary, but unfortunately lack of training during the *Eagle*'s long hunt

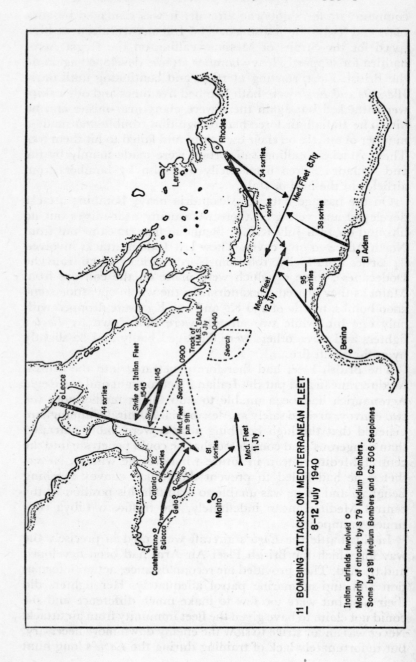

11   BOMBING ATTACKS ON MEDITERRANEAN FLEET.

8–12 July 1940

Italian airfields involved ○

Majority of attacks by S 79 Medium Bombers.

Some by S 81 Medium Bombers and Cz 506 Seaplanes.

for raiders in the Indian Ocean made all the torpedo attacks abortive. To do all these tasks with only 20 aircraft was a remarkable achievement. The old ship however was severely shaken by near-misses. The co-operation of the few R.A.F. flying-boats and the British Fleet was excellent but the same could not be said of that between the Regia Aeronautica and the Italian Fleet. Subsequently there were bitter recriminations between the Italian Navy and Air Force especially over the attacks on the Italian ships. Mussolini, however, believed the Regia Aeronautica's claims that they had inflicted serious losses on the British Fleet.

The Mediterranean Fleet had successfully confronted the Regia Aeronautica and the British were encouraged to attempt a number of further operations into the central Mediterranean. In the remaining six months of 1940, reinforcements were escorted to Malta on several occasions and the old aircraft-carrier *Argus* twice flew in Hurricane fighters to the island from Gibraltar. Naval reinforcements for Admiral Cunningham were passed through the Mediterranean, other ships were withdrawn for service in the Atlantic and on one occasion important merchant ships were sent to Greece. In all there were five incursions into the central Mediterranean and all required a full fleet operation, often from both ends at the same time. The Italians tried to counter nearly all these operations by high bombing attacks but did not make a single hit.

The most important reinforcement for the eastern Mediterranean was the new armoured aircraft-carrier *Illustrious*, which arrived in August and had an air group of 12 of the new Fairey Fulmar fighters and 22 Swordfish. By October the Italian battle-fleet consisted of six battleships, two of which were of the brand-new *Littorio* class and the remaining four had been recently modernized: and it was now superior to the eastern Mediterranean Fleet. It was therefore most important to reduce its strength and Admiral Cunningham planned to use his aircraft-carriers to make an attack on it in Taranto. It had been intended to launch 30 Swordfish from both the *Illustrious* and the *Eagle*, but the *Eagle* was suffering from defects due to the many near-misses she had received, and so five of her Swordfish were lent to *Illustrious*. Some twin-engined Glenn Martin Maryland aircraft had arrived at Malta for the R.A.F. and they were able to take photographs of Taranto so that an attack could be planned in detail. To give any chance of success with the obsolete Swordfish torpedo planes against such a heavily defended base the attack had to be made at night, and it was planned to launch the torpedoes in the light of flares. There were formidable complications, including a balloon barrage, anti-torpedo nets and shallow water.

The *Illustrious* launched 20 Swordfish in two waves after dark on 11th November from a position 170 miles to the south-east of Taranto. The attacks were a complete success and they hit the battleship *Littorio* with three torpedoes and the *Cavour* and *Duilio* with one each, for the loss of two Swordfish. The *Illustrious* then recovered her aircraft and retired at high speed. Next day the Italian Air Force made strenuous efforts to find her but her Fulmar fighters succeeded in shooting down the reconnaissance planes and she got clear away. The attack on Taranto halved the strength of the Italian battlefleet and the British Fleet again became superior. The *Cavour* became a total loss and the *Littorio* and *Duilio* were out of action for five and six months respectively. The next day the three surviving battleships of the Italian Fleet moved hastily to ports on the west coast of Italy, so removing any threat to the convoys to Greece.

Taranto was a brilliantly executed operation and the most successful air attack on ships in the history of naval warfare to date. In the past such a victory would undoubtedly have led to the gaining of complete command of the sea. In fact it made very little difference to the strategic situation in the Mediterranean. The Italians were still able to pass their convoys to Libya virtually unmolested and the British could still only penetrate into the central Mediterranean by mounting a full-scale operation with their whole fleet. Nevertheless Taranto ensured that they could still do this but there was certainly no question of restoring British merchant traffic through the Mediterranean. This was because it was still the Italian Air Force which was the dominant factor in sea power in the Mediterranean and not the Italian battle-fleet, and this had not been affected by Taranto.

At the end of November during a convoy and reinforcement operation there was a brush between Force H and the Italian Fleet in the western basin. The redeployment of the Italian battle-fleet to the west coast after Taranto had the disadvantage that reconnaissance aircraft from Malta found it more difficult to keep an eye on their movements. The Italian battleships *Vittorio Veneto* and *Cesare* with seven heavy cruisers and sixteen destroyers left Naples and Messina unobserved to oppose the passage of Force H and the convoy from Gibraltar. Their ship-borne catapult planes sighted part of Force H and they attempted to trap the *Renown* with their whole fleet. The Italians were in their turn sighted by a reconnaissance aircraft from the *Ark Royal*. As soon as the Italians realized that the *Ark Royal* was present they turned to withdraw and were attacked by eleven Swordfish but all the torpedoes missed. The surface forces made long-range contact and opened fire in what became known as the Battle of Cape Spartivento. There was no chance of overhauling the Italians unless their speed could be reduced and the *Ark Royal* flew off another seven Skuas and nine Sword-fish but they attacked some cruisers and missed and the Italian battleships got away. The Regia Aeronautica from Sardinia then counter-attacked strongly and *Ark Royal* was straddled by heavy bombs and very nearly hit but a squadron of the new Fulmar fighters succeeded in driving off some of the bombers.

The other urgent task was to dispute the passage of Italian troops and supplies to Albania and North Africa. In the first six months of the war nearly 50,000 Italian troops were sent to North Africa and over 500,000 to Albania and were kept supplied with negligible losses on the way. The key to this problem was Malta and it was essential to build up its defences so that it could be used as a base. Until this had been done, efficient reconnaissance of the central Mediterranean could not be made and air, surface and submarine striking forces could not operate from the island.

By the end of the year the Mediterranean Fleet and Force H, covered by Fulmar fighters from the *Illustrious* and *Ark Royal*, were able to penetrate with some confidence into the central Mediterranean. The Fulmar was a great improvement over the Gladiators and Skuas and fighters could be more effectively directed as a number of ships were now fitted with radar. The main reason for the failure of the Italian Air Force to prevent the British fleet movements was however that they never hit with their bombs. High-bombing had the advantage that large air-craft could be used and so heavy attacks on warships could be made far out to sea. The terminal velocity of the bombs was sufficient to penetrate armoured decks and the aircraft were practically immune from anti-aircraft fire. Nevertheless fast-moving ships were difficult targets and with a time of flight of half a minute or so they had time to take avoiding action. With a similar time of flight and size of target, naval gunfire had never expected to hit with the first salvo and this is virtually what the bombers were trying to do: it is not surprising therefore that they seldom succeeded. It was torpedo-bombers, although there were far fewer of them on both sides, that had done nearly all the damage. The Italians began to use torpedoes in October and by the end of the year, the *Liverpool*, *Kent* and *Glasgow* had all been hit by torpedoes dropped by aircraft. Three out of the four cruisers damaged in all the British sorties into the central Mediterranean in the first six months of the war were therefore the victims of torpedo-bombers. It must be admitted that the warships' anti-aircraft gunfire and the carrier-borne fighters had shot down few enemy aircraft although the guns could claim that they kept them so high that they missed. In spite of the Italian failure to hit warships, however, there were very many near-misses and the British could not stay long in the central Mediterranean or they would run out of anti-aircraft ammunition. They could not therefore interfere with the traffic to Libya which simply postponed leaving until they had gone. Nor could the passage of merchant traffic through the Mediterranean be risked and the Middle East had still to be supplied by the long route round the Cape. So, in spite of their apparent impotence, the Regia Aeronautica still denied the British the full use of the sea.

By the end of 1940, the British naval and air forces, even if they had not regained control of the sea in the central Mediterranean, were making some progress. This progress was however to be short-lived. In November, Hitler had suggested to Mussolini that German air units should be sent to the Mediterranean and at the end of the year Fliegerkorps X from Norway began to move to Sicily. This unit, which had been the opponent of the Home Fleet in the Norwegian campaign, was well trained in maritime operations. It consisted of 150 bombers, including a number of the Ju. 87 dive-bombers, 25 twin-engined long-range fighters and some reconnaissance aircraft. The bulk of Fliegerkorps X had arrived in Sicily by mid-January and was established on airfields in the Catania area. At this time the British were involved in sailing convoys in and out of Malta from east and west and in escorting an important convoy from

(Imperial War Museum Photograph)

The British armoured Aircraft-carrier *Illustrious*

(Official U.S. Navy Photograph)

An American 'Essex' Class Aircraft-carrier

PLATE V – AIRCRAFT-CARRIERS OF THE ROYAL AND U.S. NAVIES

The Short-range
Anson
(approx. radius o
action 250 miles)

The Medium-range
Lockheed Hudson
(approx. radius 500
miles)

The Long-range
Consolidated Cata
(approx. radius
700 miles)

The Very Long-range
Consolidated
Liberator
(approx radius
1000 miles)

PLATE VI — R.A.F. COASTAL COMMAND AIRCRAFT

(Imperial War Museum Photographs

Gibraltar consisting of four fast merchantmen carrying important military equipment for Malta and Greece.

On 10th January 1941 Admiral Cunningham was 60 miles west of Malta with the *Warspite, Valiant, Illustrious* and five destroyers when they were attacked by some forty Ju. 88 and Ju. 87 dive-bombers of Fliegerkorps X. The *Illustrious* at once launched more fighters but they were too late to oppose the German assault. The dive-bombers attacked with very great skill and *Illustrious* was struck by six 1000-lb bombs which put her completely out of action as an aircraft-carrier and left her out of control and heavily on fire. Her fighters succeeded in shooting down some of the German bombers as they retired but then had to land at Malta. The *Warspite* was also hit by a 1000-lb bomb, but it did little damage. In a second attack, *Illustrious* was hit again but she managed to regain control and reach Malta. Next day a dozen German dive-bombers found the cruisers *Gloucester* and *Southampton* east of Malta and hit the *Southampton* with two or three heavy bombs, causing uncontrollable fires and forcing her to be abandoned and sunk. Nevertheless all the merchant ships of the various convoys arrived at their destinations safely. Between January 16th and 23rd, Fliegerkorps X renewed its attacks on the *Illustrious* while she was under repair at Malta. Although the dockyard was badly damaged she was only hit once more and she managed to get away to Alexandria and thence to the U.S.A. for repairs. The Eastern Mediterranean now once more only had the *Eagle* but the Admiralty ordered the *Formidable*, a sister ship of the *Illustrious*, to proceed round the Cape and take her place.

The arrival of the Luftwaffe in the Mediterranean completely altered the situation: it was obvious that for warships to venture within range of the German dive-bombers was courting disaster. By the end of January, Fliegerkorps X had extended its influence: staging in the Dodecanese, it began to mine the Suez Canal, thus interfering seriously with the communications of all the forces in the Middle East. The arrival of Fliegerkorps X in the Mediterranean therefore had a far greater effect on sea power than the trebling of the size of the Italian battle-fleet during August and September or of its halving again in November. Although only a fraction of its size, Fliegerkorps X achieved more in a few minutes than the whole of the Regia Auronautica over a period of six months. Its secret was the accurate and effective dive-bomber with its special training in attacking ships at sea.

Nevertheless early in February Force H penetrated into the Mediterranean, bombarded Genoa and retired again without being attacked. This operation did not, however, come within range of the Luftwaffe in Sicily. The Italian Fleet of three battleships, three cruisers and ten destroyers attempted to intercept Force H on its way back but failed to do so because of poor co-operation between the Italian Navy and Air Force: the British were not sighted until well after the bombardment was over. Later, in March, four merchant ships were successfully convoyed into Malta from the east and although sighted by enemy aircraft escaped attack mainly because they were protected by low cloud.

During February and March the Germans had been urging the Italians to use their fleet against the British convoys to Greece. Towards the end of March it was planned to make sweeps by Italian cruisers north and south of Crete and to support them with the battleship *Vittorio Veneto*. The German and Italian air forces were to co-operate. The Luftwaffe were to fly reconnaissances of Alexandria and the central Mediterranean, to bomb Malta and to intercept British scouting aircraft from the island. The Regia Aeronautica were to bomb the British airfields in Crete, reconnoitre the area of operations north and south of Crete and give fighter protection to the fleet from Rhodes.

The Italian surface forces were sighted in the Ionian Sea by a Sunderland flying-boat from Malta and after dark Admiral Cunningham put to sea with the battleships *Warspite*, *Valiant* and *Barham* accompanied by the aircraft-carrier *Formidable*. The British had a force of 67 aircraft available; 30 of these were R.A.F. Blenheim bombers based in Greece, 5 were catapult aircraft carried in ships of the fleet and 5 were naval torpedo-bombers based ashore at Maleme in Crete. The rest were embarked in the *Formidable* whose air group consisted of 13 Fulmar fighters and 14 Albacore[12] and Swordfish torpedo-spotter-reconnaissance aircraft. Against this the German and Italian shore-based aircraft, although more numerous, were based much farther away, the nearest being in Sicily and the Dodecanese, as Cyrenaica was still in British hands. There were also some ship-borne planes with the Italian Fleet and it was one of these from the *Vittorio Veneto* which at 06.30 on

28th March sighted the British cruisers in the van of the British Fleet.

All three groups into which the Italian Fleet had been divided converged on the British cruisers in the hope of destroying them, but at 07.20 an air search from the *Formidable* located two Italian cruiser groups shortly before they became engaged. Just before nine o'clock the Italian Commander-in-Chief, Admiral Iachino, considered that he had penetrated far enough into enemy waters and ordered a general retirement. The battle thereafter turned into a stern chase in which the Italians with their superior speed were likely to be able to escape. All therefore depended on the ability of aircraft to slow them down. During the day nine separate attacks were made on various units of the Italian Fleet. Four were by torpedo-bombers from the *Formidable* and Maleme and five by the R.A.F. Blenheims from Greece. The R.A.F. bombers were unable to do better than near-misses but two torpedoes got home from F.A.A. aircraft, one on the battleship *Vittorio Veneto* and the other on the cruiser *Pola*. The *Vittorio Veneto* took on board 4000 tons of water and came to a stop but shortly afterwards she was able to get under way again and to make 20 knots. The *Pola*, however, was disabled and lay dead in the water. Some of the Italian cruisers then turned back to help the *Pola* and the British battleships came up with them during the night. They sank the *Zara*, *Fiume* and *Pola* and two destroyers. The *Vittorio Veneto* escaped as she had too great a start and her speed was not sufficiently reduced by the torpedo attack. Nevertheless the Battle of Cape Matapan was a considerable victory for the British, whose sole casualty was one Swordfish aircraft.

The Italian Navy as usual blamed their Air Force for the disaster. Co-operation was not good but they did, in fact, report Admiral Cunningham's approach accurately during the afternoon and this was discounted by Admiral Iachino. Fliegerkorps X in Sicily must receive some credit: one of the reasons that the British broke off the pursuit was to avoid being caught too close to the German air bases in daylight. In this action the aircraft from the *Formidable* finally achieved what the torpedo-bomber had been designed to do and succeeded in slowing up part of the enemy fleet so that it could be engaged. The Italians were greatly impressed by the efficiency of the British Fleet Air Arm and as a result they ordered the conversion of a fast liner to an aircraft-carrier, thus reversing the policy which they had followed since the Washington Treaty. In the meantime they decided to operate their ships only under fighter cover from the shore. Matapan decisively reasserted the superiority of the Mediterranean Fleet over the Italian Navy and made sure that it would not attempt to interfere with convoys to Greece again but it did nothing to stop the Axis traffic to North Africa and it was during this period that the German Afrika Korps was transported to Tripolitania.

Fliegerkorps X was now principally engaged in attacking Malta and on 21st April, in an attempt to hinder supplies to North Africa, the British Fleet advanced into the central Mediterranean and bombarded Tripoli during which it suffered no air attacks at all. Fliegerkorps X failed to exercise the command of the sea of which it was capable on this occasion because it was too busy doing something else. The 'flexibility of air power' can have its disadvantages.

Early in May it was considered essential to pass five fast merchantmen through the Mediterranean with important armoured reinforcements for the British Army in the Western Desert now in full retreat before General Rommel. At the same time the opportunity was to be taken to reinforce the eastern Mediterranean and to pass a convoy to Malta. The air attacks at both ends of the Mediterranean were slight and most of the merchant ships got through: only one destroyer was damaged by enemy aircraft. The successful passage of this 'Tiger' convoy, as it was called, was due to a number of factors. The first and probably the most important was that Fliegerkorps X was suffering heavy casualties over Malta and in North Africa at the time and had had to detach a number of aircraft to support General Rommel. The second was that the visibility throughout the passage was very poor and the third was that the convoy had fighter cover throughout. Fulmars under radar control from the carriers *Ark Royal* and *Formidable* protected the ships at each end of the Mediterranean and fifteen long-range twin-engined Beaufighters had been sent to Malta for the middle. Admiral Cunningham was much encouraged and now believed that the air threat had been exaggerated.

On 6th April 1941 the German Army had invaded Greece and by the end of the month the British Army sent to assist the Greeks had been evacuated. On 20th May the Germans invaded Crete, the operation being conducted almost entirely by the Luftwaffe. The unit involved was Luftflotte 4 comprising

Fliegerkorps VIII, with over 700 combat aircraft and Flieger-korps XI consisting of the equivalent of an airborne division and 500 transport planes. Fliegerkorps VIII had 430 bombers, half of which were dive-bombers, 240 fighters and 50 recon-naissance aircraft. The whole of this formidable force was based on six airfields in Greece and the Dodecanese.

The Mediterranean Fleet was determined that, whatever might be landed in Crete from the air, nothing was going to get in by sea (see Fig. 12, p. 171). To achieve this aim they had four battleships, an aircraft-carrier, eleven cruisers and twenty-seven destroyers, but the ships had practically no air support. The *Formidable* had only four Fulmar fighters left after fighting the 'Tiger' convoy through; the R.A.F. had been forced to abandon Crete altogether the day before the invasion and the R.A.F. in North Africa were a long way away and very busy in the western desert. Admiral Cunningham believed that to take Crete the Germans would have to follow up an airborne assault by a landing from the sea. If he could prevent this seaborne landing he believed that the island could be defended. To do this he divided his fleet into a number of groups. Three cruiser-destroyer striking forces were to remain south of Crete by day ready to enter the Aegean and attack any invasion convoy that was sighted: at night they would make high-speed sweeps north of Crete. These striking forces would be supported by a battleship force to the west of the island in case of interference by the Italian Navy and to provide a point on which the strik-ing forces could retire if necessary. All these forces had to be based over 400 miles away at Alexandria, to which place they had to return to fuel and replenish their ammunition. By this time Rommel had retaken Cyrenaica and was on the Egyptian frontier and the ships were liable to be attacked from this direction as well.

The Mediterranean Fleet took up its position to defend Crete in the middle of May and a few days later the German air landings took place. The *Formidable* with her depleted fighter squadrons was kept back at Alexandria, struggling to get a few more aircraft into service. The first night two cruiser-destroyer forces entered the Aegean and withdrew next day without sighting anything. Most of Fliegerkorps VIII was busy support-ing the German troops on the island but had time to sink the destroyer *Juno* and damage the cruiser *Ajax*. The Germans had intended to send in three seaborne convoys, two composed of small local craft with troops and a third of merchant ships with the heavy equipment. The first of these convoys was sighted by an R.A.F. Maryland reconnaissance aircraft near the island of Milos and that night two cruiser-destroyer striking forces closed in. The first convoy was badly mauled and the survivors turned back, the second convoy turned back before it was attacked.

Next day, the 22nd May, Fliegerkorps VIII turned on the Mediterranean Fleet in strength and in a series of dive-bombing attacks sank the cruisers *Fiji* and *Gloucester* and the destroyer *Greyhound* and damaged the battleships *Warspite* and *Valiant*, the former badly, and the cruisers *Naiad* and *Carlisle*. All ships ran short of anti-aircraft ammunition and at the end of the day Admiral Cunningham ordered them all to return to Alexandria. He did not believe that in face of this scale of air attack he could prevent seaborne landing without crippling losses. Nevertheless the Navy did its best to run in supplies to the hard-pressed forces ashore. The Germans did not attempt to land from the sea again but this did not worry them as with the capture of Maleme airfield they were able to fly in reinforcements instead. Next morning the destroyers *Kelly* and *Kashmir* were dive-bombed and sunk. Undaunted by these heavy losses a cruiser-destroyer force again entered the Aegean on the nights of 24th and 25th May. The *Formidable* which had now managed to scrape together twelve barely serviceable Fulmars, put to sea with the battleships *Queen Elizabeth* and *Barham* and attacked the airfield of Scarpanto. Her fighters became engaged in some twenty duels but only shot down two enemy aircraft. During the afternoon 25 Stukas of Fliegerkorps X from Africa, the same squadron that had attacked the *Illustrious*, struck at the *Formidable*, hitting her twice with heavy bombs and seriously damaging her. The destroyer *Nubian* was also hit and both ships had to return to Alexandria. Next day the survivors of this force were again attacked and the battleship *Barham* was seriously damaged. Shortly after this latest disaster the decision was taken to evacuate Crete.

The evacuation was carried out over a period of four nights and nearly 20,000 men were taken off, some from the north coast of Crete. The cost was high and the cruiser *Calcutta* and the destroyers *Hereward* and *Imperial* were sunk, and the cruisers *Orion*, *Perth* and *Dido* and the destroyers *Nizam* and *Kelvin* were damaged. At the end of the Battle of Crete there were only 2 battleships, 2 cruisers and 9 destroyers left in Alexandria fit for

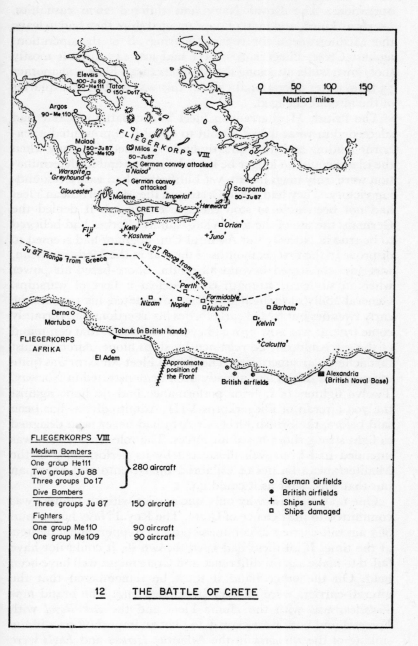

Elevsis
100-Ju 80
50-He 111 Tator
150-Do 17

Argos
90-Me 110

Molaoi
50-Ju 87
90-Me 109

Warspite
Greyhound +
+
Gloucester

F L I E G E R K O R P S   VIII

Milos
50-Ju 87
Naiad

German convoy attacked
German convoy attacked

Maleme   *Imperial*   Scarpanto
50-Ju 87
CRETE   *Hereward*

Orion

Fiji   *Kelly*
+   +*Kashmir*   +*Juno*
+

Ju 87 Range from Greece   Ju 87 Range from Africa

Perth   Formidable

Nizam   *Napier*   *Nubian*   *Barham*

Derna   *Kelvin*
Martuba

Tobruk (in British hands)   +*Calcutta*

FLIEGERKORPS
AFRIKA
El Adem

Approximate
position of
the Front   British airfields   Alexandria
(British Naval Base)

0   50   100   150
Nautical miles

FLIEGERKORPS VIII

Medium Bombers

One group He 111	
Two groups Ju 88	280 aircraft
Three groups Do 17	

Dive Bombers

Three groups Ju 87	150 aircraft

Fighters

One group Me 110	90 aircraft
One group Me 109	90 aircraft

o   German airfields
●   British airfields
+   Ships sunk
□   Ships damaged

12   THE BATTLE OF CRETE

operations. The Royal Navy had suffered 2000 casualties, nearly all killed, and most of the damaged ships then had to leave the Mediterranean for repairs. During all of the operations against Crete, Fliegerkorps VIII had lost 147 aircraft mostly shot down while attacking the Mediterranean Fleet and another 73 from other operational causes; this was well over a quarter of the aircraft engaged.

The British Mediterranean Fleet in the Battle of Crete had succeeded in what it had set out to do: it had prevented a sea-borne landing and subsequently had evacuated the Army when the island could no longer be held. Its valour and its determination were truly magnificent, yet Luftflotte 4 had won a resounding victory. They had taken Crete and the Mediterranean Fleet had not been able to stop them, even though it denied the Germans the use of the sea. What Admiral Forbes had believed to be true in Norway but Admiral Cunningham had seemed to disprove in the first six months of the war in the Mediterranean, was now confirmed beyond all doubt. Shore-based air power when in sufficient strength could defeat a fleet of warships. General 'Billy' Mitchell may have overstated his case in the early twenties but by the early forties his assertions had certainly come true. It was now proved beyond any doubt that command of the sea could not be regained by ships alone. Aircraft must first achieve air superiority. The British Fleet Air Arm was quite incapable of this as had already been demonstrated in Norway. Twelve fighters of inferior performance had no hope against the 700 aircraft of Fliegerkorps VIII. Admittedly, as has been said before, the British Fleet Air Arm had never been designed to fight strong shore-based air forces. The role for which it was intended had been well illustrated by its performance in the Mediterranean battles of Calabria, Spartivento and Matapan but that was as far as it could go.

One may well ask why only one carrier with 12 fighters was committed to the defence of Crete. The Royal Navy had a total of 7 aircraft-carriers in commission and 130 operational fighters at the time. If all these had been thrown in, it could not have failed to make a great difference and Crete might well have been held. On the other hand it must be remembered that the aircraft-carriers were all very busy elsewhere. The brand new *Victorious* was with the Home Fleet and the *Ark Royal* with Force H and both were shortly to play an important part in the sinking of the *Bismarck* in the Atlantic. *Hermes* and *Eagle* were hunting raiders in the South Atlantic and Indian Ocean: the *Furious* and *Argus* were ferrying fighters to Malta and the Middle East through Takoradi and the *Illustrious* was still under repair. The *Furious, Eagle, Hermes* and *Argus* were all old ships and were very vulnerable: how long they would have lasted off Crete is a matter of conjecture. Even the armoured aircraft-carriers *Illustrious* and *Formidable* would probably have been sunk by the Luftwaffe if it had not been for their toughness. It must also be remembered that the most numerous fighter type was still the Fairey Fulmar, of comparatively low performance. Grumman Martlets and Sea Hurricanes, which could have made a real difference, were only just coming into service.

Soon after the campaign in Crete, Luftflotte 4, with Flieger-korps VIII and XI, was withdrawn to take part in the invasion of Russia and the headquarters of Fliegerkorps X from Sicily took over all German aircraft in the Mediterranean. Its total strength was over 400 aircraft but due to maintenance diffi-culties only about 250 of these were operational at a time. It comprised Me. 109 and Me. 110 fighters, Ju. 87 dive-bombers, Ju. 88 and He. 111 medium bombers and six of the four-engined FW. 200 long-range bombers. It also had some coastal seaplanes and transport aircraft. It had two main functions: one was to support General Rommel's Army in North Africa and the other was to attack the Mediterranean Fleet and the Suez Canal area. With airfields in the Dodecanese, Crete, Greece and Cyrenaica it was in a position virtually to blockade the British Fleet in the eastern Mediterranean. Until the end of the year, when the Eighth Army retook Cyrenaica, the Fleet's activities were confined almost entirely to the supply of the besieged garrison in Tobruk and coastal support for the Syrian campaign. Having no aircraft-carrier since the departure of the *Formidable* for repairs, the fleet had to keep under cover of shore-based fighters and was unable to penetrate into the central Mediterranean. Worse still, Fliegerkorps X not only inflicted losses on the ships supplying Tobruk but bombed Alexandria and again mined the Suez Canal. Later they attacked Suez and shipping in the Red Sea, setting fire to the liner *Georgic* on 14th July, so again threatening the terminal of the vital supply route round the Cape.

Fliegerkorps X could be said to command the sea in the eastern Mediterranean except for coastal areas where R.A.F. cover could be given. Furthermore it posed a threat to the whole supply position in the Middle East. The Germans had therefore found a new way to command the sea. Without a fleet at all, they had over-run land areas on which to base air forces to operate over the sea. The British were not able to wrest control back from them until they had done the same thing and had recaptured Cyrenaica. We may be thankful that Hitler's negotiations with Franco were unsuccessful and that he was not able to apply this method of commanding the sea to the whole western basin as well.

Air reinforcements for the Middle East, especially fighters, were obviously of extreme importance if the situation was to be restored. On 15th June, 47 Hurricanes were flown off the carriers *Victorious* and *Ark Royal* in the western Mediterranean for Malta and in a series of similar operations during the summer, 269 Hurricanes were delivered, of which over 100 flew on to the Middle East. In addition the carriers *Furious* and *Argus* were employed almost entirely in ferrying fighters either to Gibraltar or Takoradi for onward passage to the Middle East. Well over half the British aircraft-carrier strength therefore spent most of its time in this auxiliary role and it was to be their contribution to the regaining of command of the sea in the Eastern Mediterranean.

There could therefore be no question of resupplying Malta from the East and it had to be done from Gibraltar by Force H. For this purpose Force H was reinforced from the Home Fleet and short sorties of the Mediterranean Fleet from Alexandria were made as diversions. Fortunately for the British they were now only opposed by the Regia Aeronautica in the western basin as Fliegerkorps X had no aircraft left in Sicily. The Italian Air Force however was stronger than before and had 200 strike aircraft in Sardinia and Sicily, 50 of which were torpedo-bombers and 30 dive-bombers. The first convoy, called operation 'Substance', got six store ships through to Malta at the end of July, the *Ark Royal*'s Fulmars providing fighter protection until the R.A.F. Beaufighters took over from the island. Nevertheless the destroyer *Fearless* was sunk and the cruiser *Manchester* damaged by torpedo-bombers, and the destroyer *Firedrake* was damaged by a near-miss from a bomb. A second convoy, of nine merchant ships, got through at the end of September (operation 'Halberd'), but one of the merchant ships was sunk and the battleship *Nelson* was damaged by torpedo-bombers. The R.A.F. supported this convoy by bombing the airfields in Sicily and Sardinia from Malta as well as providing fighter protection. The Italian Fleet in fact put to sea to oppose the passage of the 'Halberd' convoy but was hampered by poor air co-operation and the restriction that it must remain under fighter cover from the shore. It achieved nothing and it is of interest that the reason was chiefly its fear of attack by carrier-borne aircraft from the *Ark Royal*.

The recovery of Cyrenaica by the Eighth Army depended not only on the building up of its strength by the long route round the Cape but on interfering with the German and Italian sea communications across the Mediterranean. During the first half of 1941, there were only a dozen Swordfish torpedo-bombers at Malta for this purpose. Whereas submarines had sunk nineteen ships and a destroyer striking force had sunk nine, aircraft had only disposed of four. In April, however, Blenheim bombers from the North Sea joined the Swordfish at Malta and by August there were 20 of each with 12 Wellingtons to bomb the ports by night and 10 Marylands for reconnaissance. The Swordfish worked at night and their effort was partly devoted to the attack of ships at sea with torpedoes and partly to the laying of mines. The targets were located by a leading Swordfish fitted with radar which then dropped flares in the light of which the others launched their torpedoes. The Blenheims worked by day and attacked ships at sea by masthead level bombing. In the last half of the year, aircraft headed the list by sinking 26 ships to 17 by submarine and 11 by Force K, the surface striking force that operated from Malta in November and December. The sinkings by surface ships, however, were only possible with the aid of air reconnaissance and the gaining of air superiority over Malta by the fighters. From July onwards the total sinkings represented about 20 per cent of the Axis traffic to North Africa but in November it rose to 62 per cent and this was a considerable factor in the reconquest of Cyrenaica at the end of the year. In November a Swordfish from Malta had a success against a warship, torpedoing the Italian cruiser *Abruzzi*.

In October, in the eastern Mediterranean, the R.A.F. aircraft for co-operation with the fleet were organized into No. 201 (Naval Co-operation) Group which consisted of two flying-boat squadrons, two General Reconnaissance Squadrons and two long-range fighter squadrons. The R.A.F. had hitherto opposed this move as they believed that it was contrary to the principle of the flexibility of aircraft. They wished to retain the option to transfer squadrons, for instance, from fleet co-operation to support of the army or air defence as the situation demanded. The change, however, which was simply the institution of a small Coastal Command in the Mediterranean, was a decided advance from the maritime point of view and made for better and closer co-operation.

The year 1941 ended in the Mediterranean with a series of disasters for the British. The battleship *Barham* was sunk by a U-boat and the battleships *Queen Elizabeth* and *Valiant* were severely damaged by Italian human torpedoes in Alexandria harbour. Most serious of all was the loss of the aircraft-carrier *Ark Royal*, torpedoed by the German *U.81* east of Gibraltar as she returned from flying fighters to Malta. This left no aircraft-carrier in the Mediterranean at all and, with the beginning of the war in the Far East, little prospect of one being sent.

VII

Aircraft in the Attack and Defence of Trade

1940 – 1941

WHILST THE EVENTS related in the last chapter were taking place in the Mediterranean, the war at sea continued unabated off the coasts of western Europe and on the oceans of the world. The Luftwaffe, after its victory in Norway, was able to make the British Home Fleet keep its distance; although two carrier strikes were made by the *Furious* against Tromsö and shipping on the Norwegian coast in September they did little damage. Nevertheless the Luftwaffe could not break the blockade and so alter fundamentally the maritime stranglehold which Great Britain had upon Germany. The over-running of Norway, the Low Countries and France however did alter the maritime strategic position considerably. The Germans now had access to the oceans of the world and there could no longer be any hope of confining them to the North Sea.

After the Norwegian campaign, and during the summer of 1940, the Germans turned increasingly to an attack on British commerce and on 17th August Hitler ordered a total blockade of the British Isles. To enforce this blockade, U-boats, aircraft, surface raiders and mines were all used. Aircraft not only made direct attacks on shipping but laid mines and helped the U-boats and surface raiders: they were also used to counter all three methods of attack. Of these campaigns, the most important to our subject is the German air attack upon shipping which was the first of its kind on a large scale in naval warfare. The campaign had started early in the war and continued both during and after the Battle of Britain. With bases available in the Low

Countries and Northern France air attacks against coastal shipping were now waged with greater effect. Fighter Command had been very busy with the Battle of Britain up to October and could spare little for defence of coastal shipping, so German aircraft caused a steady casualty rate amongst merchant ships, fishing vessels and small auxiliary warships, which began to rival that caused by mines. In November some heavy attacks on shipping in the Thames Estuary and the Straits of Dover were defeated by Fighter Command but not until eleven ships had been lost. In December only four ships were sunk within fighter range of the coast but minelaying aircraft, operating at night at low altitude, were very hard to intercept and mines continued to be laid from the air.

In July, German Bv138 flying-boats, which had a range of a thousand miles, arrived at Brest and began to operate south of Ireland and as far as longitude 9° west. In August, four-engined F.W. 200 land-planes of Gruppe 40 arrived at Bordeaux and with double the endurance began to range to the west and even north-west of Ireland. Although the original intention was that they should scout for the U-boats, they waged an independent anti-shipping offensive with considerable success and in October they set the liner *Empress of Britain* on fire 70 miles off the north-west coast of Ireland. Coastal Command attempted to counter this with long-range Blenheim fighters patrolling to the south-west of Ireland but it was found very difficult to make interceptions at random and in attempts to respond to distress calls from merchant ships they found that they generally arrived too late. The Blenheim had in any case an insufficient margin of speed to catch a F.W. 200 in a stern chase and they failed to shoot any of them down. In all, German aircraft disposed of 95 ships of 269,884 tons by the end of the year. Mines laid by ships and U-boats as well as aircraft destroyed another 98 ships of 188,941 tons.

Successful as the German air attacks proved to be, they were only disposing of a fraction of the tonnage sunk by the U-boats. The U-boats resumed the Battle of the Atlantic in June 1940 after the Norwegian campaign and soon began to operate from bases on the Atlantic coast of France. As a result their time on passage was substantially decreased and they were able to spend longer in the operational area which they were also able to extend to 25° W. Convoys were at this time only escorted to and from 17° W but this was extended to 19° W in October. As this was still not enough, the U-boats were able to operate against shipping that was both unescorted and beyond the range of most Coastal Command aircraft. The U-boats had also developed new tactics: they now attacked on the surface at night and concentrated in 'wolf-packs' against the convoys. These measures achieved considerable success and by the end of the year they had sunk 343 ships of 1,754,501 tons.

Coastal Command had increased in strength since the beginning of the war from 265 aircraft at the outset to 490 by 1st July 1940. The main increase is accounted for by the addition of 93 long-range fighters and some 50 torpedo-bombers. There were, however, 64 more general reconnaissance machines and a greater proportion of this type were now medium-ranged Hudsons and Sunderlands instead of the very short-ranged Ansons. During the autumn, Coastal Command was very busy in the Narrow Seas with anti-invasion patrols so that aircraft for the Battle of the Atlantic were few and far between. The effort of the few aircraft available was expended in escorting convoys but as they could go only 250 miles to seawards and the U-boats operated farther out, the majority of the convoys which they escorted were not even threatened. Only a few Sunderlands could reach the areas where most of the sinkings were taking place and the U-boats were not unduly worried by them. The Sunderland flying-boats could always be seen a long way off and the U-boats could generally submerge in ample time to avoid attack. Even if they could not, the small 100-lb A/S bomb, with which the aircraft were armed, needed a direct hit to give any chance of a kill and so was not particularly lethal. Those aircraft which had radar found it difficult, with the primitive sets available, to pick up U-boats on the surface at night and they had no means of detecting or attacking them submerged at all. All the same, U-boats were sighted fairly frequently on the route round the north of Scotland but, by the end of 1940, of the 32 German U-boats that had been sunk since the beginning of the war, Coastal Command could only claim a share in two of them. Nevertheless air patrols inconvenienced the U-boats as they had to dive whenever an aircraft approached and were therefore denied their surface mobility to a certain extent. They needed surface mobility to concentrate against convoys, and so the U-boats moved steadily seawards so that they could carry out their chosen surface tactics unmolested. The farther the U-boats went to the westwards, however, the harder it became to find the convoys. It was to help them in this that the long-range F.W. 200 aircraft had originally been sent to operate in the Atlantic. There were, at this time, however, only enough F.W. 200 aircraft to make one scouting flight each day.

The third and, in damage done, the smallest threat to British commerce, was the German surface raider. Six armed merchant raiders left Germany during the summer of 1940 and broke out into the Atlantic, the Indian Ocean and the Pacific and they were followed in October and November by two warship raiders. Between them they sank another 68 ships of 436,363 tons, and the air and surface measures which had to be taken to counter them were extensive.

The merchant raiders *Atlantis* and *Orion* left Germany in March and April 1940, before the Norwegian campaign began: they passed northwards in disguise and broke out into the Atlantic through the Denmark Strait without being detected by the Coastal Command North Sea patrols. The *Widder*, *Thor* and *Pinguin* followed in May and June and the *Komet*, with Russian connivance, navigated the Arctic sea route to the Pacific in July. At the end of October the pocket battleship *Scheer* also broke out by the Norwegian coast and Denmark Strait without being sighted and attacked convoy HX84, sinking the armed merchant cruiser *Jervis Bay*. In early December she was followed by the heavy cruiser *Hipper* also without being seen. She was driven off by the escort of convoy WS5A and she entered Brest on 27th December.

Once out on the wide oceans, finding these ships was a very difficult problem. A hunting group was formed round the new aircraft-carrier *Formidable* on her way round the Cape to replace the *Illustrious*, and the *Hermes* and the seaplane-carrier *Albatross* did their best in the South Atlantic. A great deal of trouble and many casualties would have been saved if these ships could have been intercepted as they left Germany. This was however by no means easy, especially as the German ships made the best use of the weather to avoid being sighted. Coastal Command had been able to resume patrols across the North Sea in the autumn of 1940 but they now had to compete with German fighters off Norway. In November, Coastal Command was able to start to patrol the Denmark Strait and the Iceland–Faeroes gap using short-range aircraft based in Iceland, and to patrol off the Biscay ports from bases in the south-west of England. The Home Fleet, which shared the responsibility with Coastal

German Junkers 87
Dive Bomber
(used in Norway and
Crete)

man Focke-Wulf
200 Condor
(used over the
Atlantic)

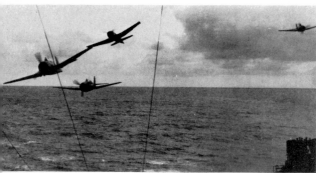

American Grumman
F6F Hellcats
(the victors of the
Battle of the
Phillipine Sea)

(Imperial War Museum Photographs)

merican Douglas
SBD Dauntless
Dive Bomber
(the victor of the
attle of Midway)

(Official U.S. Navy Photograph)

PLATE VII — TYPES OF AIRCRAFT USED AT SEA 1939–45

The British Carr
Force in Operati
'Pedestal'
(H.M. Ships
Indomitable and *E*
taken from *Victor*

H.M.S. *Furious*
Operating Fairey
Swordfish Torpedo-
Spotter
Reconnaissance
aircraft

Hawker Hurrican
Fighter flying off
CAM ship

PLATE VIII

(Imperial War Museum Photograpl

Command, was reduced to the *Furious* after the *Ark Royal* had been sent to join Force H and in November she had to be employed ferrying fighters to the Middle East. At the time of the break-out of the *Scheer* and the *Hipper*, therefore, the Home Fleet had no aircraft-carrier at all.

These defensive measures to protect trade took up most of the British naval and maritime air effort, but it was still found possible to attack the enemy's sea communications. Apart from a few submarines and coastal forces, this was done almost entirely by aircraft. The targets were principally the iron-ore ships from Narvik to Germany and the German industrial coastal traffic. In the second half of 1940, 914 mines were laid in enemy waters mostly by Bomber Command aircraft, sinking 62 ships of 49,348 tons and damaging another eight, at a cost of 21 aircraft. Some direct air attacks were also made on ships at sea in this period, originally by three Fleet Air Arm squadrons working from the shore and then by two Blenheim squadrons of Bomber Command. In September the Blenheims were joined by two squadrons of the new twin-engined Beaufort torpedo-bombers of Coastal Command. In the second half of 1940 they sank six ships of 5,091 tons and damaged another fourteen, but lost 82 aircraft. These results were unspectacular and somewhat expensive, but they were attacks on the enemy's commerce that could not be made by any other means.

During 1941 the U-boat held its place as the chief destroyer of commerce although in January there were still only twenty-two of them operational. During the year they sank 432 ships of 2,171,754 tons and as early as 6th March the Prime Minister issued his famous directive giving the Battle of the Atlantic a very high priority. In general, the year 1941 was a period of great expansion of the area of U-boat operations. They spread right across the North Atlantic to Newfoundland and the area south of Greenland and penetrated into the South Atlantic as far south as St Helena. The aim of the U-boats was to find areas where shipping was unescorted and where their sinking-rate would be high and above all to find areas where there were no air patrols and they would be able to operate on the surface.

Coastal Command had grown to a total of 564 aircraft by the beginning of 1941 with a slightly higher proportion of aircraft of medium range than before. At the time the greatest need of the Command was to extend its operations to follow and harass the U-boats in their new areas. In March Sunderland flying-boats were sent to Freetown in West Africa and in April Sunderlands and Hudsons were sent to Iceland, the Sunderlands operating from the R.A.F. seaplane tender *Manela*. In May the short-range London flying-boats at Gibraltar were replaced by Catalinas of well over double the range and in October, although they were not yet at war, U.S. Flying Fortresses and Catalinas began to work with the Royal Canadian Air Force from Newfoundland and Iceland. By July, 15 Group of Coastal Command, which covered most of the North Atlantic, had three squadrons of Catalinas and some Whitleys and Wellingtons which gave them more long-range aircraft. The extreme radius of all these types, however, still did not exceed 600 miles and there was a large gap in the centre of the Atlantic which could not be covered. There were two ways of filling this gap, the first was to use much longer-range aircraft and the second was to use aircraft-carriers. There were no aircraft-carriers available but the first very long-range squadron of B.24 Liberators was formed in June and began to operate in the autumn. By December, however, it still only consisted of nine aircraft. By the end of the year Coastal Command had expanded to 633 aircraft and the short-range Anson and London types had practically all been replaced.

On 15th April 1941 the Admiralty were given 'operational control' of Coastal Command, by which was meant that they were able to order broad requirements and priorities but the method of meeting them was left entirely to the R.A.F. In September a new directive was given to Coastal Command jointly by the Admiralty and Air Ministry in which reconnaissance for enemy ships and U-boats at sea and in harbour had first priority. Next came offensive measures against enemy warships, U-boats and merchant ships which included mine-laying, and lastly defensive measures such as the escort of convoys.

Up to this time aircraft had not been at all successful in taking the offensive against the U-boats: an airborne depth charge was now in service, but little was as yet being achieved. With the decision to give the offensive a higher priority, patrols were stepped up in the Bay of Biscay and round the north of Scotland and sightings became much more frequent. In January an aircraft of 210 Squadron sank the Italian submarine *Marcello* west of the Hebrides and in August a Hudson of 269 Squadron working from Iceland so badly damaged *U.570* that she surrendered. On 30th November the 'Bay Offensive' had its first success when a radar-fitted Whitley of 502 Squadron sank *U.206*. Nevertheless of the 33 German and Italian U-boats sunk in the Atlantic during the year these were the only three to be dispatched by aircraft, three more were destroyed jointly by ships and aircraft whilst 26 were sunk by ships alone.

Admiral Dönitz, commanding the German U-boats, was still keen to use air reconnaissance to help the U-boats find the convoys. In January, whilst Göring was away on leave, he succeeded in getting Gruppe 40 at Bordeaux placed under his operational control. In order to range as far as possible to the westwards a scheme to land at Stavanger was brought in. The aircraft's reports, however, were found to be very inaccurate and they could not stay long enough to 'home' the U-boats on to the convoys by radio. Such tactics required the U-boats to make long fast passages on the surface and this was found to be increasingly difficult as Coastal Command patrols became more numerous and could operate for extended periods with the very short summer nights in northern latitudes. The U-boats, therefore, found it better to move even farther westwards out of range of Coastal Command where they could pursue their surface tactics undisturbed and incidentally find more independent merchant ships.

The F.W. 200 aircraft, now some thirty strong, being unable to reach the U-boat operating areas, continued throughout 1941 to wage an independent campaign against shipping. In January they sank 20 ships and in February another 27 and damaged others. Convoys were then routed farther north towards Iceland to try to avoid these attacks and Coastal Command sent long-range Blenheim fighters from the east coast to Northern Ireland. Interception from the shore was, however, still mainly a matter of chance and it was apparent that fighters must be operated from ships at sea if success was to be achieved. The Admiralty's first move was to convert the old seaplane carrier *Pegasus* to carry three Fulmar fighters which could be launched by catapult. These were, however, 'one-shot' aircraft that would have to ditch in the sea if out of range of a shore base. Three merchant ships, the *Maplin*, *Springbank* and *Ariguani* were also converted to Fighter Catapult Ships and all four were ready in April. It was not until August, however, that a Hurricane launched from the

Maplin shot down a F.W. 200 but in the two following months both the *Springbank* and *Ariguani* were torpedoed by U-boats.

It was clear that the Fighter Catapult Ships could only provide part of the answer to the F.W. 200, whose depredations continued throughout the summer. The final solution had already been decided upon and it was to convert suitable merchant ships to small aircraft-carriers which could carry fighters. A German prize, the *Hannover*, was already in hand and in May it was decided to convert five more ships and to order six in the U.S.A. under Lend-Lease. It was hoped that these ships would also be able to carry some anti-submarine aircraft and so help in the war against the U-boats, especially in the mid-Atlantic gap. The first escort carrier, renamed *Audacity*, entered service in June and in September, with six Grumman Martlet fighters, she was sent to escort the Gibraltar convoys. She achieved success almost at once and her aircraft shot down a F.W. 200 whilst escorting convoy OG74 on 20th September. The escort carrier was however a fairly long-term solution and it had already been decided in April to fit 50 merchant ships with a catapult to carry a single Hurricane 'one-shot' fighter so that every convoy would have a degree of protection. These CAM ships, as they were called, remained cargo-carrying merchant ships and the aircraft were provided by two squadrons of R.A.F. Fighter Command.

The Gibraltar convoys were the most threatened by the F.W. 200 and on this route they were able to work with some success in co-operation with U-boats. In December the *Audacity* accompanied HG76 which had a heavy and efficient radar-fitted escort of anti-submarine vessels, during its passage home. It was covered for the first part of its route by Coastal Command aircraft from Gibraltar and for the last part by 19 Group aircraft from England, the *Audacity*'s fighters being used in the anti-submarine role while the convoy was out of range of shore-based aircraft. The convoy was heavily attacked for five days but no fewer than five U-boats were sunk by the ships and two F.W. 200 were shot down by the Martlets. Two merchant ships and one of the escorts were sunk and then the *Audacity* whilst operating aircraft outside the screen was torpedoed and sunk by a U-boat. Nevertheless the fighting through of this convoy was a considerable success and in spite of its vulnerability the worth of the escort aircraft-carrier was proved beyond doubt.

The German air attacks on coastal shipping continued

throughout 1941. There were only two shipping casualties in January but these had risen to 21 in March. Partly as a result of the Directive on the Battle of the Atlantic and partly because of an increase in strength, Fighter Command was able to increase its protection of shipping by day from under 10 per cent of its total effort in February to 49 per cent in April and to keep this up for the rest of the year. The Germans inflicted considerable casualties at night in May and June but the number of attacks fell off sharply as the Luftwaffe was withdrawn for the Russian campaign. Radar cover now extended farther to seawards and co-operation between ships and aircraft was much improved. As a result, casualties by day fell and five ships were sunk at night for every one by day. By December only one merchant ship, one fishing vessel and two minesweeping trawlers were sunk for nearly 500 sorties by German aircraft. In the second half of 1941, a total of 68 ships were sunk in coastal waters by air attack as well as 18 fishing vessels and 26 minor war vessels. For the whole of 1941, Axis aircraft in all areas sank over a million tons of shipping which was half the amount sunk by U-boats but over twice that disposed of by all the surface raiders on the oceans.

At the beginning of 1941, six armed merchant cruisers and the pocket battleship *Scheer* were still at large on the trade routes and all efforts to find them had so far proved abortive. They were all, of course, out of range of any shore-based aircraft and the sea was simply too big to be covered by the few carrier and shipborne aircraft that were deployed. At last on 22nd February the cruiser *Glasgow*'s aircraft sighted the *Scheer* north of Madagascar but lost touch before she could be brought to action.

At the end of January the German battle-cruisers *Scharnhorst* and *Gneisenau* broke out into the North Atlantic. They sailed from Kiel on 23rd and when they passed through the Great Belt intelligence of this movement reached London. Extra air patrols were organized and the Home Fleet took up an intercepting position south of Iceland. The Home Fleet at this time consisted of three capital ships, eight cruisers and eleven destroyers but no aircraft-carrier: the *Furious* was still being used to ferry aircraft for the Middle East. The two German battle-cruisers succeeded in reaching the Atlantic without being seen by Coastal Command. They were sighted momentarily between Iceland and the Faeroes by the cruiser *Naiad* of the Home Fleet's screen but they doubled back, refuelled in the Arctic and later broke out by the Denmark Strait. The cruise of the German battle-cruisers lasted nearly two months, they sank 22 ships and during this time aircraft only sighted them three times. On 8th March, a shipborne aircraft from the *Malaya* saw them north of the Cape Verde Islands and on 20th March a Swordfish from the *Ark Royal* picked them up 600 miles off Cape Finisterre but they were too far away and the weather was unsuitable for a striking force to be flown off. Finally on 21st March a Hudson of Coastal Command sighted them 200 miles from Brest, which they entered on 22nd March. Aircraft could not therefore claim to have inconvenienced the two ships very much. Indeed the fact that the two raiders had not caused more damage was due more to the battleships *Ramillies*, *Malaya* and *Rodney* which by their presence prevented them from destroying three convoys. Aircraft were no more successful in preventing the return of the *Scheer* to Germany at the end of March or a short sortie of the *Hipper* from Brest and her subsequent return to Germany at about the same time. Nor did they succeed against the armed merchant raiders, and in April the *Thor* returned to Germany through the Channel.

Once in Brest the German battle-cruisers were heavily bombed from the air. Some 13 per cent of Bomber Command's strength was used in the four months following their arrival, dropping 2000 tons of bombs. In the same period Bomber and Coastal Commands laid 275 mines off Brest. In spite of this weight of attack only four hits were secured on the *Gneisenau* in April and one on the *Prinz Eugen* after her arrival there in July: furthermore 31 aircraft were lost. On 6th April however a Beaufort of No. 22 Squadron of Coastal Command torpedoed the *Gneisenau* in the stern but was immediately shot down. On 22nd July the *Scharnhorst* moved to La Pallice and on 24th July fifteen of the new four-engined Halifax bombers made a daylight raid and secured five hits, damaging her so badly that she had to return to Brest for repairs. The effect of this considerable air effort was disappointing but it did prevent the ships leaving their bases in France to co-operate with the sortie of the *Bismarck*. At the end of July, all three ships having been hit, the main weight of Bomber Command was switched back to Germany.

The new and powerful battleship *Bismarck* with the heavy cruiser *Prinz Eugen* left Gdynia on 18th May and their passage through the Belts was duly reported to London. Intensive air patrols were flown by both sides but the weather was very bad.

The British failed to sight the *Bismarck* and the Germans failed to detect that the Home Fleet had left Scapa Flow. A naval aircraft from the Orkneys however was able to establish that the *Bismarck* had left Kors Fjord near Bergen and so could be assumed to be at sea. The *Bismarck* was sighted by the cruisers *Norfolk* and *Suffolk* in the Denmark Strait and was subsequently brought to action by the *Hood* and *Prince of Wales*. The *Hood* was sunk and the German ships pressed on to the southwards shadowed by the two cruisers and Coastal Command aircraft from Iceland. The Home Fleet, which was over 300 miles away at the time, included the new aircraft-carrier *Victorious* but she had only just joined the fleet; she was about to take Hurricanes to Malta and so only had a small air group of nine Swordfish and six Fulmars on board. She was detached at once from the fleet with an escort of four cruisers to close the *Bismarck* and launch an air strike to try to slow her down. The strike was flown off when 120 miles away, whilst *Bismarck* was still being shadowed, and one torpedo hit was secured amidships. Unfortunately it did little damage and two Fulmars which were with the Swordfish were lost. Shortly afterwards the shadowing cruisers lost touch and an air search from the *Victorious* at dawn 25th May failed to regain contact.

Coastal Command had now to establish patrols to cover the *Bismarck*'s return to Germany as well as her course to the French Atlantic coast. It was a Catalina of 209 Squadron of Coastal Command which found the *Bismarck* again when she was nearly 700 miles west of Brest, but the Home Fleet was too far astern to catch up and the *Bismarck* would soon be covered by the Luftwaffe based in France. Force H from Gibraltar with the *Ark Royal* who had been brought north from Gibraltar to co-operate, was however in a position to intercept. An air search from the *Ark Royal* soon gained contact and a strike of fourteen Swordfish was flown off. Unfortunately they attacked the cruiser *Sheffield* by mistake but a second strike of fifteen Swordfish five hours later secured two torpedo hits on the *Bismarck*. The first was on the armour belt and again did no damage but the second disabled her rudder and proved mortal. The Home Fleet though very short of fuel was able to overhaul and sink her next day. On the following day as the fleet retired up the west coast of Ireland, Luftwaffe bombers from France sank the destroyer *Mashona*.

At last a German surface raider had been intercepted and

sunk before she could reach the trade routes. Again Coastal
Command air reconnaissance had initially failed to sight her
and it was the cruiser patrols which, using radar, had made
contact. The chase then moved out of range of shore-based
aircraft but the *Bismarck* was re-located by a Catalina of Coastal
Command as soon as she was in range again. The successful
outcome of the action was largely due to the presence of two
aircraft-carriers, which succeeded in slowing her down for the
battle-fleet to engage in the way the British had always intended
that their Fleet Air Arm should be used. It was a salutary lesson
for air reconnaissance, however, that without the cruisers in
the Denmark Strait she might never have been found at all.

In the middle of June, Coastal Command secured their first
real success against a raider. Intelligence was received that the
pocket battleship *Lützow* was about to break out on to the trade
routes and a Coastal Command strike of fourteen Beauforts was
sent to the Skagerrak, one of them hitting her with a torpedo.
The *Lützow* got back to Germany but she was out of action for
six months. At the time of this success there were still four
armed merchant raiders at large but the cruiser *Cornwall* had
sunk the *Pinguin* in May and the *Devonshire* the *Atlantis* in June.
In both cases they had used their shipborne aircraft to find the
raiders and it was in these operations on the oceans that they
proved really useful. The German raiders all carried seaplanes
as well and made very good use of them. The raiders *Orion* and
Komet both got safely back to occupied France in August and
November without being brought to action although the *Komet*
was in fact sighted by a Coastal Command aircraft.

In September more bomber raids were made on Brest, but
the *Scharnhorst* and *Gneisenau* were not made a primary target
again until they were thought to be ready to sail in December.
On 17th–18th of that month, over a hundred aircraft attacked
at night followed by forty next day escorted by ten squadrons of
fighters, but although they hit the gates of the dock in which the
Scharnhorst was lying only the *Gneisenau* was slightly damaged.
Heavy raids continued until the end of the year and in December
nearly 50 per cent of Bomber Command effort was involved.
Although they damaged the dockyard they failed to hit the
ships and eleven aircraft were lost. In all, from August to the
end of December, nearly 1300 tons of bombs were dropped
in 851 sorties which represented some 10 per cent of Bomber
Command's effort during that period.

The British attack on German shipping continued throughout 1941 along the whole coast of occupied Europe. With the concentration of practically all the British submarines in the Mediterranean this was mainly done by aircraft. Aircraft were, in any case, able to go where shallow water prevented submarine operations and in inland waters such as the Baltic and the Norwegian leads where it was practically impossible for naval forces to operate at all; they had, however, to face considerable opposition from the Luftwaffe and casualties were high. The German attack upon Russia made it important to attack the supply lines of the German armies in North Norway and Finland, which ran from Germany up the Norwegian coast, and also made such attacks easier because of the withdrawal of much of the Luftwaffe's strength to fight in that campaign.

Of the three main types of air attack on shipping, the first and by far the most successful was minelaying by night. In the last six months of 1941, Bomber Command made 843 sorties laying 677 mines which sank 39 ships of 30,343 tons for the loss of 36 aircraft. It therefore needed some twenty sorties and the loss of an aircraft to sink a ship. The air minelaying campaign was an excellent example of inter-service co-operation: the mines were supplied by the Admiralty who also specified the settings and said where they should be laid; the actual laying of the mines then became entirely an R.A.F. operation. Over this period the air minelaying campaign represented some 4 per cent of Bomber Command's effort.

The second type of attack on shipping was a direct attack on ships at sea, generally by day with bombs and torpedoes. The aircraft used were twin-engined types: Hudsons for reconnaissance, Blenheims for low-level bombing, Beauforts with torpedoes and other Blenheims as fighter escort. In the Channel, Hurricane fighter-bombers were also used under radar control from the shore. In the six months from June to December 1941, 3461 sorties were made, sinking 41 ships of 53,575 tons, but 127 aircraft were lost. With this method, therefore, it needed four times as many sorties to sink a ship and three aircraft were lost for every ship sunk. The losses from German fighters and anti-aircraft fire from the ships and their escorts, were therefore heavy and in November the Blenheims were withdrawn and returned to Bomber Command.

The third type of attack was by carrier-borne aircraft, which with the departure of much of the Luftwaffe's strength for

Russia, became possible again. At the end of July, in response to appeals from Russia, the Home Fleet made the first of these carrier strikes on shipping in North Norway and Finland. The *Furious*, after spending nine months continuously transporting aircraft for the Middle East, joined the *Victorious* and, escorted by the cruisers *Devonshire* and *Suffolk* and ten destroyers, took part in the operation. The force approached its objectives in fog and low visibility and was not sighted until just before launching its strike aircraft. A total of 38 Albacores, mostly armed with torpedoes, escorted by 15 Fulmars were flown off on 30th July and attacked Petsamo and Kirkenes in daylight. They were opposed by Me. 110 and Me. 109 fighters and some Ju. 87 dive-bombers, and 12 Albacores and 4 Fulmars were lost. The damage done was small as there were few ships in the two ports, so the attack was undoubtedly a sharp reverse and showed the folly of using such low performance aircraft by day against fighter opposition. The Germans however did not counter-attack the carriers and the four Sea Hurricanes and three Fulmars they had kept back for defence were not required. The *Victorious* struck again at coastal shipping at Tromsö in August and in the Vestfjord in October; on these occasions no aircraft were lost and there was no reaction from the Luftwaffe but the damage done was small.

The decrease in the Luftwaffe's strength in Western Europe also made it possible to supply Russia with much needed arms and equipment through the northern ports of Murmansk and Archangel. In August the *Argus* took Hurricane fighters to North Russia and in September the North Russian convoys were started; by the end of 1941 six of these PQ convoys had sailed and they were practically unopposed by the Germans.

Although, during the period from mid-1940 to the end of 1941 covered by this chapter, aircraft took a very large part in the war at sea in the Atlantic, their influence was not nearly so great as it had been in the Mediterranean. The Luftwaffe still had a considerable effect off the coasts of Europe and the British Home Fleet was reluctant to try conclusions with it until after the beginning of the Russian campaign. German aircraft joined the U-boats and surface raiders in their attack on British commerce and while achieving considerably more success than the British attacks on German shipping, failed to stop the traffic in either the North Sea or the Channel.

British aircraft from the R.A.F. Coastal, Bomber and Fighter Commands as well as the Fleet Air Arm all took part in maritime operations. They achieved some successes but also had some failures, the greatest of which was probably that on twenty-one occasions they allowed warships and armed merchant raiders to leave or return to Germany without even sighting them and they failed to attack three enemy vessels which they did sight. Their sole successes were the torpedoing of the *Lützow* and the re-location of the *Bismarck* that led to her destruction by the Home Fleet. The principal reasons for the failures were the bad weather in northern latitudes, of which the Germans were adept in taking advantage; the poor endurance of most of the aircraft; their lack of suitable radar; the activity of German fighters off the enemy coasts; and a lack of any long-range striking power. Part of the blame must also fall upon the Home Fleet and it is of interest that they seldom had an aircraft-carrier with them until the *Bismarck* operation.

Throughout the period there was a serious shortage of British aircraft-carriers and the loss of the *Glorious* and the *Courageous* early in the war was keenly felt. The loss of the *Ark Royal* right at the end of the period was a greater blow still. Five new aircraft-carriers had by then been delivered but the *Illustrious* and *Formidable* were still under repair and the *Indomitable* had only just been completed. The *Furious* and *Argus* were used from the end of 1940 almost entirely to transport aircraft to the Middle East and the *Eagle* and *Hermes*, being slow and elderly and with small aircraft complements, were used to hunt for raiders on the oceans. The *Ark Royal* up to the time she was sunk remained with Force H at Gibraltar and, until the *Victorious* was commissioned in 1941, the Home Fleet was virtually without a carrier. It was fortunate for the British that the *Bismarck* was not accompanied by the aircraft-carrier *Graf Zeppelin*; work on this ship had been suspended in April 1940 in order to build more U-boats. It had subsequently been decided to complete her but there had been little progress and much difficulty was experienced with the Luftwaffe over suitable aircraft for her. The shortage of aircraft-carriers also put the British at a disadvantage when searching for raiders on the oceans but here shipborne aircraft, which came into their own on both sides, had a considerable share in the destruction of the two raiders sunk.

Of much interest is the minor campaign by Bomber Command against the *Scharnhorst* and *Gneisenau* and other warships which attempted to use Brest as a base to raid Atlantic shipping. The greater part of their effort was expended in night bombing and this proved almost totally ineffective against the ships although it did considerable damage to the dockyard. Most of the damage that was done was in a daylight bombing raid and in an air torpedo attack. Nevertheless it was aircraft which prevented these ships joining the *Bismarck* or making any subsequent forays. An enemy battle-fleet had, in fact, been neutralized, not by a blockade as of old or even a threat of being brought to action by a superior fleet as in the First World War, but by air attack in harbour.

Aircraft of Coastal Command proved a useful auxiliary to ships in the war against the U-boats, but could not claim to be more than that. Their greatest achievement was to persuade the U-boats to operate farther to seawards where they had great difficulty in finding the convoys. It would be too strong to say that these aircraft 'drove them out of the focal areas': the U-boats could have remained if they had wished but would not have been able to operate, in their chosen way, on the surface. As U-boat killers, aircraft were unimportant at this stage and only destroyed a fraction of the number sunk by ships using asdics and depth charges. Coastal Command, using their long-range fighters, were no more successful against the F.W. 200 aircraft in the Atlantic. It was only towards the end of the period that the escort carrier with fighters embarked and to a lesser extent the Fighter Catapult Ships and the CAM Ships began to get the measure of them. The German air attacks on coastal shipping by day were defeated by Fighter Command but were continued at night.

Of the British air attacks on German shipping, minelaying by night proved to be by far the most successful. Direct attacks were very expensive for relatively small results and the few carrier strikes achieved very little. Nevertheless aircraft were able to attack shipping when ships were unable to do so and if it had not been for them the German traffic would have been practically unimpeded. The indecisive nature of both the British and German air campaigns against shipping in coastal waters is a striking contrast to the victory of air power over the sea in Crete. The reason was that in both of the former campaigns, the attacking aircraft were opposed by comparable numbers of defending aircraft: whereas in the latter campaign German air superiority was complete and their attacks were opposed only by ships' anti-aircraft gunfire.

VIII

Japanese Naval Air Power in the Pacific

1941 — 1942

IN THE AUTUMN of 1941, Japan found her plans for the establishment of the 'Greater East Asia Co-prosperity Sphere' stopped dead by the oil embargo imposed by the United States as a counter-move to her invasion of southern Indo-China. The only alternative to complete accession to American demands for no further aggression and the evacuation of China seemed therefore for her to go to war. Although Japan was vastly superior to the American forces in the Far East and could easily take the Philippines, war with the U.S.A. involved a very great risk. American resources were enormous, she was as yet uncommitted in Europe and there was the danger of a subsequent American advance across the Pacific in which Japan could lose all she had gained. The Japanese believed, therefore, that it was of paramount importance to remove the threat of the U.S. Pacific Fleet at the outset. Provided this could be done the chance of a successful outcome to a war seemed better in the autumn of 1941 than it had ever been before and was ever likely to be in the future.

The problem was how to destroy the U.S. Pacific Fleet. The Japanese Navy, since the American withdrawal of three battleships and an aircraft carrier to the Atlantic in 1941, was just superior to the U.S. Pacific Fleet at Pearl Harbor. In surface ships Japan had an overwhelming superiority over the combined fleets of America, the British Commonwealth and Holland stationed in the Far East, but there was not very much difference on paper between the strength of the Japanese Navy and all the combined Allied navies in the whole Pacific Ocean. In air power it was a very different matter. The Japanese naval air force by itself outnumbered all the Allied air forces deployed in the Far East by three to one and, when the available Japanese Army Air Forces were added, by over four to one. Even when the American air forces stationed in the central Pacific were taken into account, the Japanese still had a two-to-one superiority. It was therefore to the air forces that Japan looked for a decisive blow against the U.S. Pacific Fleet.

The Japanese Naval Air Force had some 1500 planes available for operations outside Japan, and half of these, roughly, were carried in ships and half were based ashore. Since 1936, when the Japanese aircraft-carrier force came third in the world with four ships of some 68,000 tons, it had been expanded until it was the largest with ten carriers of 178,000 tons. Against this the United States now had eight aircraft-carriers and the British nine. However, there were only three U.S. aircraft-carriers stationed in the Pacific and one small British one in the Indian Ocean. Since 1938 the Japanese carriers had all been grouped together in the First Air Fleet and it was the intention that this force should be employed as an independent offensive strike weapon. The Japanese admit that they obtained the idea for such a use for aircraft-carriers from the U.S. Navy; their particular achievement was to exploit the principle to a far greater extent than their mentors. The Japanese First Air Fleet with its 500 aircraft was the most powerful ship-borne air force that had ever put to sea.

The Japanese First Air Fleet had the endurance, if it refuelled from tankers at sea, to steam the three thousand miles to the American Fleet base at Pearl Harbor and attack it. Admiral Yamamoto, the Japanese Commander-in-Chief, began planning such an operation early in 1941, a month or two after the successful British attack on the Italian Fleet at Taranto. If the American Pacific Fleet could be knocked out in one blow it would give the Japanese ample time to over-run and consolidate the Greater East Asia Co-prosperity Sphere before any counterattack could be mounted. The Americans however had a substantial number of aircraft in the central Pacific. There were 230 U.S. Army planes on Oahu as well as over 150 Naval and Marine Corps machines with detachments at Wake and Midway Islands; in addition there were some 220 planes in the three American aircraft-carriers. This meant that the 500 aircraft of the First Air Fleet could have to face over 600 American planes if they were alert and ready. The Japanese therefore considered it essential that the attack of the First Air Fleet should come as a complete surprise and so they decided to carry it out without a declaration of war.

For the attack on Pearl Harbor the Japanese decided to use only their six largest and fastest carriers but to augment their air group to a total of 450 planes. The six carriers were to be escorted by two fast battleships and two heavy cruisers, which carried another twelve seaplanes, and a screen of a light cruiser and nine destroyers. Three submarines went ahead to give warning of any merchant ships, and eight tankers and supply ships accompanied the force to replenish it. This fleet, after intensive training, was assembled by the 27th November under the command of Vice-Admiral Nagumo, in an unfrequented bay in the Kurile Islands. The greatest care was taken to preserve wireless silence and to give the impression that the carriers were still in the Inland Sea.

On the 26th November the First Air Fleet sailed and followed a northerly route for seven days in fog and heavy weather (see Figs. 13 and 14, pp. 197 and 198). This route had been chosen partly because there were practically no merchant ships in this area and partly because bad visibility was normally encountered. Course was set to pass Midway Island out of range of the American Catalina flying-boats which were stationed there. On 3rd December after over a week's steaming at moderate speed to allow the tankers to keep up, the fleet refuelled. So far only one merchant ship had been sighted and she was Japanese. On the evening of the 6th December the fleet reached a point some 500 miles north of Oahu and information that the American Fleet was in harbour, together with its berthing plan, was received through the Japanese Consul at Honolulu. Speed was then increased to 26 knots, the tankers were left behind and course was set for a launching point 275 miles north of Pearl Harbor. The cruisers *Tone* and *Chikuma* were sent ahead so that their float-planes might provide final confirmation that the American Fleet was still there. At dawn on the 7th December the carriers turned into the north-east trade wind and launched their first strike of 190 aircraft. Forty of these were torpedo-bombers, fifty high-level bombers, fifty dive-bombers and fifty were fighters.

The Americans were taken completely by surprise. Eight of the nine battleships of the Pacific Fleet were berthed in Pearl Harbor with eight cruisers, thirty-eight destroyers and a

number of auxiliaries. As it was Sunday morning a large proportion of the crews were ashore. Only a few air patrols had
been flown and these were to the south of the island: an effective
radar watch was not being kept and nearly all of the 420-odd
aircraft on Oahu were on the ground on the six airfields or
moored at the flying-boat base at Kanoehe. Few anti-aircraft
guns either ashore or afloat were manned or had ready-use
ammunition available.

The first wave attacked just before 08.00 and within a very

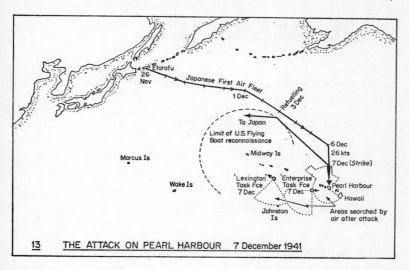

13   THE ATTACK ON PEARL HARBOUR   7 December 1941

few minutes over twenty torpedo hits had been scored by the
torpedo-bombers followed by a similar number of hits with
heavy bombs from the dive-bombers. The forward magazine of
the battleship *Arizona* exploded and she sank with over a
thousand casualties. The *Oklahoma* capsized and the *West
Virginia* sank bodily on to the bottom. The *Nevada* and *California*
were torpedoed and struggling to keep afloat. The *Tennessee*
and *Maryland* were in the inside berths and so screened by other
ships from torpedo attack but both were hit by bombs, the
former being badly set on fire. Torpedoes also hit the cruisers
Helena and *Raleigh* which were seriously damaged, and two
more were wasted on the old target battleship *Utah* which
capsized. All the airfields except one were attacked and large
numbers of aircraft were destroyed on the ground. Only ten
U.S. fighters managed to scramble during the whole attack,

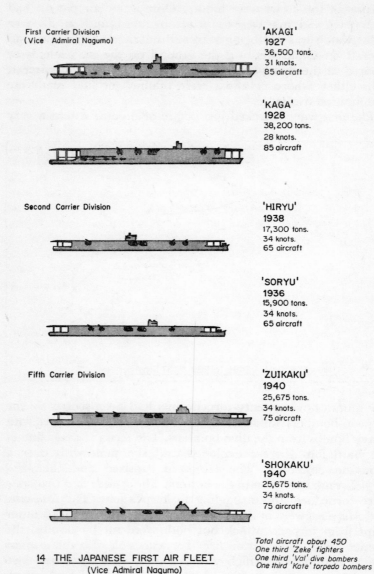

First Carrier Division
(Vice Admiral Nagumo)

'AKAGI'
1927
36,500 tons.
31 knots.
85 aircraft

'KAGA'
1928
38,200 tons.
28 knots.
85 aircraft

Second Carrier Division

'HIRYU'
1938
17,300 tons.
34 knots.
65 aircraft

'SORYU'
1936
15,900 tons.
34 knots.
65 aircraft

Fifth Carrier Division

'ZUIKAKU'
1940
25,675 tons.
34 knots.
75 aircraft

'SHOKAKU'
1940
25,675 tons.
34 knots.
75 aircraft

14   THE JAPANESE FIRST AIR FLEET
(Vice Admiral Nagumo)

Total aircraft about 450
One third 'Zeke' fighters
One third 'Val' dive bombers
One third 'Kate' torpedo bombers

and these were from the one small airfield in the north of the island which the Japanese had overlooked.

Whilst the first wave was attacking, the Japanese carriers were launching a second strike consisting of another fifty high-bombers, eighty dive-bombers and forty fighters. They kept back fifty fighters for their own defence and used the battle-ship and cruiser float-planes to patrol in the vicinity of the fleet. The second strike again attacked the airfields and the ships in Pearl Harbor hitting the battleship *Nevada* which had just got under way and the flagship *Pennsylvania* which was in dock. The *Nevada* had to be beached and subsequently the *California*, torpedoed by the first wave, slowly settled on to the bottom. Soon after midday, when Admiral Nagumo had landed on his planes, he withdrew to the north-westwards at 25 knots. The total Japanese loss had been 29 aircraft.

The Battle Force of the Pacific Fleet was decimated. Of its nine battleships, five were resting on the bottom of Pearl Harbor. Three others had all been hit with varying degrees of damage. Only the *Colorado*, which was refitting on the west coast of the U.S.A., escaped completely. It was two weeks before the three survivors, the *Tennessee, Maryland* and *Pennsylvania*, after temporary repairs, crawled away to the west coast. The *Nevada, West Virginia* and *California* were refloated during the next few months and left for the west coast for repairs but the *Oklahoma* and *Arizona* were a total loss. The cruisers *Helena* and *Raleigh* were completely put out of action but were later repaired; the *New Orleans* and *Honolulu* were damaged to a smaller degree. Three destroyers which had been bombed in the dockyard were completely knocked out, a small minelayer capsized and sank and the seaplane carrier *Curtiss* was badly damaged. The shore-based air forces were halved, the actual number of aircraft destroyed by the Japanese being 188, including 27 of the 36 reconnaissance flying-boats.

Fortunately for the U.S. Navy the destroyer and submarine forces at Pearl Harbor were practically intact as well as the Naval Base itself and the fuel tanks. Considering the complete-ness of Admiral Nagumo's victory over the shore-based air forces, it is just as well for the Americans that he did not re-arm his planes and strike again at these targets. By far the most important among the Pacific Fleet survivors were, however, all three of their aircraft-carriers with their air groups which had been absent from Pearl Harbor at the time. The *Lexington* with three heavy cruisers and five destroyers was on her way to Midway to deliver some Marine Corps aircraft: the *Enterprise* with three more heavy cruisers and nine destroyers was returning from a similar mission to Wake Island and the *Saratoga* was off San Diego on the west coast. Three other heavy cruisers were also at sea, one exercising and two escorting transports to and from the Philippines, and two older light cruisers were off Panama and the coast of Peru. Substantial forces therefore survived but the elimination of the Battle Force of the Pacific Fleet should have meant, by all the old standards, the complete loss of the command of the sea.

In spite of the magnitude of the disaster, the American main thought was to counter-attack the enemy. During and immediately following the attack, the light cruisers *Detroit*, *St Louis* and *Phoenix* with some destroyers had managed to put to sea and they joined the *Minneapolis* which was the cruiser which had been exercising. The Army Air Force had fourteen bombers operational, the Navy three Catalina flying-boats at Oahu and some others at Midway and the Marines a dozen scout-bombers. Admiral Kimmel, the U.S. Commander-in-Chief in the Pacific, ordered the *Saratoga* to join him at Pearl Harbor as soon as possible. No one at Pearl Harbor, however, knew where the attack had come from. A large number of inaccurate and panicky enemy reports were received, often about American ships. The indication seemed to be that the enemy had attacked from the south and were now on their way back to the Marshall Islands. The *Enterprise* task group was ordered to intercept and attack the enemy and her aircraft narrowly avoided sinking the *Minneapolis* group. The *Lexington* task group turned to the south, which was away from the enemy, and searched in that direction. The Japanese were never sighted and passed 500 miles north of Midway and so out of range of reconnaissance aircraft; they had however taken the additional precaution of sending two destroyers to bombard the Midway seaplane base to make their withdrawal more certain but they only succeeded in knocking out one flying-boat. In fact both the *Lexington* and *Enterprise* groups had been in a position to intercept the Japanese had they searched to the northwards. It is perhaps as well, as the American official historian remarks, that they did not do so since they had only 130 aircraft between them and Admiral Nagumo still had 360 or so.

On the 15th December, on his way home, Admiral Nagumo detached a force consisting of the carriers *Soryu* and *Hiryu* with two cruisers and some destroyers to support a second attack on Wake Island after a first had been repulsed. At the same time Admiral Kimmel decided to use his three carriers to attempt to relieve the island. The intention was that the *Lexington* group should make a diversionary raid on the Marshall Islands while the *Enterprise* group covered Pearl Harbor from the westwards. The group formed round *Saratoga*, which had now arrived from the west coast, would then attempt to relieve the island. Wake Island, however, fell on 23rd December when *Saratoga* was still 425 miles away. Although neither side knew it, the *Saratoga* group was the same distance from the *Soryu* and *Hiryu* and this was the nearest to a carrier battle that occurred during the Pearl Harbor attack. On the fall of Wake Island however both forces returned to base.

The Japanese attack on Pearl Harbor was an outstanding naval victory but it was not decisive. 'No General is defeated until he thinks he is'[13] and the Americans certainly didn't think they had lost the war. In fact this treacherous attack in time of peace united them as never before and made them determined to fight until Japan was utterly defeated. This they were sure they could do with their immense resources and manpower. The Americans had a very large naval construction programme already under way and could replace their losses. Above all they had six more aircraft-carriers of the new *Essex* class already laid down and five more authorized. Immediate steps were taken in Washington to try and make good the losses of the Pacific Fleet and the carrier *Yorktown*, the battleships *New Mexico*, *Mississippi* amd *Idaho* and some light forces were at once ordered through the Panama Canal.

Battleship protagonists could claim that the disaster would never have happened if the air forces had been ready and intelligence better. They would not have been caught in harbour with guns unmanned and in a low state of damage control. Nevertheless aircraft had for the first time destroyed a battle-fleet, and in spite of the fact that the ships were all elderly, this could not do other than support the view that the day of the battleship was past. In fact the battle-fleet destroyed at Pearl Harbor could scarcely have prevented the subsequent advance of the Japanese in Asia. In the following months their absence was to show that aircraft and their mobile bases the aircraft-carriers were well able to compete without them.

With their First Air Fleet engaged in attacking Pearl Harbor, air support for the Japanese southern advance was to be given by their Eleventh Air Fleet (see back endpaper). This force of some 500 aircraft was entirely shore-based and was composed of bombers, torpedo-bombers and fighters with a few reconnais-sance flying-boats. It was a mobile offensive air force capable of advancing from airfield to airfield and was fully trained in maritime operations. It was, in fact, manned by naval personnel and was part of the Japanese Navy. The Eleventh Air Fleet was supported by another 200 seaplanes carried in 8 seaplane-carriers or tenders and in battleships and cruisers. To this was added the small aircraft-carrier *Ryujo* which had been detached from the First Air Fleet.

The Eleventh Air Fleet was initially stationed in Indo-China and Formosa and was divided into three Air Flotillas. The 21st Air Flotilla of 190 planes and the 23rd Air Flotilla of 200 were mixed forces of bombers and fighters with a few flying-boats. Their primary function was to knock out the American Air Force in the Philippines. In this they were to be assisted by 144 planes of the Japanese Army's 5th Air Division. The 22nd Air Flotilla was stationed in Indo-China and it consisted of 138 bombers, 36 fighters and 6 reconnaissance planes making 180 in all. Their duty was to protect the forces for the invasion of Malaya, support the landing and co-operate with the Japanese Second Fleet which was covering the landing. In this they were assisted by 3 seaplane-tenders with 30 seaplanes.

The first move in the Japanese southern advance had to start well before the attack on Pearl Harbor. Convoys carrying the Japanese Army for the invasion of Malaya left Hainan and Saigon on the 4th and 5th December. They kept close inshore with air protection from the 22nd Naval Air Flotilla in Indo-China. Nevertheless on 6th December they were sighted, first by an American Catalina flying-boat from the Philippines and later by an Australian Hudson reconnaissance aircraft from Malaya. The Japanese landings in Malaya and Siam therefore came as no surprise.

The Japanese were unable to achieve complete surprise

against the American Air Force in the Philippines either, but this was because of the time difference, dawn being some hours later than at Pearl Harbor. Nevertheless it was planned to attack the American airfields in Luzon as soon as possible in the hope that news of the Pearl Harbor attack would not by then have got through. The attack was further delayed by fog over the airfields in Formosa but a force of 192 planes taking off at 10.00 on 8th December caught most of the American planes on the ground. They destroyed 12 B.17 bombers and 30 fighters, leaving only 17 bombers, 5 of which were damaged, and less than 40 fighters. Next day, the 9th December, bad weather prevented any attacks at all but on the 10th, 54 bombers attacked the airfields again and practically wiped out the naval base at Cavite. Thirty-five American fighters opposed the attack but they were outnumbered by three to one by the Japanese fighters which escorted their bombers. The American surface forces stationed in the Philippines, consisting of three cruisers and eight destroyers, escaped damage: they had wisely been kept in the East Indies well to the south. Nevertheless the air attacks on their base and the complete loss of air superiority over the Philippines made it quite impossible for them to return and oppose the subsequent Japanese landings. The Eleventh Air Fleet was therefore more responsible than the Japanese Second Fleet, which was covering the landings, for gaining command of the sea.

Pre-war British defence plans against Japan had always envisaged the stationing of a battle-fleet at Singapore and this was the keystone of their strategy. Commitments in the war in Europe, however, made this almost impossible but in the end two capital ships, the *Prince of Wales* and *Repulse*, were sent with the primary aim of deterring Japanese aggression. No aircraft-carrier was available to accompany them as the new *Indomitable*, which had been earmarked for this duty, had run aground in the West Indies and was under repair; and the small carrier *Hermes* was refitting at Durban.

When France fell in 1940, the fact had to be faced that no battle-fleet would be available for the Far East whilst the war continued with Germany and Italy. After anxious deliberations, the Chiefs of Staff considered that if a large enough air force could be assembled in Malaya, they would probably be able to defend the naval base against a seaborne attack. An estimate by the Chiefs of Staff was that 336 aircraft would be required but the local commanders estimated that 582 would be needed: that is, a force of roughly the same size as the Japanese Eleventh Air Fleet. The war in Europe, of course, prevented anything like this scale of reinforcement and when war with Japan broke out the British air forces in Malaya, which included some Australian and New Zealand units, were only 158 planes strong. Their immediate opponents, the 22nd Air Flotilla and its 180 planes, was not on paper much stronger, but it must be remembered that they also had to compete with the Japanese 3rd Army Air Division of 354 planes. Furthermore the aircraft allocated to Malaya were nearly all of obsolete types. They consisted of 60 Brewster Buffalo day-fighters which were inferior to the modern Japanese types and 12 Blenheim night-fighters. The 35 Blenheim bombers were the most modern part of the force but the 24 Vildebeeste biplane torpedo-bombers were hopelessly out of date. Reconnaissance was supplied by 24 Lockheed Hudsons and 3 Catalina flying-boats. The force was spread on nine airfields from the north to the south of Malaya and they had to provide air support for the Army and air defence for the whole peninsula as well as to try to co-operate with naval forces and oppose an invasion from the sea.

The Japanese convoys which had been seen entering the Gulf of Siam on 6th December landed their troops practically unopposed at Singora in Siam without a declaration of war. Other forces which landed at Khota Bharu in northern Malaya were at once resisted. The Japanese quickly established themselves on captured airfields from which the 3rd Air Division began to operate. Blenheim bombers sent from Singapore to attack the landings encountered heavy fighter opposition and suffered losses without doing much damage. At the same time the Japanese began to raid all the British forward airfields in northern Malaya in strength.

At the time of the attack, the new battleship *Prince of Wales* and the battle-cruiser *Repulse* were at Singapore and Admiral Phillips decided to put to sea and attack the landings at Singora and Khota Bharu at dawn on 10th December (see back endpaper). He asked the R.A.F. for air reconnaissance to the north of his force during the 9th and fighter protection as well as reconnaissance over Singora at dawn on the 10th. The two ships with four destroyers, which were all that were available, then left Singapore on the evening of the 8th December.

The Japanese invasion was being covered by sizeable surface

forces. Besides destroyer escorts for the transports, four heavy cruisers under Admiral Kurita were in close support. Covering the operation was Admiral Kondo with the two battleships *Kongo* and *Haruna* and the heavy cruisers *Takao* and *Atago* and a number of destroyers. Mines had been laid between the Anamba Islands and Pulo Tioman before war began and twelve submarines were on a patrol line just to the northwards. The battleships and cruisers all carried float-planes and the whole area was within range of the aircraft of the 22nd Air Flotilla in Indo-China.

Although Admiral Phillips knew that a battleship of the *Kongo* class was somewhere in the area, he did not realize how strong the air opposition from Indo-China was likely to be. He believed that if he could achieve surprise and was given the air support that he had asked for he had a chance to destroy the landing forces at Singora. Admiral Phillips set a course to pass to the east of the Anamba Islands to avoid the shallow water, where he rightly suspected that mines had been laid, and then turned to the northwards. Unknown to him he was sighted by one of the Japanese submarines on patrol soon after midday. Admiral Kondo ordered Admiral Kurita's cruisers to fly off float-planes to find the British and then to rendezvous with him for a surface action. One of the float-planes sighted the *Prince of Wales* and *Repulse* just before dusk. The 22nd Air Flotilla, whose bombers were about to raid Singapore, at once re-armed with torpedoes and took off but by this time it was dark and they failed to make contact.

During this time the R.A.F. in north Malaya had suffered very severely: they had lost half their aircraft, many of which were destroyed on the ground. To make matters worse the airfields in the Khota Bharu area were directly threatened by the Japanese Army. As a result it was impossible to provide fighter protection for Admiral Phillips off Singora, and he was told so, but it was still thought possible that the Hudsons and Catalinas could give him the reconnaissance he required. At 8.15 p.m., however, realizing that no fighters would be available and that he had been seen by Japanese reconnaissance planes, he reluctantly abandoned the operation and turned to the south. Soon after midnight he received reports, false as it turned out, of another landing at Kuantan and as this was well to the south of him he turned to attack at daylight. He was again sighted by a Japanese submarine in the early hours of the morning and on receipt of this report Admiral Kondo realized that there was no chance of a surface action and ordered the 22nd Air Flotilla to attack at dawn. The 22nd Air Flotilla flew off a search of twelve planes before it was light, followed by a striking force of 34 bombers and 51 torpedo planes. At dawn the *Prince of Wales* flew off her own aircraft to scout and on arrival off Kuantan could find no landing.

The ships had just altered away from Kuantan when they were sighted by one of the Japanese scouting planes. The Japanese air striking force had, by this time, passed well to the south and was on its way back. The British detected the Japanese planes by radar at 11.00 and the Japanese attacked in six waves spread over a period of an hour and a half. The first attack on the *Repulse* by high-bombers secured a single hit that failed to pierce the armoured deck. The second attack, by torpedo planes, missed the *Repulse* but hit the *Prince of Wales* with two torpedoes, one near her rudder and propellers, reducing her speed to 15 knots and leaving her out of control. After this attack the *Repulse*, having heard nothing from the flagship, radioed to Singapore that the ships were under air attack. A simultaneous assault by high- and torpedo-bombers then again missed *Repulse* but the next attack, by torpedo-planes, scored three more hits on the *Prince of Wales* and no less than five on the *Repulse*. The *Repulse* turned over and sank in a few minutes as eleven Buffalo fighters took off from Sembawang airfield near Singapore in answer to her signal. More high-bombers then attacked the *Prince of Wales* and secured another hit: fifty minutes after the loss of *Repulse* she too capsized and sank just as the British fighters arrived overhead. The total losses of the 22nd Air Flotilla in these attacks were three aircraft.

The Eleventh Air Fleet, by the destruction of these two ships, had secured an even greater victory than their colleagues in the First Air Fleet at Pearl Harbor. This time the success was achieved against battleships at sea fully alerted and ready for action and one of them was a brand new ship. Much has been written about the disaster and its causes, of which the failure to provide fighter protection for the two ships is generally assessed the greatest. Experience off Norway and in the Mediterranean had shown that battleships could not operate against shore-based aircraft without fighter protection and Admiral Phillips was perfectly aware of this. Indeed he called off the operation when he was definitely informed that it could not be provided at

Singora. Why he did not ask for fighters off Kuantan, where they could have been sent, is a mystery. However, even if fighters had been available in north Malaya it now seems doubtful whether the attack by the two ships was a viable operation. There were never more than twelve day-fighters in North Malaya and even if these had been reinforced from Singapore, they would almost certainly have been overwhelmed by the many Japanese Army fighters operating from airfields in the Singora area. Furthermore, the 22nd Air Flotilla used only half of its strength to sink the two ships and could have brought a heavier attack to bear. The damage the two ships could have done would have been of little use as the troops had already landed from the transports well before the morning of the 10th December.

The R.A.F. did their best to compete with their many responsibilities which included naval co-operation, air defence and support for the Army, with far too few aircraft. Shore-based air reconnaissance however did not provide Admiral Phillips with a picture of the situation on 9th December and he was in consequence practically blind. He knew nothing of Admiral Kondo's fleet, except that it was in the South China Sea, or of Admiral Kurita's cruisers, which on the evening of the 9th, must have been only just over the horizon. Admirals Kondo and Kurita, on the other hand, knew exactly where Admiral Phillips was. Submarines and ship-borne float-planes were more responsible for this than the 22nd Air Flotilla but one cannot help admiring the close co-operation of all Admiral Kondo's forces including his shore-based naval air force. It is significant that all of these were naval units and under his full command. The R.A.F.'s real problem, however, was that they were outnumbered by three to one and it is doubtful, even if they had had the 336 aircraft which the Chiefs of Staff thought necessary, that they could have competed; the 582 which the local commanders had asked for were clearly nearer the mark.

The disaster has often been attributed to the fact that there was no aircraft-carrier with the force. The *Indomitable*, which should have been there, carried 9 Hurricane and 11 Fulmar fighters which would have been outnumbered by the attackers by four to one. Nevertheless they would undoubtedly have shot down a number of them and spoilt the aim of others especially as the attacks were spread out over a long period and each wave was comparatively small. The presence of a carrier could almost certainly have prevented the sinking of the two ships off Kuantan, but it seems doubtful whether she could still have done so off Singora where the 22nd Air Flotilla could have had a heavy escort of Japanese Army fighters. If the *Indomitable* had been with the ships, it is probable that her aircraft would have sighted Admiral Kondo's fleet and events might have taken a very different turn. It is also probable, however, that the best use of *Indomitable*, if she had been there, would not have been to protect the two battleships in the South China Sea but to strike at the enemy landings at an earlier stage across the Malay peninsula from the Malacca Straits.

In many people's minds the subsequent loss of Singapore was mainly attributable to this disaster, but it is very doubtful whether it made very much difference except to morale. It was a great shock to those who still believed that sea power depended upon the battleship but it was not the cause of the loss of Malaya. What had been suspected for many years was now proved beyond doubt. Even modern battleships fitted with the latest anti-aircraft gunnery systems could be sunk by aircraft at sea. On the Japanese side there is little doubt that it was the 22nd Air Flotilla which covered the landings at Singora rather than Admiral Kondo's battleship force. It was now clear to the most reactionary of the nautical faction that the battleship was no longer the only counter to the battleship and a battle-fleet by itself could no longer claim to be the arbiter of sea power. Furthermore it was doubtful if this was what was needed even if it could have had full air support. What had really been required in Malaya was an air force which could compete with the 500-odd aircraft of the Japanese 22nd Air Flotilla and 3rd Army Air Division and gain air superiority over the sea as well as the land. Whether this was shore-based or carrier-borne was immaterial provided a proportion of it was well-trained in working over the sea. Admiral Kondo had no aircraft-carrier and relied upon support by an air force based ashore with an indisputably successful outcome. Nevertheless a carrier force would not have been tied to the defence of Malaya and would have been more flexible and more mobile. If, in addition to the air forces that were actually in Malaya, all four of the existing British armoured aircraft-carriers, with modern air groups and proper escort, had been available, as they were two years later, Malaya could probably have been defended. It was this which

was required rather than a so-called balanced fleet based on the
battleship.

The Japanese surface forces amassed for the southern advance
outnumbered those of the Allies stationed in the Far East for
the defence of the area. The Japanese had 2 battleships, 12
heavy and 7 light cruisers and 54 destroyers against the Allies' 2
heavy and 10 light cruisers and 25 destroyers. Although all the
Japanese battleships and cruisers carried float-planes and they
had the light carrier *Ryujo* and four seaplane tenders, they
depended for air support almost entirely upon the Eleventh Air
Fleet. They did not venture outside its cover and a pillar of
Japanese strategy was to seize airfields so as to gain air
superiority ahead of the main amphibious operations and thus
provide the air component of the air–sea team required to
command the sea. The main amphibious operations were
mounted to capture territory and raw materials, especially oil.
Many of these also yielded airfields for a further advance but
some amphibious operations had to be carried out solely to
obtain airfields.

On 20th December, a force from Palau, covered by the aircraft of the *Ryujo* and the seaplane-carrier *Chitose*, landed at
Davao at the southern end of the Philippines (see back endpaper).
From there it moved on to the island of Jolo, capturing the
airfields in both places. In early January the 21st Air Flotilla
moved from Formosa to Davao and the 23rd Air Flotilla to
Jolo. At the same time, in the South China Sea, an amphibious
operation took the airfields in the neighbourhood of Kuching
for the 22nd Air Flotilla. These bases put the Eleventh Air
Fleet within range of their next objectives at Tarakan in
Borneo and Menado at the northern end of Celebes. Further
amphibious operations were then mounted under their cover to
take these places, landing in early January 1942. The two Air
Flotillas then moved forward again in their wake to establish
themselves on the captured airfields. These new positions in
their turn put them within range of Balikpapan in Borneo and
Kendari at the south-eastern point of Celebes, and other
amphibious steps were taken to these places towards the end of
January.

Up to now the Japanese air superiority was so complete that

the Allied surface forces in this area had not made contact with any Japanese surface forces. The only Allied ripostes had been by American and Dutch shore-based aircraft but these had been priority targets for the Eleventh Air Fleet planes and, except for some Dutch bombers operating from secret bases, had already been driven back to Java and Australia. They were too weak to do more than cause a few casualties. An American task force consisting of the cruisers *Boise* and *Marblehead* with four destroyers had retired to Timor before the Japanese advance and Admiral Glassford, its Commander, now saw a chance to attack the landing force at Balikpapan. The task force sailed on 20th January but the cruiser *Boise* ran aground when passing through the Sape Strait and only the destroyers were able to go on. They crossed the Flores Sea in daylight without being seen by the 23rd Air Flotilla which had not yet moved down from Tarakan, and in a night attack sank four transports and retired without loss. Two weeks later, on 4th February, Allied reconnaissance planes indicated further activity off Balikpapan as though the next amphibious operation was about to sail. On 4th February, Admiral Doorman of the Royal Netherlands Navy, now in command of a combined Dutch and American striking force consisting of the cruisers *de Rujter, Houston, Marblehead, Tromp* with eight U.S. and Dutch destroyers, sailed from Surabaya to attack. By this time, however, the 21st Air Flotilla was well established at Kendari. Sixty bombers took off and severely damaged the *Marblehead, Houston* and *de Rujter.* The *Marblehead* was damaged so badly that she had to retire to Ceylon. The task force then abandoned the strike and retreated to Tjilatjap on the south coast of Java as Surabaya was now within range of the Eleventh Air Fleet.

The sea route to Singapore by the Malacca Straits had been rendered unusable early in the campaign by the proximity of the airfields captured by the Japanese in northern Malaya, and this also cut the air reinforcement route from India. The British naval forces were engaged mainly in escorting convoys by the alternative route to Singapore through the Sunda Straits. The 22nd Air Flotilla seems to have spent the majority of its effort on bombing Singapore itself. It found time however to attack many of the convoys but seven of them were escorted in with the loss of only one ship.

Well before the fall of Singapore the British bomber aircraft in Malaya had to retire to Sumatra. Here, from airfields in the vicinity of Palembang, they did their best to oppose the Japanese landings on the island. There were however too few of them and they never looked like even gaining local air superiority with the result that the Japanese southern movements were unchecked. On 4th February a striking force was formed of the cruisers *Exeter* and *Hobart* to attack a Japanese force which had been reported, falsely as it turned out, making towards southern Sumatra. They were heavily attacked by bombers from western Borneo and were forced to retire; miraculously they escaped damage. The whole Allied striking force, now consisting of *de Rujter*, *Tromp*, *Java*, *Exeter* and *Hobart*, with four Dutch and six American destroyers under Admiral Doorman, was then sailed through the Sunda and Gaspar Straits to attack from north of Banka Island. It was sighted by Japanese reconnaissance planes and attacked by aircraft from the carrier *Ryujo* and subsequently by waves of high-level bombers from western Borneo. These did not succeed in making a single hit but there were many near misses and the attacks achieved their object for Admiral Doorman, having no fighter protection, turned and retired through the Sunda Straits. At the same time that the Allied striking force had been under attack, an attempt was being made to reinforce Timor with troops from Darwin in a convoy escorted by the *Houston*. They were attacked by 35 bombers from Kendari and also had to turn back.

At the end of the year the new aircraft-carrier *Indomitable* rounded the Cape and entered the Indian Ocean but the need for fighters at Singapore was so great that she was diverted to Port Sudan to embark 48 Hurricanes from the Middle East. She launched them from south of Java at the end of January and they flew on to Singapore. She then had to return to pick up her own air group and so this valuable ship was not available to work with the Allied surface striking force in the East Indies. Up to this time, except for the American destroyer attack on Balikpapan, the Japanese Eleventh Air Fleet had warded off all surface opposition to the southern advance. By the middle of February when Singapore fell and the Japanese had landed at Palembang, the Eleventh Air Fleet had established a ring of airfields round Java, stretching from Sumatra through Borneo and Celebes to Bali and Timor. The greater part of their effort was thereafter used to bomb Java in preparation for the invasion.

The remaining Allied cruisers and destroyers based at Tjilat-jap tried hard to interfere with the invasion convoys. The allied striking force was forced to remain south of the Malay barrier for most of the time and was only able to make occasional raids, under constant threat of air attack in daylight, to the north-wards. As the distances were too great for these raids to be completed in darkness they were in peril for a great deal of the time. The Japanese generally sighted them in plenty of time, diverted their convoys and sent aircraft and superior surface forces to deal with them. In this way they were completely defeated at the Battle of the Java Sea and subsequent minor actions and of the whole force only four American destroyers got away.

The Eleventh Air Fleet was not directly responsible for the final sinking of the allied ships, which were dispatched mainly by surface forces, but its operations contributed substantially. They had bombed the Allied force out of its base at Surabaya: many of the ships were damaged, all the ships were short of fuel, ammunition and stores and the crews were worn out. In the Battle of the Java Sea and the subsequent efforts to escape, Japanese aircraft, which included float-planes from their cruisers, told their ships exactly where the allied forces were and the Japanese were able to use them for spotting. At the end of February the veteran American aircraft-carrier *Langley*, long reduced to the status of a seaplane tender, tried to run 32 modern American fighters into Tjilatjap but she was sunk by aircraft of the Eleventh Air Fleet 75 miles south of the port.

Allied co-operation with what was left of their air force was weak and ineffective and such reports of the enemy as there were took so long to get through that they were useless. The Eleventh Air Fleet however never repeated against the Allied cruiser and destroyer striking force in the East Indies the success it had achieved by sinking the *Prince of Wales* and *Repulse*. In fact they obtained far fewer hits probably because for some reason they used high-bombers rather than dive-bombers or torpedo planes.

At this time Admiral Nagumo's First Air Fleet reinforced by Admiral Kondo's battleships appeared south of Java. After Pearl Harbor Admiral Nagumo did not tarry long in Japan and his carriers were sent south to assist the 24th Air Flotilla with the capture of Rabaul and then to support the 21st Air Flotilla in the capture of Ambon in the East Indies. On 18th

February they left an anchorage near Kendari in Celebes, crossed the Banda Sea at night and at dawn flew off 152 bombers escorted by 36 fighters to attack Darwin. Supported by bombers of the 21st Air Flotilla they sank eleven ships and the American destroyer *Peary* and damaged others. They also destroyed 23 aircraft and put the base out of action for some months, all for the loss of 5 aircraft.

On 9th March Java surrendered and the Japanese had conquered the whole of the Greater East Asia Co-prosperity Sphere except Burma in the remarkable short period of three months and with trifling losses. The southern advance was achieved by a very skilful use of air power over the sea. Until the end carriers were hardly employed at all and the air support for the amphibious operations and for the warships escorting them was provided by shore bases. The Japanese operations were so successful that it would be true to say, not that ships commanded the sea with the help of the Eleventh Air Fleet, but that the Eleventh Air Fleet commanded the sea, ships and amphibious operations being used to assist in the acquisition of the airfields needed.

After the fall of Singapore and Java, the British sea communications and possessions in the Indian Ocean were clearly threatened. Ceylon and even India might be invaded and the sea routes included not only the communications to these two places but also the vital oil traffic from the Persian Gulf and the communications of all the British forces fighting in the Middle East. The British had done their best to raise an Eastern Fleet and the ships that had been hastily assembled in Ceylon looked a formidable force on paper. They consisted of five battleships, three aircraft-carriers, seven cruisers and sixteen destroyers. Their new Commander-in-Chief, Admiral Sir James Somerville, had just arrived to take command when intelligence was received that the Japanese First Air Fleet was about to enter the Indian Ocean and attack them.

The information was correct. Admiral Nagumo left Celebes on 28th March with the carriers *Akagi*, *Soryu*, *Hiryu*, *Shokaku* and *Zuikaku* escorted by four battleships of the Kongo class, the cruisers *Tone* and *Chikuma* and a screen of a light cruiser and eleven destroyers. Their plan was to try and catch the British in

P

harbour at Colombo on Sunday 5th April. At the same time a second force consisting of the light carrier *Ryujo*, seven cruisers and eleven destroyers was to enter the Bay of Bengal through the Malacca Straits to attack shipping.

The disparity of force at first sight does not look very great but in fact the British were very much the weaker side. Admittedly the *Indomitable* and *Formidable* were new armoured aircraft-carriers, although the *Hermes* was small and nearly twenty years old. The British inferiority lay in their air groups rather than the ships: whereas the Japanese fleet carried over 300 planes, the British had no more than ninety. The shore-based air force in Ceylon, No. 222 Group R.A.F., had another ninety aircraft but even so the Japanese could pit 300 planes against the British total of 180. When the quality of the aircraft is compared the disparity will be seen to have been greater still. Of the British carrier-based planes, only 33 were fighters and of these the Fulmars and Martlets were inferior to the hundred or so Japanese Zero fighters and only the 9 Sea Hurricanes were their equals. Against this the British carriers were fitted with radar and so should have been able to direct their fighters more efficiently. The rest of the British air groups were composed of Albacore and Swordfish torpedo-spotter-reconnaissance machines. These were obsolescent biplanes with a maximum speed of 160 m.p.h. and a range of only just over 500 miles. Compared to these the hundred or so Kate torpedo-bombers had a speed of over 220 m.p.h. and more than twice the range: moreover they carried a 21-inch instead of an 18-inch torpedo. But these were not the only Japanese strike aircraft: there were a hundred Val dive-bombers embarked capable of carrying 1000 lbs of bombs for over 1000 miles.

Sixty-five of the shore-based aircraft in Ceylon were fighters, mostly Hurricanes, but with a few Fulmars. There were 6 long-range Catalina flying-boats but only 14 Blenheim bombers and 6 shore-based Swordfish torpedo-bombers as a striking force. The British carrier-borne torpedo-bombers had, however, one advantage and that was that they could operate at night. A proportion of them were fitted with radar and so could find the enemy in the dark and drop flares in the light of which the remainder could make an unseen and unopposed torpedo attack. This technique had proved extremely effective against Italian convoys in the Mediterranean. Nevertheless the problem of getting close enough to the enemy to use it was a very difficult one as all three types of enemy aircraft had double the strike range of the British torpedo planes. By day the British carriers would obviously be at a hopeless disadvantage: their defending fighters would probably be outnumbered and outclassed by the Japanese strike escort and their own strike planes would be shot to ribbons by at least an equal number of defending Zeros. The difficulty was to keep out of strike range of the Japanese by day whilst getting close enough to attack at night.

The British actually had one more battleship than the enemy and a substantial superiority in gunpower of forty 15-inch against thirty-two 14-inch. Furthermore they had radar for ranging while the Japanese had not. All the battleships on both sides were of First World War vintage and the *Warspite* and *Revenge* had actually fought at Jutland. All four of the Kongo class had been modernized whereas of the British ships only the *Warspite* had been brought up to date; she now had an armoured deck and a speed of 25 knots. The four other battleships, *Revenge, Royal Sovereign, Resolution* and *Ramillies*, the 'wobbly R's' as they were called, were very slow, had no armoured decks and their underwater protection was poor. They were not afraid of the Japanese battleships but had little chance against the highly efficient Japanese strike aircraft which would almost certainly dispatch them even more speedily than the *Prince of Wales* and *Repulse*.

Admiral Somerville was inclined to try to avoid action and keep his 'fleet in being'. He considered that even if the Japanese landed and took Ceylon all would not be lost. On the other hand if he fought and his fleet was destroyed, the sea communications in the Indian Ocean would be cut and the whole of the British forces in the Middle East as well as in India would be in jeopardy. He was however prepared to try for a night torpedo-bomber attack on the enemy. He appreciated that the Japanese would approach Ceylon from the south-east and launch their air groups at dawn to attack Colombo or Trincomalee (see Fig. 15, p. 218). Relying on the six Catalina flying-boats to detect the approach of the enemy, he took up a position to the south-east of Ceylon from which his carriers would be close enough to strike at the enemy during the night as they approached their launching point. In daylight he would keep the fleet close to Ceylon where he hoped his ships could benefit from fighter protection from the shore as well as from his carriers. If his torpedo-bombers succeeded in damaging the enemy, then his battleships, slow as they were, might well be able to close in and complete their destruction.

Admiral Somerville took up this position on the 31st March and divided his fleet into a fast division with the *Indomitable*, *Formidable* and *Warspite* and a slow division with the 'R' class battleships and the *Hermes*. He held this position for three days, by day trying to keep out of the line of possible Japanese reconnaissance aircraft on their way to Colombo or Trincomalee and at night advancing in the hope that he could make an air strike if the enemy were sighted. When nothing had happened by the evening of the 2nd April, he assumed that the intelligence was at fault and, as his ships needed fuel and water and there was a considerable danger from submarines in such a restricted area, he abandoned the operation and set a course for his secret base at Addu Atoll at the southern end of the Maldive Islands. For various good reasons some of his ships were sent into Colombo and Trincomalee. This move was premature: Admiral Nagumo was already in the Indian Ocean and was fuelling from his tankers to the south-west of Java.

The British Eastern Fleet had just arrived at Addu Atoll when a Catalina sighted the enemy 360 miles from Ceylon and heading towards it. It was clear that they could now strike Ceylon without the Eastern Fleet being able to do anything about it. Admiral Somerville therefore decided to complete fuelling and then sail for a position 250 miles south of Ceylon in the hope of engaging any ships which might be damaged by the R.A.F. He still hoped that he might get in a night torpedo-bomber attack. At dawn on 5th April, the Japanese First Air Fleet launched 53 bombers and 38 dive-bombers escorted by 36 fighters to attack Colombo from a position 300 miles to the south. The British were ready and had sent as many ships to sea as possible. Forty-two fighters opposed the strike but the destroyer *Tenedos* and an armed merchant cruiser were sunk, the submarine depot ship *Lucia* was hit and considerable damage was done to the port. The Japanese only lost 7 aircraft and the British, in addition to 2 flying-boats, lost 19 fighters and all 6 Swordfish of the shore-based striking force which ascended with torpedoes in the hope that they might get a shot at the enemy. Fourteen Blenheim bombers also took off to counter-attack the Japanese carriers but were unable to find them.

Two of the ships which had been detached to Colombo by

Admiral Somerville were the heavy cruisers *Dorsetshire* and *Cornwall* and they were at once ordered to rejoin the fleet. Their course took them within 150 miles of the enemy carriers and they were sighted by a float-plane from the Japanese cruiser *Tone*. A strike of 53 aircraft was flown off and both cruisers were dive-bombed and sunk. The fast division of the Eastern Fleet was, in fact, only 80 miles or so to the south-west at the time and had been flying searches to the eastwards but had found nothing. Night air searches were then flown but they also drew blank, the Japanese having recovered their planes and withdrawn to the south-east. So vanished the only chance of engaging the First Air Fleet and, in fact, the only chance the British had during the Second World War of a battle between carriers.

After the raid on Colombo and the loss of the two cruisers, the Admiralty, nervous for the safety of the 'R' class battleships, gave Admiral Somerville discretion to withdraw them to East Africa. Admiral Sommerville did not hesitate. It had been just worth the risk when there was a good chance to get in a night torpedo-bomber attack, but to steam about aimlessly in the same general area as the Japanese who had twice the strike range was courting disaster. With the loss of the *Dorsetshire* and *Cornwall* to remind him of what had happened to the *Prince of Wales* and *Repulse*, discretion clearly came before valour. He therefore, after refuelling them at Addu Atoll, sailed the slow division of the fleet to Kilindini to protect the sea communications to the Middle East. Admiral Somerville decided to keep the fast division in Indian waters but to avoid the vicinity of Ceylon whilst the Japanese were about.

Admiral Nagumo, keeping more than 500 miles from Ceylon, to try to avoid being seen by the flying-boats turned into the Bay of Bengal to attack Trincomalee. He was sighted and reported by a Catalina 470 miles from the port and steering towards it; the Catalina was then shot down. This gave time however for the port, as at Colombo, to be cleared of all ships including the *Hermes* whose air group contained no fighters. The First Air Fleet flew off a large striking force which attacked Trincomalee and did considerable damage to the airfield but caught no ships in harbour. Twenty-three fighters opposed the raid and nine of them were lost. Nine Blenheims took off to attack the enemy fleet and bombed the *Akagi*, scoring three near misses. They were attacked by fighters and five Blenheims were

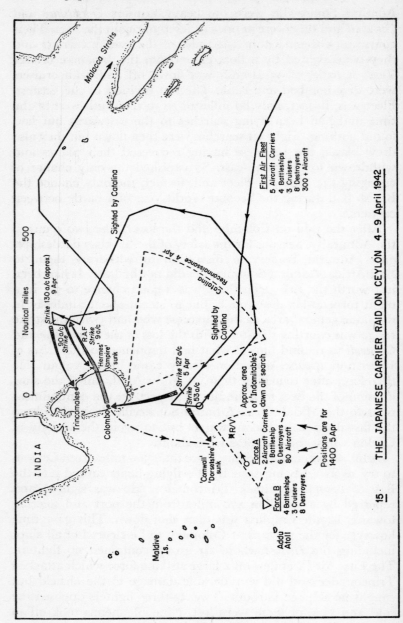

shot down. This strike, unsuccessful though it was, is notable as the only one made on the Japanese First Air Fleet in the first six months of the war in the Pacific.

The *Hermes* with the Australian destroyer *Vampire* was 65 miles to the south of Trincomalee when she was sighted by one of the Japanese aircraft. She proceeded at full speed towards Trincomalee in the hope that she could obtain fighter protection but owing to a communication failure no fighters arrived. A strike of 90 planes from the Japanese carriers found her and sank her by dive-bombing in under twenty minutes. They then completed the destruction of the *Vampire* and one or two other auxiliaries in the vicinity. The First Air Fleet then recovered its aircraft and retired through the Malacca Straits to Japan. While the First Air Fleet had been operating off Ceylon the Japanese cruiser force in the Bay of Bengal sank 19 ships of nearly 100,000 tons and stopped traffic on the east coast of India for some time.

Although the First Air Fleet had failed to destroy the British Eastern Fleet, it had sunk two heavy cruisers, a small carrier and two destroyers and had brought down 39 British aircraft for the loss of only 17 of its own. The result was inevitable: the hastily scraped together Eastern Fleet, notwithstanding its modern carriers, was no match for them at all. The vulnerability of the 'wobbly R's' to Japanese attack was one reason for this weakness but the great inferiority in numbers and in performance of the British Fleet Air Arm aircraft was of great significance. The Japanese air groups were essentially a striking force of great range and power and fully capable of competing with shore-based aircraft. The British air groups were the result of a policy of a fleet air arm designed to give air support to a battle-fleet. Admiral Somerville certainly maintained his fleet 'in being': the fact that the Japanese never knew of the existence of his secret base at Addu Atoll was a major contribution towards this totally negative success. His policy was very similar to that of Torrington after Beachy Head in 1690 and it succeeded for the same reason, that was that Nagumo, as Tourville before him, failed to follow up his advantage.

The American carriers in the Pacific had been very busy since Pearl Harbor. Early in January 1942 they were reinforced by the *Yorktown* from the Atlantic but on the 11th the *Saratoga* was torpedoed by a Japanese submarine and had to return to the west coast for extensive repairs. Shortly after this, the *Lexington* set out to strike at Wake Island, but her only replenishment tanker was torpedoed by a submarine and she had to return for lack of fuel. *Enterprise* and *Yorktown* were at first used to cover an important convoy of reinforcements for Samoa, and then early in February to strike at the Gilbert and Marshall Islands. Their aircraft took part in some air fighting and did not do very much damage. Nevertheless this was the first American offensive operation of the Pacific war; valuable experience was gained and it was good for morale.

On 23rd January the Japanese had taken Rabaul and established the 25th Air Flotilla on two airfields in New Britain. It was believed that they had further offensive intentions and so the *Lexington* was sent down to reinforce the surface ships already in the Anzac area; in the middle of February she planned a strike at Rabaul but became involved in a considerable air battle with the 25th Air Flotilla and called it off. A number of Japanese planes were shot down for small U.S. losses and the operation indicated that American carrier-borne aircraft were quite capable of competing with Japanese shore-based aircraft. In the meantime the *Enterprise* crossed the central Pacific and struck at Wake Island on 24th February and Marcus Island on 4th March. These strikes were designed to divert the attention of the Japanese from the southern area. Very little damage was done and there is no indication that they diverted any attention, nevertheless these two strikes too were good for morale and good practice. In mid-February the *Yorktown* was sent down to join the *Lexington* in the south and together they planned to renew a strike on Rabaul. During their approach, the Japanese landed at Lae and Salamaua in New Guinea and the target was shifted to these two places. On 10th March both carriers flew off a heavy strike from across the tail of New Guinea. They met no air resistance and sank an armed merchant cruiser, a minesweeper and a supply ship. *Lexington* then returned to Pearl Harbor leaving *Yorktown* on guard in the south.

In March, the new carrier *Hornet* joined the American Pacific Fleet. It had already been planned that she should take sixteen Army Mitchell bombers right across the Pacific to attack Tokyo. This was to be a one-way mission as the twin-engined bombers could not land on again. She left San Francisco with the Mitchells on board on 2nd April and after meeting the *Enterprise* as escort, she flew off the bombers on 18th April. All the planes attacked their targets in Tokyo and three other Japanese cities and then flew on to China. One plane was interned at Vladivostok, one fell in the sea and the others either crash-landed or their crews baled out in China. All the bombers were therefore lost but most of the crews were saved. Little damage was done by this raid but to have attacked the Japanese capital was a tremendous fillip to morale.

The three battleships *Idaho, Mississippi* and *New Mexico* had by this time arrived from the Atlantic and joined the *Colorado* and the two least damaged survivors of Pearl Harbor, the *Maryland* and *Pennsylvania*, to reform the Battle Force of the Pacific Fleet. Although it was considered that they might be used for active operations, they were too slow to work with carrier task forces and there were insufficient tankers to refuel them and so they were left on the west coast. It was already clear that the new carrier task forces were able to operate by themselves successfully and even to go over to a limited offensive without the support of a battle-fleet. Indeed no real use could be found for the reconstituted battle force and it was feared that it would, like the 'wobbly R's' in the Indian Ocean, prove more of a liability than an asset in the operational area.

The main Japanese First Fleet under Admiral Yamamoto now consisted of seven battleships including the immense 64,000-ton *Yamato* which was, with nine 18·1-inch guns, the most powerful battleship in the world and which had joined soon after war broke out. These ships, with two small cruisers, a seaplane carrier and eight destroyers, took no part either in the strike on Pearl Harbor or against the British in the Indian Ocean or indeed in the southern advance. They remained throughout in Japanese waters and their sole function was to give battle if either the British or the American battle-fleets challenged them in their own waters. With the British battle-fleet retired to Kilindini and the American at San Pedro in California, this was a somewhat remote possibility. One is tempted to suggest that it would have made little difference if none of these three battle-fleets had been there at all. However it is enough for the moment to remark that none of them was exerting any influence, even indirectly, on the course of events.

Students of sea power have become used to battle-fleet in-activity in war, but have accepted in the past that their role has nevertheless been paramount. Their supremacy has relied on the fact that they could only be destroyed by other battle-fleets, and battle or the threat of battle has been the key to their power. The first two days of the Pacific war showed in no un-certain terms that no longer were battleships the only answer to battleships. Threat of Japanese battle-fleet attack certainly did not deter the U.S. carriers from their operations any more than the 'wobbly R's' would have deterred Nagumo from operating against the Middle East sea communications on the east coast of Africa if he had wanted to do so. On the other hand the mere threat of air attack was enough to deter any battle-fleet from its purpose. The Japanese First Air Fleet had shown, on the other hand, both at Pearl Harbor and off Ceylon that it was happy to accept the challenge of shore air power. The series of U.S. carrier strikes, notably the recent repulse of the 25th Air Flotilla by the *Lexington*, had begun to show that the U.S. carriers could compete with shore-based aircraft too. They were able to do this because their weapons were also aircraft. Already therefore it was becoming clear that the real arbiters of sea power were now aircraft operating either from the shore or from carriers and that their role was not merely the support of ships.

The Allied disasters in the Far East in 1941–2 have been blamed on many things, among them the surprise obtained at Pearl Harbor, British pre-occupation in Europe and a general underestimation of the Japanese. A very great deal was how-ever due to the fact that the Japanese were far and away the first to understand that in the war at sea it was now aircraft, whether shore-based or carrier-based, that came first and ships that came second. To the Japanese, command of the sea automatically meant air superiority over it first, a concept very different from the earlier one whereby air support and fighter protection was provided for ships so that they could command the sea.

IX

Air Power Over the Sea
in European Waters
January – July 1942

AT THE END OF 1941, Hitler decided that the *Scharnhorst*, *Gneisenau* and *Prinz Eugen* were achieving nothing in Brest and must be transferred to northern waters for the defence of Norway against a possible British landing. It was Hitler's personal decision that they should return through the Channel and that they should sail from Brest immediately after dark which meant that they would pass the Straits of Dover in daylight. The British Admiralty appreciated that they would use the Channel route and forecast their movements accurately with the exception that they believed they would aim to pass the Straits of Dover in darkness. They informed Coastal, Bomber and Fighter Commands of the R.A.F. and arranged for 98 mines in five fields to be laid by Bomber Command on the anticipated route. Although they produced a plan to attack the German ships in the Channel they provided a force of no more than six elderly destroyers, seven motor-torpedo-boats, and six Swordfish torpedo-bombers which was obviously too weak to stop them. In retrospect it is clear that they expected the R.A.F. to do the job for them, which considering what had happened at Pearl Harbor and to the *Prince of Wales* and *Repulse* seemed reasonable to them at the time.

Coastal Command flew air patrols by radar-fitted Hudsons off Brest and along the northern coast of France and these were backed up by day by Fighter Command's usual anti-shipping sweeps. Coastal Command had a striking force of 33 Beaufort torpedo-bombers which they concentrated in the south of England. Bomber Command had about 240 aircraft able to attack by day out of its total operational strength of some 400, while Fighter Command had 550 fighters in the south of England.

The German plan was to sweep a passage clear of mines and for the three ships to sail immediately after dark and proceed at full speed through the Straits of Dover to Germany. A night of no moon and a flood tide up the Channel was chosen and arrangements were made for strong fighter support from the French coast during the day. A force of 252 F.W. 190 and Me. 109 fighters of Jagdgeschwaders 1, 2 and 26 under General Galland was allocated, and they planned to keep a minimum of 16 fighters airborne over the ships. A Geschwader of bombers was also kept ready to attack any British warships which opposed the movement.

The German squadron sailed from Brest at 22.45 on 11th February, a little behind schedule because of an air raid, but was extremely lucky and avoided detection by the Coastal Command patrols. Both the Hudsons on patrol had radar trouble and had to return to base. It was a Fighter Command anti-shipping patrol which eventually sighted the German squadron off Le Touquet next morning but it took some time to 'de-brief' the pilots and it was not until this had been done that it was realized that the German squadron was at sea.

The various British naval and air forces were then thrown into action piecemeal as fast as they could be made ready. The six Swordfish of 825 Squadron of the Fleet Air Arm took off from Manston and, in spite of a Spitfire escort, were all shot down as they attacked ten miles north of Calais. Of the Coastal Command Beauforts only 14 attacked during the afternoon in ones and twos in bad visibility: all their torpedoes missed and three of them were lost. Bomber Command attacked in three waves: the first consisting of 73 aircraft secured no hits; of the second of 134 planes few found the targets in the mist; and the 35 of the last wave failed to find the target at all. They lost fifteen aircraft altogether and achieved nothing and the six destroyers and seven motor-torpedo-boats which attacked had no more success. Fighter Command committed nearly 400 aircraft which fought the German fighters until dark, inflicting heavy losses on them and losing 17 of their number. In the late afternoon *Scharnhorst* detonated a mine and had to stop but she got under way again after half an hour. Shortly afterwards the *Gneisenau* also exploded a mine and as the *Scharnhorst* arrived at the same position off Terschelling, she was damaged by another.

Both ships, however, were able to reach German ports during the night. In a raid on Kiel on 26–27th February, Bomber Command hit the *Gneisenau* and blew up her forward magazine. After a year in dockyard hands, repairs were eventually abandoned and she took no further part in the war.

The escape of the German ships caused a public outcry in Britain. How was it that our capital ships could be sunk with ease by Japanese aircraft, while the Germans could pass close to the main British metropolitan air force with impunity? Undoubtedly the co-ordination of operations by the British was not good; three R.A.F. and three Naval Commands were involved as well as the Admiralty; but this was not the real cause of the failure. Certainly the results would have been different if the Coastal Command patrols had detected the Germans leaving Brest as the British attacks would then have been better co-ordinated. The root cause, however, was that the Royal Navy, having predicted the enemy's intentions accurately, virtually left the whole operation to the R.A.F. and the warships they provided were quite incapable of stopping the German ships.

There were, of course, essential differences between the air situation in the Channel in February 1942 and that of two months earlier in the South China Sea when the *Prince of Wales* and *Repulse* had been sunk. Bomber Command aircraft were trained for night bombing operations over the land and were totally unable to hit fast moving warships in poor weather conditions or of competing by day against well-organized and strong fighter opposition. The torpedo-bombers of Coastal Command and the Fleet Air Arm which were trained in maritime warfare, would undoubtedly have done better if it had proved possible for them to attack *en masse* but they too were very vulnerable to the fighters. The attack on the *Prince of Wales* and *Repulse* by the Japanese 22nd Air Flotilla was in good weather, by pilots well-trained in maritime operations and there was no fighter opposition, and this was a very different proposition. There is no doubt that the flexibility of British air power was overestimated and the effect of efficient German fighter support was underestimated, and that this was why the R.A.F. were unable to prevent the escape of the German ships. Nevertheless it was Bomber Command's mines which did the only damage to the ships and subsequently Bomber Command made one of its most successful raids of the war in which the *Gneisenau* was disabled. If one good thing came out of the fiasco it was a greater demand for the replacement of the Navy's obsolete strike aircraft. Yet the Fairey Barracuda which was already under development did not come into service to replace the Albacore and the Swordfish for over a year.

We now return to the Mediterranean where at the beginning of 1942, the British Navy had lost the *Ark Royal* and the *Barham*, and the *Queen Elizabeth* and *Valiant* had been seriously damaged. At the eastern end, the fleet was reduced to 5 light cruisers and 26 destroyers, but the Eighth Army had relieved Tobruk and captured Benghazi with the result that ships could now be supported by shore-based aircraft in the central Mediterranean. Force H at Gibraltar was reduced to the battleship *Malaya*, the very elderly aircraft-carrier *Argus* and a light cruiser and 8 destroyers. Against this the Italians had 4 battleships, 8 cruisers and 25 destroyers at Taranto and their air forces were as strong as ever.

The heavy losses of Rommel's supplies on the way to North Africa towards the end of 1941 had prompted Hitler to send Fliegerkorps II from Russia back to Sicily. The intention was that it should bomb Malta so heavily that the island could no longer be used as a base for surface, air or submarine striking forces and so enable the Axis to regain control of the central Mediterranean. Fliegerkorps II could also be used to prevent any reinforcements arriving at Malta or passing through the Mediterranean. The Luftwaffe began its attack in January and in that month dropped 669 tons of bombs on Malta, damaging all three airfields and sinking the destroyer *Maori*. In the same month the British found it possible to escort six merchant ships from Alexandria to supply Malta and only one ship, which had to turn back, was lost. This success was possible because the ships had fighter protection from either Cyrenaica or Malta for the whole voyage. In spite of heavy air attacks by Fliegerkorps II after their arrival, all their cargo was landed safely.

So great was the need to get supplies from Italy to the Axis armies in Africa that the Italian battle-fleet put to sea during January and escorted two important convoys. Although British torpedo-bombers sank the large troopship *Vittoria*, 66,000 tons of stores were landed. As a result General Rommel was able to advance and take the airfields in western Cyrenaica. In

February, therefore, a convoy of three ships for Malta from Alexandria could not be given fighter protection in the Ionian Sea; two of the merchant ships were sunk by air attack and one was damaged and had to turn back. To make matters worse, Fliegerkorps II stepped up its attack on Malta in February to 1020 tons of bombs: the forces based on the island were unable to interfere with the Axis supplies to North Africa which consequently flowed freely and with little loss.

In March it became essential to try to get another British convoy through to Malta, not only to reinforce the defences and to try to keep the striking forces in action, but to meet civilian needs. The defending fighters were already reduced to 32 Hurricanes and in three trips in March, the aircraft-carriers *Argus* and *Eagle* from Gibraltar flew in a total of 47 Spitfires. On 20th March, an attempt was again made to send four supply ships through from the east. The R.A.F. 201 (Naval Co-operation) Group in the Middle East now had 119 aircraft to help, 40 of which were long-range fighters and 40 torpedo-bombers. They were based in Egypt, however, which was too far away for effective support to the convoy. In the event, Axis air attacks on the convoy were light, partly because the weather was very bad, partly because the Luftwaffe was very busy in the western desert and partly because the R.A.F. had heavily bombed their bases. The convoy was also opposed by the Italian battleship *Littorio* with supporting cruisers and destroyers but Admiral Vian, with an escort of light cruisers and destroyers succeeded, very skilfully, in getting the convoy past. All was, however, in vain as Fliegerkorps II sank one of the ships as she approached Malta and damaged another which had to run itself ashore. The remaining two ships reached harbour but were sunk before they could be unloaded and only about a quarter of the cargo that left Alexandria was saved. Fliegerkorps II now had 335 aircraft and dropped over 2,000 tons on the island during March. By the end of the month the surface striking force of cruisers and destroyers had to leave the island and the air striking force was reduced almost to impotence. As a result Axis traffic to North Africa continued with negligible loss. British aircraft had, in fact, only sunk five ships during the first three months of 1942 while British submarines, less affected by the bombing, had sunk sixteen.

In April Fliegerkorps II redoubled its efforts and dropped 6700 tons of bombs on Malta putting the dockyard almost completely out of action and reducing the fighter strength at times to half a dozen operational machines. Towards the end of the month the submarines after losing four of their number in harbour also had to leave. There were only four torpedo-bombers left on the island and British aircraft only sank one ship in April and May. It was obvious that there was no point in trying to send another convoy to Malta until the fighters could guarantee that it could be unloaded and on 20th April, the U.S. aircraft-carrier *Wasp* flew off 47 Spitfires to the island. Many of these were lost on the ground and by the 23rd they were down to half a dozen operational machines again. *Wasp* and *Eagle* together flew in 81 more Spitfires during the month and this time they were able to regain air superiority over the island. This was partly due to the departure of much of the strength of Fliegerkorps II to Russia and North Africa as the Germans believed, prematurely as it turned out, that they had neutralized Malta. By the end of the month the Germans had lost 40 aircraft and only had 91 planes left. As a result, the weight of bombs dropped in May fell to 520 tons. Axis command of the central Mediterranean however was still firm. On 11th May the destroyers *Jackal*, *Kipling* and *Lively* were all sunk by dive-bombers from Crete when they attempted to interfere with an Italian convoy to Cyrenaica.

By mid-June it was imperative to run another convoy to Malta. As a preliminary the *Eagle* flew in 59 more Spitfires in two operations in early June, bringing the total up to 95, and there was now a squadron of long range Beaufighters on the island as well. It was difficult to know whether a convoy had a better chance from the east or the west. The enemy air forces in Sardinia, Crete and Sicily could be switched to either end as could the superior Italian battle-fleet at Taranto. In the end it was decided to run convoys from both east and west simultaneously with the hope that the enemy's forces would be split and one or both would get through. The convoy from the west (Operation 'Harpoon') of five supply ships and a tanker was to be escorted as far as the Sicilian narrows by the *Eagle* and *Argus* with 16 Sea Hurricanes, 6 Fulmars and 18 Swordfish embarked and by the *Malaya*, 4 cruisers and 17 destroyers. One anti-aircraft cruiser and 9 destroyers were to go through with the convoy to Malta. The convoy from the east (Operation 'Vigorous') was to consist of eleven merchant ships escorted by 8 cruisers and 26 destroyers, half of which had come through the canal from the Eastern Fleet. Air support was to come entirely from the shore. Fighter protection was to be given from Egypt at the beginning of the voyage and from Malta at the end, but the convoy would have to rely on its guns in the Ionian Sea. The R.A.F. however intended to bomb the enemy airfields in Italy, Sicily, Cyrenaica and Crete. If the Italian battle-fleet put to sea, the plan was to use shore-based aircraft and nine British submarines to attack and try to stop it reaching the convoy. There were some thirty Wellington and Beaufort torpedo-bombers available and some of these were to be based at Malta as well as in Egypt. There were also nine American Liberators at Fayid in the Canal zone and about thirty reconnaissance aircraft, some of which were fitted with radar.

The first few days of the passage of both convoys went well. The 'Harpoon' convoy from the west was attacked on 14th May by four or five groups of German and Italian aircraft from Sardinia and Sicily. The heaviest attack consisted of 28 torpedo-bombers and 10 high-bombers escorted by 20 fighters. One merchant ship was sunk and the cruiser *Liverpool* was damaged and had to turn back for Gibraltar. Most of the attacks were aimed at the *Argus* and *Eagle* but they were not hit and their fighters shot down 13 and damaged others for the loss of 7 of their number. In the late evening, as the convoy neared Cape Bon, the carriers and the main escort turned back and Beaufighters from Malta arrived to give protection to the remaining five ships of the convoy. Meanwhile the 'Vigorous' convoy from the east had already had one merchant ship damaged by dive-bombers and had had to send another back as she could not keep up. During the 14th May it was subjected to seven attacks by sixty to seventy dive-bombers but only lost one merchant ship. Another ship was found to be too slow and was sunk by 40 dive-bombers after she had been detached. By sunset on the 14th, the seven surviving merchant ships were south-west of Crete but still 400 miles from Malta.

On the 14th May, the Italian battleships *Littorio* and *Vittorio Veneto* with 4 cruisers and 12 destroyers put to sea from Taranto to attack the 'Vigorous' convoy and the cruisers *Savoia* and *Montecuccoli* and five destroyers sailed from Palermo to attack the 'Harpoon' convoy. All now depended on whether the submarines and air striking forces could stop these powerful warships. The Italian battle-fleet was attacked with torpedoes by four Wellingtons during the night and by nine Beauforts at dawn and they succeeded in hitting the cruiser *Trento*. The Italian fleet then ran through the British submarine line unscathed and was attacked by American Liberators from Fayid and twelve Beauforts from Sidi Barrani, the Liberators scoring a single bomb hit on the *Littorio*. The 'Vigorous' convoy escort had lost the destroyer *Hasty* and had the cruiser *Newcastle* damaged by motor-torpedo-boats and, in the circumstances, the convoy had to turn back to save itself. On its way back to Alexandria, it lost the destroyers *Nestor* and *Airedale* to air attack and the cruiser *Hermione* to a U-boat. Meanwhile the escort of the 'Harpoon' convoy west of Malta was able to beat off the Italian cruisers but the destroyer *Bedouin* was sunk and the *Partridge* was badly damaged. While the fighter protection from Malta was still very thin, three more merchant ships were sunk by air attack and the surviving two ships were all that reached Malta of the total of seventeen that had sailed in the two convoys.

The operation had therefore proved very expensive. Six merchant ships had been sunk by air attack and two damaged. A cruiser and five destroyers had been sunk and three cruisers and two destroyers damaged, roughly half of them by air attack. The Italians had lost the *Trento* which had been finished off by a submarine, and the *Littorio* was damaged when she was hit by a torpedo from a Wellington from Malta as she retired to Taranto. In addition the destroyer *Vivaldi* was disabled and they lost 22 aircraft attacking the 'Harpoon' convoy alone. Nevertheless the Axis had won the battle and they had done it by a combination of air and surface forces, which at this time commanded the sea in the central Mediterranean. Without the strong air support given by the R.A.F. from Malta and Egypt, however, the operation would have gone even less well for the British. That the Axis attacks were not more effective was in large measure due to the R.A.F. bombing of Axis airfields and their interception of striking forces as they rook off from Cyrenaica. Shore-based fighter protection at long range proved uneconomical and it was difficult to keep more than a few fighters over the ships at a time; without air bases in Cyrenaica the R.A.F. were in any case unable to provide fighter protection for the whole route. The *Argus* and *Eagle* were far more effective but their fighter complement was weak so that, although they were on the spot, they had difficulty in competing with the enemy escorting fighters, let alone the strike aircraft. The Allied shore-based air striking forces were unlucky and made only two hits out of the 32 torpedoes they carried into action; they made contact successfully with the enemy at long range but there were not enough of them to stop a battle-fleet. Yet the convoy, although it had to turn back, was saved and 201 Group can take some of the credit for this, as well as for the infliction of further damage on the Italian battle-fleet on its way home.

Battlefleet protagonists could, of course, say that the failure of the 'Vigorous' convoy to get through was simply because there were no British capital ships in the eastern Mediterranean. Certainly battleships would have been able to protect the convoy from the Italian fleet but whether they could have competed with the Luftwaffe in the Ionian Sea without fighter protection is quite another question. It was a pity that the three British armoured aircraft-carriers in the Indian Ocean were not available as they would have been able to provide fighter protection in the Ionian Sea and also a striking force to attack the Italian battle-fleet. Strong though the shore-based air forces provided for the two convoys were, they were too far away to be able to provide the support required at sea and could not make up for the lack of aircraft-carriers. As a result the situation in the Mediterranean got worse. Malta, upon which an attack on the Axis sea communications in the central Mediterranean depended, had fuel only for fighter aircraft and could not play its part as an offensive base. Tobruk fell and by the end of June the Eighth Army was back at El Alamein and Rommel was threatening the naval and air bases in Egypt.

The first four convoys to North Russia in 1942 totalled 42 ships and they arrived with the loss of only one of them. In the middle of January, Hitler, fearing British designs in Norway, transferred the battleship *Tirpitz* from Germany to Trondheim where she arrived without being sighted by Coastal Command patrols. She was followed in the middle of February by the *Prinz Eugen* and the *Scheer*. As a precaution, therefore, the convoy PQ12, which left Iceland for Russia on 1st March, was covered by the whole Home Fleet. On 6th March, the submarine *Seawolf* sighted the *Tirpitz* leaving Trondheim to attack the convoy. PQ12 had been sighted earlier by a F.W.200 reconnaissance aircraft but the *Tirpitz* failed to find it in bad weather. On her way back she was attacked with torpedoes by twelve

Albacores from the *Victorious* but they failed to hit her and two of them were shot down. Nevertheless the Germans were alarmed by this attack and had never known that the Home Fleet was at sea. Hitler subsequently ordered the aircraft-carrier *Graf Zeppelin* to be completed, which shows the moral effect an aircraft-carrier could have on a fleet which had not got one.

Hitler also ordered the North Russian convoys to be attacked with more vigour by air, submarine and surface forces. As a result convoys PQ13, 14 and 15 totalling 69 ships which sailed in March and April lost 10 ships, 5 of them to air attack and the damaged cruiser *Trinidad* returning from Russia in May was also sunk by aircraft. The North Russian convoys could get little air support as the North Cape was 1,000 miles from the nearest bases in the Shetlands and in Iceland and the Russians would not at first permit the use of their airfields.

In May the *Prinz Eugen*, which had been damaged by a submarine, sailed from Trondheim for Germany. This time she was intercepted and attacked by a Coastal Command striking force of 27 Beaufort torpedo-bombers and 13 Hudsons escorted by 14 Beaufighter and Blenheim fighters. Little damage was done however, and this powerful striking force suffered heavy losses which was a severe set-back for Coastal Command.

The Germans now had a force of 264 aircraft based in North Norway at Banak, Bardufoss and Kirkenes, of which 133 were dive-bombers and 57 were torpedo-bombers. By the time PQ16 of 35 ships sailed on 21st May the Luftwaffe were ready for a trial of strength. Convoy PQ16 had had its normal anti-submarine escort augmented by the anti-aircraft ship *Alynbank* and the CAM ship *Empire Lawrence* carrying a single Hurricane fighter. It was escorted as far as Bear Island by four cruisers: the main Home Fleet was in a covering position to the westwards. Over a period of five days, the convoy was attacked by roughly a thousand aircraft sorties. The single Hurricane from the CAM ship shot down one aircraft and damaged another but six ships were sunk by air attack and one by a U-boat. With so little fighter support against such a powerful air force it says much for the convoy's A.A. fire that casualties were not higher. Nevertheless they were as heavy as in the 'Harpoon' – 'Vigorous' convoys in the Mediterranean. It was clear that they would increase unless air support could be improved, but in spite of this it was decided to sail PQ17 of similar size and escort from Iceland at the end of June.

On 1st July, PQ17 was found by a U-boat north-east of Iceland and after an abortive air attack was shielded for two days by fog. The convoy was east of Bear Island before it was again attacked from the air when it lost three ships. It was making good progress when, out of the blue, it was ordered by the Admiralty to scatter. The *Tirpitz* with the *Hipper*, *Scheer* and six destroyers had left Trondheim for Alten Fjord to be in a position to attack PQ17. Their movements had been detected by air reconnaissance and the Admiralty believed surface attack on the convoy to be imminent. The *Tirpitz* group, after a short sortie, decided to leave the destruction of the dispersed ships to U-boats and aircraft and returned to Alten Fjord. Another 20 ships were then sunk, half by air attack and half by U-boats, and only 11 of the original 36 reached Russian ports. The order to scatter, although normally justified when a convoy is actually attacked by a powerful surface ship, was fatal when it was given in U-boat infested waters and in range of enemy airfields. The real reason for the disaster however was deeper than an error of judgement in the Admiralty. The Home Fleet was stronger than the *Tirpitz* and her consorts and would have been happy to engage them; if it had escorted the convoy the whole way, the convoy would never have had to scatter, but the Home Fleet was by no means happy to accompany the convoy into waters off North Norway dominated by the Luftwaffe, even with the *Victorious* in company. This is understandable as the odds were numerically 264 aircraft to 33 and qualitatively much greater. It was therefore primarily the strengthening of the Luftwaffe in Northern waters rather than the arrival of the German heavy ships which was responsible for the disaster to PQ17, and which, for three months afterwards, stopped the North Russian convoys.

With the entry of the United States into the war, the U-boats seized the opportunity to cross the Atlantic and massacre the dense unescorted shipping off the American coast before counter measures could be instituted. In the first seven months of 1942 they sank 681 ships of 3½ million tons. The campaign thus moved out of the British zone of operations. In March Coastal Command had 553 aircraft working in Home Waters and the Atlantic and the R.C.A.F. had another 85. Only 36 of all these were of the very long-range type, but at last a system of detecting and attacking U-boats on the surface at night had been perfected. Patrolling aircraft could now locate a U-boat by radar and could secure a point of aim for the actual attack by using a searchlight, known as a Leigh light, by which a stick of torpex depth charges could be dropped. This system only came into service in the spring of 1942 and its effect was immediate. In May two U-boats were damaged and in July *U.502* and *U.751* were both sunk in the Bay of Biscay. The Germans then reversed their normal procedure on passage and began to dive at night and surface by day. Coastal Command's 'Bay Offensive' which had begun early in 1941 and which by the end of April 1942 had only sunk one U-boat and damaged two others, at last began to take effect. In the first seven months of 1942, aircraft showed a general improvement in air anti-submarine operations in other areas as well. Of the grand total of 51 German, Italian and Japanese U-boats sunk in this period, aircraft had been responsible for 13 and had helped ships to sink 4 more. Ships were still the most important U-boat killers with 21 to their credit while Allied submarines had sunk 10.

In early 1942, there were no German warship raiders at sea but the merchant raiders *Thor, Stier* and *Michel* and the minelayer *Doggerbank* put to sea from ports in Occupied France and avoided detection by proceeding along the north coast of Spain in disguise. They only sank twenty ships of some 140,000 tons which was a very small result compared to the depredations of the U-boats: nevertheless they were a nuisance. Using the same route, eight blockade runners arrived from Japan and four left for the same destination without being detected. It was difficult for Coastal Command to find these ships without aircraft of greater range; the very long-range aircraft that they possessed were all very busy with the U-boats.

The need for much longer ranged reconnaissance aircraft was the cause of some dispute at this time between the Admiralty and the Air Ministry. For this type of aircraft, Coastal and Bomber Commands were in direct competition. The Air Ministry, understandably, did not wish to divert any of them from the bomber offensive against Germany, but the Admiralty pointed out that they did not require very many and that the whole campaign against the U-boats might be lost if they were not provided. There is little doubt that the Air Ministry's argument that they could best influence the war at sea by strategic bombing was fallacious. In spite of the heavy attacks on German industry, U-boats were now being produced at the rate of 20 a month and the building rate was increasing; no German U-boats were sunk in harbour by bombing in this period whereas there was a distinct chance that the numbers sunk by aircraft at sea could be substantially increased if aircraft could reach the U-boat operating areas in mid Atlantic.

In the first half of 1942, the British direct air attacks on German coastal shipping continued at about the same rate as before with much the same results. Some four times as many mines were now being laid from the air and they caused about twice the casualties of 1941; minelaying proved, as before, to be by far the best way to attack the German shipping in coastal waters. The total casualties caused by both these methods was 114 ships of 144,197 tons which was far greater than those achieved by the German air attacks on Allied shipping. German aircraft in small numbers continued to attack east coast shipping, generally at night, and to lay mines but the British defences were much improved, especially the low radar cover and the co-operation of ships and aircraft. They only succeeded in sinking sixteen ships of under 20,000 tons in this period; this was no more than an irritant.

The first half of 1942 had been a frustrating one for British air forces working over the sea. They failed to stop the *Scharnhorst* and *Gneisenau* in the Channel; they did not stop a single merchant raider or blockade runner arriving at or leaving the French coast and they detected few of the German warships on their way to Norway. Attacks on German warships such as that of the Fleet Air Arm on the *Tirpitz* or Coastal Command on the *Prinz Eugen* were failures and when the British strike aircraft in the Mediterranean tried to stop the Italian battle-fleet from attacking the 'Vigorous' convoy they failed also. On the other hand German air power over the sea seemed invincible; working with surface ships it had made the Malta and North Russia convoys prohibitively expensive. The Japanese had done better still: they had sunk two battleships sent against them; they had driven the Allies out of the Far East and the main British carrier force had had to retreat before them from Ceylon to the east coast of Africa. The only hopeful trends were that

Fighter Command had the German air attack on coastal shipping well in hand, Bomber Command was having some success with mining, and Coastal Command was just beginning to get the measure of the U-boats.

In searching for the reasons for these apparent inconsistencies, no simple solution such as the colour of the aviator's uniform or that carrier-based aircraft were better than shore-based aircraft, can be found. It is, however, possible to discern some principles. The first is sheer strength: for example, 264 Luftwaffe planes in north Norway against 33 in the *Victorious* were enough to prevent the Home Fleet escorting PQ17; and the 90 aircraft in the British carriers in the Indian Ocean were no match for the 350 aircraft of Admiral Nagumo's First Air Fleet. Conversely 40 R.A.F. strike aircraft in the Mediterranean were too few to stop the Italian Fleet. An important feature, also independent of the colour of the uniform, was training in maritime air warfare. Fliegerkorps X and the Japanese 22nd Air Flotilla could really hit ships at sea whereas Bomber Command and the high-level bombers of the Regia Aeronautica could not. Training in maritime air warfare explains to a great extent why the principle of the flexibility of air forces did not apply to Bomber Command so much as to Fliegerkorps II, VIII, and X of the Luftwaffe. Although whether the planes were carrier-borne or shore-based supplies no general rule, air power could not fail to be influenced by the distance of its bases from the scene of action. Sometimes the shore bases were close, as in the Battle of Crete, and sometimes more distant as in the German operations against the Russian convoys. With aircraft-carriers, however, the bases were always close: 22 fighters in the *Argus* and *Eagle* were more effective in defending the 'Harpoon' convoy than the 40 long range fighters of 201 Group in defending the 'Vigorous' convoy. The value of aircraft-carriers was, probably best illustrated by the way that both the German and Italian Fleets felt blind without them and by the fact that both had taken steps to provide themselves with them.

The type of aircraft already discussed in many places is also, of course, very relevant. The effectiveness of aircraft over the sea was therefore an amalgam of strength, training, equipment and proximity of base. The fact is that in the first part of 1942 the Germans and Japanese did better because they fulfilled these conditions to a greater degree than the British.

X

The Great Carrier Battles of the Pacific in 1942

THE JAPANESE CONQUEST of the Greater East Asia Co-prosperity Sphere had been completed in half the time that they had allowed for it. At the same time they felt a need to push their defence further out to the eastwards. They wanted to prevent any repetition of the carrier raid on Tokyo; to complete the capture of New Guinea and to try to drive a wedge south-eastwards from Rabaul between the Americans and the Australians. They decided to undertake the first instalment of this plan in the south-west Pacific in early May by seizing seaplane bases at Tulagi and Deboyne Islands and then making an amphibious assault on Port Moresby. As the majority of the Japanese Navy was required for a subsequent thrust to Midway and the Aleutians which they had planned for June, the forces available were moderate. They consisted of four heavy and three light cruisers and nine destroyers with the new light aircraft carrier *Shoho* carrying 21 aircraft. The shore-based 25th Air Flotilla in the area consisted of 41 bombers, 18 fighters and 12 seaplanes and it was reinforced by 45 bombers and 45 fighters, bringing it to a total of 161 aircraft. It was also decided to use the aircraft-carriers *Zuikaku* and *Shokaku* from Admiral Nagumo's fleet which had just returned from the Indian Ocean; these two fleet carriers had 125 aircraft on board and they were escorted by the heavy cruisers *Myoko* and *Haguro* and 6 destroyers.

The plan was first for a small force to seize Tulagi and establish a base for six seaplanes (see fig. 16(a), p. 240). Then eleven transports escorted and covered by several independent groups including the light carrier *Shoho* and the 25th Air Flotilla would sail from Rabaul for Port Moresby. This would be supported by the carrier striking force from Truk which, keeping out of range of American search aircraft, would round the end of the Solomon Islands and enter the Coral Sea. It was thought that the Americans had only one aircraft-carrier in the area, which they hoped to eliminate at this stage. Subsequently the carrier striking force would make a heavy attack on the U.S. Army air force bases in Queensland, so that the Port Moresby landing would not be molested. The Port Moresby invasion force on its way would establish the seaplane tender *Kamikawa Maru* at Deboyne Island.

The Allied forces in the area were partly American and partly Australian and consisted of 2 heavy and 1 light cruisers and 2 destroyers, with 12 Catalina flying-boats at Noumea and 11 submarines at Brisbane. On the Queensland airfields and at Port Moresby there was a strong U.S. Army air force of 100 fighters and 192 bombers, 48 of which were Flying Fortresses. The Americans had already received intelligence that a thrust was to be made into the Coral Sea and guessed that Port Moresby was its objective. There was therefore a task force from the U.S. Pacific Fleet in the area which consisted of the carriers *Yorktown* and *Lexington* with 143 aircraft between them. These were supported by 5 heavy cruisers and 11 destroyers. The carriers *Hornet* and *Enterprise* had only just returned to Pearl Harbor from the Tokyo raid and although it was hardly expected that they could arrive in time, they were also sent south at full speed.

The forces available to the two sides were therefore fairly even. The Japanese had three carriers to the American two but the number of aircraft embarked was approximately the same. The total surface forces committed were almost exactly equal. The Americans, however, had a grand total of very nearly 450 aircraft against 315 for the Japanese. The American superiority in numbers was, however, offset by the fact that the U.S. Army Air Force in Australia had many planes out of action awaiting spares; they were not trained to operate over the sea and were not under naval command. The Japanese planes on the other hand were all naval and trained in maritime operations and were all under the same command as the ships.

The first Japanese move to seize Tulagi and establish the seaplane base was made on 3rd May without opposition but it was detected by reconnaissance planes from Australia (see fig. 16(a), p. 240). Admiral Fletcher, who commanded the Allied naval task forces, had already concentrated his force in the

Coral Sea and was at the time engaged in re-fuelling from his tankers. The *Yorktown* was the first to complete fuelling and she was sent at full speed towards Tulagi and during 4th May launched three successive strikes of 40, 38 and 21 planes. The invasion group had already left but the planes badly damaged the destroyer *Kikuzaki* which eventually sank, and destroyed four minor warships and five seaplanes. At the time the Japanese carrier force under Admiral Takagi was fuelling well out of range to the north. *Yorktown* then turned south and rejoined the *Lexington* group early on 5th May and continued fuelling. Meanwhile the Port Moresby invasion group had left Rabaul.

At dawn on 6th May the Japanese carrier striking force was already in the Coral Sea but the searches flown by the American carriers just missed them on two occasions (see fig. 16(b) p. 240). Admiral Takagi was at one time only seventy miles away and neither side knew that the other was there. The Japanese for some reason, probably as they did not want to give their position away, flew no long range searches on 5th or 6th May. Oddly enough both these carrier forces were sighted by the other side's shore-based planes but none of the reports got through to the opposing carrier commanders. The Australia-based army planes, however, had sighted the Port Moresby invasion group and Flying Fortresses had attacked the *Shoho*. The *Kamikawa Maru* arrived at Deboyne Island ready to use her seaplanes to scout on 7th May. Admiral Fletcher, with no information about the Japanese carriers, moved westwards ready to strike at the Port Moresby invasion group next day.

On 7th May the Japanese carriers flew a search which found the American tanker *Neosho* (see fig. 16(c), p. 241) and her escort the destroyer *Sims*, which were reported as a 'carrier and a cruiser' and the *Shokaku* and *Zuikaku* launched a heavy strike which sank the *Sims* and disabled the *Neosho*. Admiral Fletcher was well to the westward at the time busy with the Port Moresby invasion group: he had already detached a surface force under Rear Admiral Crace, consisting of the Australian cruisers *Australia* and *Hobart* and the U.S. cruiser *Chicago*, to block the Jomard Passage. At the same time, search planes from the *Yorktown* made contact with the support group of the Port Moresby invasion group which they erroneously reported as two aircraft-carriers. Admiral Fletcher took this to be the main Japanese carrier force and launched a heavy strike of ninety-three planes from both *Yorktown* and *Lexington* keeping back

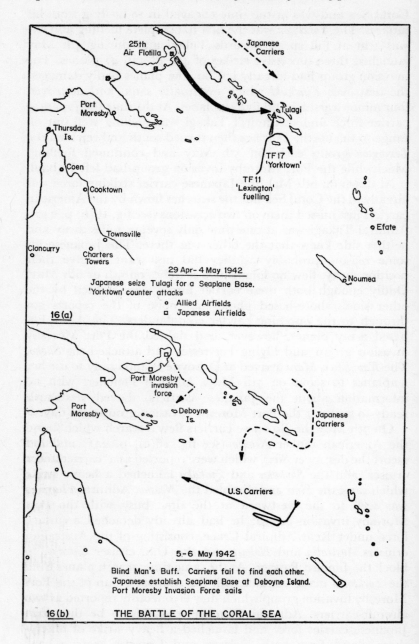

25th
Air Flotilla

Japanese
Carriers

Port
Moresby

Tulagi

Thursday
Is.

Coen

TF 17
'Yorktown'

TF 11
'Lexington'
fuelling

Cooktown

Efate

Townsville

Cloncurry

Charters
Towers

29 Apr- 4 May 1942

Japanese seize Tulagi for a Seaplane Base.
'Yorktown' counter attacks

Noumea

○ Allied Airfields
□ Japanese Airfields

16 (a)

Port Moresby
invasion
force

Port
Moresby

Deboyne
Is.

Japanese
Carriers

U.S. Carriers

5-6 May 1942

Blind Man's Buff. Carriers fail to find each other.
Japanese establish Seaplane Base at Deboyne Island.
Port Moresby Invasion Force sails

16 (b)   **THE BATTLE OF THE CORAL SEA**

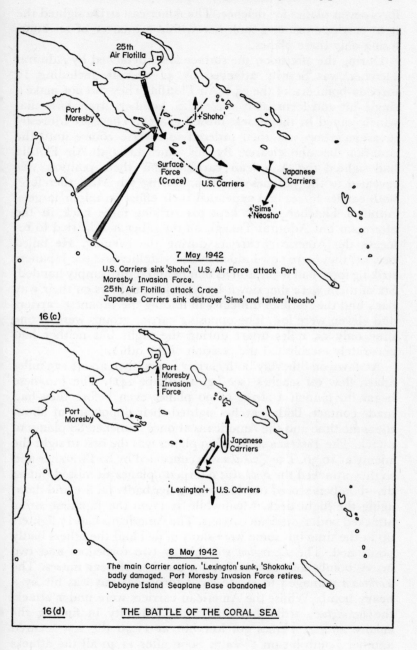

25th Air Flotilla

Port Moresby

+ 'Shoho'

Surface Force (Crace)   **U.S. Carriers**   **Japanese Carriers**

+ 'Sims'
+ 'Neosho'

7 May 1942

U.S. Carriers sink 'Shoho', U.S. Air Force attack Port
Moresby Invasion Force.
25th Air Flotilla attack Crace
Japanese Carriers sink destroyer 'Sims' and tanker 'Neosho'

16 (c)

Port Moresby Invasion Force

Port Moresby

Japanese Carriers

'Lexington'+ **U.S. Carriers**

8 May 1942

The main Carrier action. 'Lexington' sunk, 'Shokaku'
badly damaged. Port Moresby Invasion Force retires.
Deboyne Island Seaplane Base abandoned

16 (d)   THE BATTLE OF THE CORAL SEA

forty-seven planes for defence. The American strike sighted the light carrier *Shoho* and sank her within a quarter of an hour, losing only three planes.

During the afternoon the surface force detached by Admiral Fletcher was heavily attacked by 42 aircraft, including 12 torpedo-bombers, of the 25th Air Flotilla. They did not make a single hit and fortunately three U.S. bombers from Australia, which joined in the attack, missed as well. The Port Moresby invasion group was then ordered to reverse course until the situation became clearer. By this time the 25th Air Flotilla had sighted the American carriers and their position was revealed to the Japanese carriers. During 7th May, therefore, both carrier forces had expended their effort on minor targets. Admiral Fletcher wisely kept his striking force back in the afternoon but Admiral Takagi, on the other hand, tried to re-locate the American carriers during the evening. He failed because they were now hidden in bad visibility and the Japanese striking force had to jettison its bombs and return empty handed. Six of them were shot down by American fighters on their way back and the rest had difficulty in finding the Japanese carriers and eleven were lost. The opposing carrier groups were at one time only 95 miles apart during the night but neither had accurately established the position of the other.

At dawn on 8th May both carrier forces, then about 175 miles apart, flew off seaches (see fig. 16(d), p. 241). The Japanese began to launch a strike of 90 planes even before they had made contact. Both searches sighted their opponents at about the same time and the Americans at once launched 88 planes to attack. The *Yorktown* strike of 49 planes was the first to sight the enemy at 10.30. The *Zuikaku* was concealed by bad weather and so they attacked the *Shokaku*: 9 torpedo planes all missed but 24 dive-bombers scored two hits setting her badly on fire and dam-aging the flight deck. Meanwhile at 11.20 the Japanese strike attacked both American carriers. The Americans had 17 fighters up at the time but some were short of fuel and the others badly positioned. The *Lexington* was hit by two torpedoes and two heavy bombs and also suffered a number of near misses. The *Yorktown* managed to evade eight torpedoes but was hit by a heavy bomb. Whilst the American carriers were under attack, the *Lexington's* strike had had some difficulty in finding the enemy and 21 planes got lost but at 11.40 the rest secured another bomb hit on *Shokaku*. Soon after 11.40 all the attacks were over and the planes of both sides were returning to what was left of their carriers.

The *Shokaku* had lost over a hundred men killed and although she managed to extinguish the fires, she was in no state to continue flying operations and had to withdraw from the battle. She was out of action for two months but the undamaged *Zuikaku* was able to take some of her planes. *Yorktown* had 66 men killed but also brought her fires under control and was able to land-on her planes. *Lexington* made a remarkable recovery; she corrected her list and extinguished fires and began to land-on her aircraft. At 12.47, however, there was a heavy internal explosion caused by petrol vapour and the fires broke out all over again. Two hours later there was another explosion, the fires got out of control and she had to be abandoned. Later in the day she had to be sunk by an American destroyer. Thirty-six planes went down with her but nineteen had landed on *Yorktown*.

Both forces now retired from the battle. The Americans had lost another 33 planes in the fighting and the *Yorktown* ended the day with 49 ready for action. The Japanese lost 43 planes during the day and some 50 others were damaged or carried off in *Shokaku* and so *Zuikaku* had only 39 planes available at the end of the action. Next morning the U.S. Army aircraft from Australia heavily bombed the seaplane base at Deboyne Island and the Japanese were forced to withdraw; they also continued their attacks on the retiring Port Moresby invasion group and the Japanese had to call off the operation altogether.

The Battle of the Coral Sea was the first action between aircraft-carriers in the history of naval warfare. Surface ships made no contact with each other and all the damage was done by aircraft. By counting the casualties the Japanese got the better of it although it was really a case of mutual mutilation: the Americans sank a light aircraft-carrier and a destroyer but lost a very large fleet aircraft-carrier as well as a tanker and a destroyer. Strategically it was a very different matter. The U.S. and Australian forces had prevented the invasion of Port Moresby and had forced the transport to turn back and had neutralized the Japanese seaplane bases that they had set up to cover the Coral Sea. The morale of the Allies received a great lift. At last, after six months of retreat, the Japanese had been stopped.

.   .   .   .   .   .

At the end of May, the Japanese were ready to undertake the next part of their plan which was to advance into the central Pacific and take Midway Island and at the same time to seize Kiska, Attu and Adak in the Aleutian Islands. The setback in the Coral Sea did not worry them unduly: they thought that they had sunk both the American carriers and hoped before long to be able to renew their attack on Port Moresby. Admiral Yamamoto believed that Midway was sufficiently important to the Americans to force them to defend it with the whole of their Pacific Fleet. He hoped for a fleet action in which he could complete the destruction begun at Pearl Harbor. Admiral Yamamoto therefore intended to use practically the whole of the Japanese Navy for this operation. He was able to do this as the British in the Indian Ocean had been thrown back on the defensive; they were busy occupying Madagascar and were no threat to his rear.

The Japanese surface forces consisted of 11 battleships, 13 heavy and 8 light cruisers and 59 destroyers. There were 8 aircraft-carriers and 4 seaplane carriers with a total, counting the float-planes in the battleships and cruisers, of well over 500 aircraft. In addition there were 22 submarines of the Advanced Expeditionary Force, 18 tankers to refuel the fleet and 16 transports to carry the troops. The plan was a complicated one and the ships were divided into no fewer than thirteen separate groups with different functions. In general however there were to be two carrier striking forces, Admiral Nagumo's four carriers of the First Air Fleet to support the Midway operation and the two light carriers *Junyo* and *Ryujo* for the Aleutians. The main body of the Japanese Fleet, under Admiral Yamamoto himself, would be in support and there would be invasion groups to take Midway and the Aleutian Islands. Two other small aircraft-carriers and the seaplane carriers were embodied in the invasion support forces. The Japanese intended to strike at the Aleutians first in the hope of drawing some American forces to their defence. Midway would then be taken; it was expected that this operation would come as a complete surprise to the Americans and that their fleet would not have time to interfere. Subsequently, when the Pacific Fleet moved up to counter-attack, the Japanese main body would close in for a decisive action.

To oppose this huge armada, the Americans had just over half the number of ships of all types. With the loss of *Lexington*

A Japanese plane shot down by an American Carrier Task Force

The Japanese aircraft-carrier *Zuiho* at the Battle of Leyte Gulf

PLATE IX — ACTION SHOTS IN THE PACIFIC
(Official U.S. Navy Photographs)

H.M.A.S. *S*
off Korea
Sea Furies a
Fireflies

H.M.S. *Albion* in Suez Operation
Seahawk and Sea Venom fight
and a Whirlwind helicopter

H.M.S. *Victorious*
A fully angled deck carrier. Scimi
Sea Vixen, Gannet aircraft
Wessex helicopters

PLATE X
THREE 'GENERATIONS' OF
THE FLEET AIR ARM
(*Imperial War Museum Photographs*)

at the Coral Sea, they were again reduced to four carriers in the Pacific Fleet and of these the *Saratoga* with repairs only just completed was still working up at San Diego and the *Yorktown* had not yet been repaired after the Battle of the Coral Sea. They had a total of 9 heavy and 4 light cruisers, 32 destroyers and 25 submarines in the central Pacific and the Aleutians but Admiral Pye's battle force, now of 7 battleships, was at San Francisco. In aircraft their inferiority was not so marked. The air groups of the carriers totalled 233 planes and they had managed to pack 115 into Midway Island. At the same time there were 173, mostly U.S. Army fighters, in the Aleutians area.

The Americans, however, had one very great advantage and that was the efficiency of their intelligence. While the Battle of the Coral Sea was in progress, Admiral Nimitz was informed of the main Japanese intentions. He recalled *Enterprise* and *Hornet* which were on their way to the south-west Pacific and ordered the *Yorktown* and *Saratoga* to Pearl Harbor as soon as possible. A second advantage was that Midway itself could function as an immobile but unsinkable aircraft-carrier and the 115 aircraft there were as many as the airfield would hold. Nineteen U.S submarines were deployed defensively in the area and the Aleutians were reinforced with two cruisers and two destroyers but their air defence was left to the Army Air Forces. *Enterprise* and *Hornet* were sent to wait north-east of Midway where they were unlikely to be discovered by Japanese scouting planes but where they were well placed to counter attack if the Japanese carriers launched air strikes against the island. The *Yorktown* was patched up at Pearl Harbor in two days, her air group was brought up to strength and she made all speed to join the *Enterprise* and *Hornet*. *Saratoga* left San Diego at full speed on 1st June but was not able to arrive in time.

The Japanese put to sea in eight groups from the Inland Sea, the north of Japan, and the Marianas during the last week in May (see fig. 17(a), p. 248). On 2nd June, twenty-four hours before the Japanese arrived, the three American carriers were in position just over 300 miles north-east of Midway. Very early next morning the Japanese carriers *Ryujo* and *Junyo* launched a strike to attack Dutch Harbour in the Aleutians. The weather was very bad and many of the planes got lost but 17 of them from the *Ryujo* did considerable damage. A few hours after this attack a Catalina flying-boat from Midway sighted the Japanese transport group some 600 miles away approaching the island. Nine army B17 bombers took off to attack but their bombs all missed. Later, after dark, four Catalinas armed with torpedoes secured a hit on a tanker. While these attacks were in progress, Admiral Nagumo, who was sheltered by poor visibility, was closing Midway at 25 knots and preparing to strike the island at dawn. Admiral Fletcher, commanding the American carriers, guessed that this was what he was doing and moved to a position 200 miles north of Midway ready to counter-attack the Japanese carriers as soon as they were located.

At 04.30 4th June, Admiral Nagumo launched a powerful strike of 108 planes to attack Midway Island from a position 240 miles away (See fig. 17(b), p. 248). He kept back 93 planes with armour-piercing bombs and torpedoes in case enemy ships were sighted, and some 50 fighters for his defence. At the same time he flew off a search to the eastwards with the float-planes from his cruisers. No sooner was the strike launched than he was seen by a Catalina flying-boat from Midway and the report was received by the American carriers. At the time the *Yorktown* was about to recover a dawn search so Admiral Fletcher ordered *Hornet* and *Enterprise* under Admiral Spruance to close and strike at the Japanese carriers.

At 06.30 the Japanese strike attacked Midway and were opposed by Marine Corps Buffalo and Wildcat fighters. The strike did a great deal of damage but did not put the airfield out of action. The American defences shot down some 30 aircraft but 17 of their fighters, which was over half the number they had, were lost. Midway then flew-off ten torpedo planes which found and attacked the Japanese carriers but were decisively repulsed without scoring a single hit; only three returned and they were damaged.

The *Enterprise* and *Hornet* began to launch a strike of 116 planes soon after 07.00. Sixty-nine of these were dive-bombers and 29 torpedo-bombers escorted by 20 fighters. They kept back 8 scout-bombers and 36 fighters for defence. In the middle of launching, the American carriers were sighted by a float-plane from the cruiser *Tone* but before its report could get through, Admiral Nagumo was persuaded that a second strike was needed against Midway. He therefore ordered the 93 stand-by planes to be re-armed to attack shore targets. After half an hour, when this work was only half completed, the report from his search plane came in and he then, after some hesitation, reversed the order and told his air groups to be prepared to attack ships again.

At this point 16 Marine Corps dive-bombers from Midway attacked but all their bombs missed and half of them were shot down by guns and fighters. These were followed by a very high bombing attack by 15 U.S. Army B17's from Midway and these missed too but suffered no casualties. Yet another Marine bombing attack of 11 planes then came in and yet again they missed. In the four attacks they had made on the Japanese carriers, the Midway air group was halved and had caused no damage at all.

The 116 planes launched by the *Hornet* and *Enterprise* set off soon after 08.00 and 17 dive-bombers, 12 torpedo planes and 6 fighters from the *Yorktown* followed them (See fig. 17(c), p. 249). While the strikes were on their way, the Japanese carriers were recovering their own striking force from Midway and as soon as they had them on board, Admiral Nagumo turned towards the American carriers to close. Of the first wave of the American striking force, the 35 dive-bombers and some fighters from the *Hornet* failed to find the enemy and had to land on Midway. The 41 torpedo planes from all three U.S. carriers were the first to attack and with disastrous results; 35 of them were shot down and those torpedoes they succeeded in dropping either missed or were evaded. Nevertheless they played their part for they drew the Japanese defending fighters down to a low level and left the way clear for the American dive-bombers. The 49 dive-bombers from *Enterprise* and *Yorktown* caught the Japanese carriers with nearly all their planes on board in the middle of refuelling and re-arming. Attacking immediately after the torpedo bombers, they scored three hits with 1000-lb bombs on both the *Akagi* and the *Soryu* and four hits on the *Kaga* and all three carriers burst into flames. They were put completely out of action and were, in fact, finished although they remained afloat for some time. In these attacks the American air groups lost 69 planes. Admiral Yamamoto with the battle-fleet was several hundred miles to the westwards and when he heard of the disaster he ordered the *Junyo* and *Ryujo* from the Aleutians to join the *Hiryu*, which was the sole survivor of Admiral Nagumo's carriers.

Hiryu was still intact and she launched 18 dive-bombers and 6 fighters at 11.00 and they were led by two of *Chikuma's* float-planes to the *Yorktown* which was about to land on her victorious strike. Twelve defending fighters did their best but six of the Japanese dive-bombers made three hits causing serious fires and

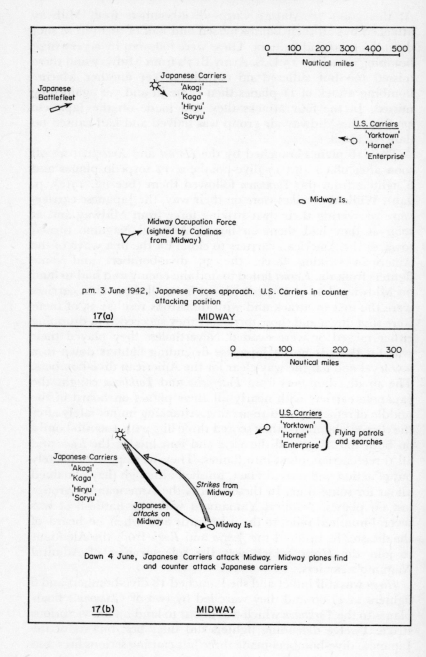

0 100 200 300 400 500
Nautical miles

Japanese
Battlefleet

Japanese Carriers
'Akagi'
'Kaga'
'Hiryu'
'Soryu'

U.S. Carriers
'Yorktown'
'Hornet'
'Enterprise'

Midway Is.

Midway Occupation Force
(sighted by Catalinas
from Midway)

p.m. 3 June 1942, Japanese Forces approach. U.S. Carriers in counter
attacking position

17(a)   **MIDWAY**

0   100   200   300
Nautical miles

U.S. Carriers
'Yorktown'
'Hornet'   } Flying patrols
'Enterprise'   and searches

Japanese Carriers
'Akagi'
'Kaga'
'Hiryu'
'Soryu'

Strikes from
Midway

Japanese
attacks on
Midway

Midway Is.

Dawn 4 June, Japanese Carriers attack Midway. Midway planes find
and counter attack Japanese carriers

17(b)   **MIDWAY**

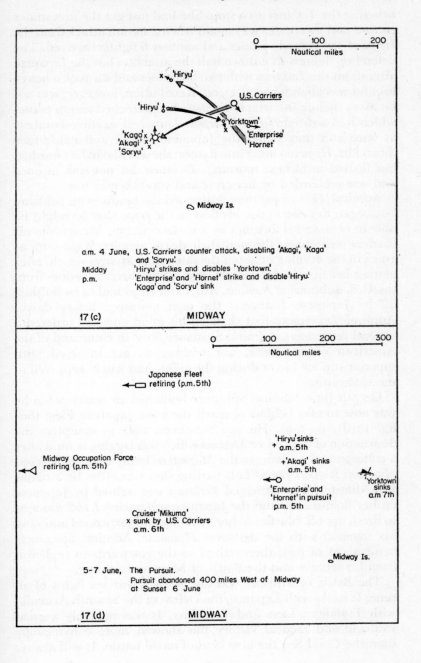

0   100   200
Nautical miles

× 'Hiryu'

'Hiryu'   U.S. Carriers

'Kaga' ×   Yorktown
'Akagi' ×   'Enterprise'
'Soryu'   'Hornet'

◇ Midway Is.

a.m. 4 June,   U.S. Carriers counter attack, disabling 'Akagi', 'Kaga'
                    and 'Soryu'.
Midday   'Hiryu' strikes and disables 'Yorktown'.
p.m.   'Enterprise' and 'Hornet' strike and disable 'Hiryu'.
                    'Kaga' and 'Soryu' sink.

17 (c)   MIDWAY

0   100   200   300
Nautical miles

Japanese Fleet
◁☐ retiring (p.m. 5th)

'Hiryu' sinks
+ a.m. 5th

Midway Occupation Force   +'Akagi' sinks
retiring (p.m. 5th)   a.m. 5th
◁   Yorktown
                              ◄─○   sinks
                         'Enterprise' and   a.m 7th
                         'Hornet' in pursuit
                            p.m. 5th

        Cruiser 'Mikuma'
        × sunk by U.S. Carriers
           a.m. 6th

                                                      ○ Midway Is.

5-7 June,   The Pursuit.
            Pursuit abandoned 400 miles West of Midway
            at Sunset 6 June

17 (d)   MIDWAY

bringing the *Yorktown* to a stop. She had just got the fires under control and was under way again when a second attack from the *Hiryu* of ten torpedo planes and another 6 fighters arrived. The defending fighters shot down half the attackers but the Japanese aircraft hit the *Yorktown* with two torpedoes and she took a heavy list and was abandoned at 15.00. Retribution, however, was not far away. Before the attack, *Yorktown* had launched search planes which had sighted *Hiryu*. *Enterprise* launched 24 dive-bombers at once and they found the Japanese carrier and made four direct hits. *Hiryu* too burst into flames: she was also in fact finished but floated until next morning. *Yorktown* did not sink at once and was reboarded by her crew and was taken in tow.

Admiral Yamamoto then postponed the landings on Midway but began to concentrate his fleet in the hope that he might be able to retrieve his fortunes by a surface action. Meanwhile his carriers were going through their death agonies. *Kaga* sank at 19.25 in the evening at about the same time as *Soryu* which, after getting her fires under control, was hit by three torpedoes from the U.S. Submarine *Nautilus*. *Akagi* and *Hiryu* had to be finished off by Japanese destroyers the next morning. Before dawn, Admiral Yamamoto had changed his mind and had ordered a general retirement. Admiral Spruance, now in command of the American carrier force, not wishing to get involved with superior surface forces during the night, had wisely kept well to the eastwards.

On 5th June Admiral Spruance launched air searches but he was now too far behind to catch the main Japanese Fleet (See fig. 17(d), p. 149). He was, however, able to complete the destruction of the cruiser *Mikuma* which was lagging behind after a collision and to damage the *Mogami* so badly that she was out of action for two years. This parting shot was offset by a major U.S. disaster. The damaged *Yorktown* was sighted by Japanese cruiser float-planes and the Japanese submarine *I.168* was sent to finish her off. She found her next day and torpedoed and sank her together with the destroyer *Hammann*. Admiral Spruance, being short of fuel, then retired to the eastwards to replenish from his tankers and the Battle of Midway was over.

The Battle of Midway was one of the great sea fights of all time. It ranks with Lepanto, the Defeat of the Spanish Armada, with Trafalgar, Lissa and Tsushima. It was not only a great strategic and tactical victory but showed more convincingly than the Coral Sea the new kind of naval battle. It will always be taken as an illustration of the final replacement of the battleship as the unit of sea power by aircraft, especially when operated from aircraft-carriers. The spectacle of a fleet of eleven battleships, including the most powerful ship in the world, retiring before two aircraft-carriers with somewhat depleted air groups leaves no doubt that this had now come about. If any confirmation is needed, the impotence of the American Battle Force of seven battleships, which made a timid lunge into the Pacific from San Francisco but never came within two thousand miles of the enemy, must surely provide it.

In the whole of the Battle of Midway, as at the Coral Sea, no surface ships made contact and the whole action was won by aircraft assisted in a minor way by submarines. All the damage was in fact done by carrier-based aircraft and those based ashore, whether they belonged to the U.S. Navy, Marine Corps or Army, achieved very little. The reasons, however, seem to have more to do with training, marksmanship and the method of attack than with where the aircraft were based. High-bombing by the U.S. Army B17's was not accurate enough at this stage of the war to hit ships at sea. The American torpedo-bombers, whether carrier-borne or shore-based, proved a complete failure, in spite of the fact that six of them were of a brand new type. The U.S. Marine Corps dive-bombers were of an older type and they failed to hit anything; nevertheless they exhausted the Japanese resistance before the carrier planes attacked. All the damage was done by dive-bombers, using 1000-lb bombs, and they were all carrier-based planes. The four Japanese carriers were destroyed by fire rather than loss of buoyancy and all of them took a long time to sink.

The rout of a powerful aggressor by a smaller defensive force always fascinates the student of war. Probably the most important reason for the Americans' success was the excellence of their intelligence and the skill with which they concentrated their carriers in very much the right place at the right time. On the other side, the certainty by the Japanese that they would surprise the Americans and the way Admiral Nagumo allowed his force to be used to support an amphibious operation before he was certain that he had the command of the sea, were equally to blame. Admiral Yamamoto, in fact, got his fleet action but it was of a very different kind with a very different result from the one he had intended.

Strategically this heavy repulse of the main Japanese fleet and the destruction of the *Akagi*, *Kaga*, *Soryu* and *Hiryu* of the heretofore invincible First Air Fleet, finally put an end to any further Japanese expansion. The loss of the four fleet carriers was bad enough but the loss of their 250 first line aircraft, and a high proportion of their pilots was a greater disaster still.

Grievous though the Japanese losses had been at Midway, they still had a respectable carrier force. The repairs to the *Shokaku* after the Coral Sea were nearly completed and *Zuikaku* was training a new air group. They had the three light carriers *Ryujo*, *Hosho* and *Zuiho*, one of which was required for training, and the *Junyo*, converted from a fast liner, was in service carrying fifty-three aircraft and her sister ship the *Hiyo* was nearing completion. Including the escort carrier *Taiyo*, this totalled six carriers ready for operations and two more nearly so. With the loss of *Yorktown*, the Americans were again down to three fleet carriers. However the *Wasp*, which had been operating with the British in European waters, passed through the Panama Canal soon after Midway.

On 5th July, U.S. reconnaissance aircraft found that the Japanese were building an airfield on Guadalcanal. The American Chiefs of Staff reckoned that it was essential to prevent this airfield, which would extend the 25th Air Flotilla's operations into the Coral Sea, from being brought into use. They therefore decided to mount an amphibious operation to seize the island without delay. The *Saratoga*, *Enterprise* and *Wasp* were all sent down to support the operation: they covered and supported the landings on 7th August and their 99 fighters helped to repulse the counter-attacks by the 25th Air Flotilla from Rabaul. On 8th August they shot down 17 out of 26 torpedo-bombers but one transport and a destroyer were lost. Admiral Fletcher, who was in command, was nervous about remaining for too long in a position which was within range of the 25th Air Flotilla. On the evening of the 8th August he started to withdraw and that night a Japanese cruiser force surprised and heavily defeated American and Australian cruisers at the Battle of Savo Island. Although Admiral Fletcher knew that the action was in progress he did not use the *Wasp's* air group, which was trained in night operations, and continued his withdrawal next day when he could have been within easy range of the victorious Japanese cruisers. There is little doubt that the defeat at Savo Island could have been prevented by the powerful American carrier force. If the right searches had been flown the American carriers could have struck the Japanese force the day before and have prevented it ever getting as far as Savo Island. The withdrawal of the American carriers left the Marines on Guadalcanal without air support until 20th August when the escort carrier *Long Island* was able to bring 19 Marine fighters and 12 dive-bombers to operate from the captured airfield on Guadalcanal which the Americans had completed and renamed Henderson Field.

By day the sea area round Guadalcanal had become one in which there was heavy air fighting between the Henderson Field planes and the 25th Air Flotilla and neither side had established air superiority. At night, ships of both sides moved in to try to support and reinforce or supply the troops ashore, withdrawing before daylight. The Japanese ships would withdraw north-westwards seeking air protection from the 25th Air Flotilla and the Americans south-eastwards towards their carriers which were still at sea.

It was two weeks before the Japanese managed to muster a carrier force to send to the Solomons area. The plan was ostensibly to throw another 1500 troops into Guadalcanal under cover of strong carrier and surface forces. The real aim, however, was to destroy the American carriers, so that the U.S. surface forces could be driven out of the Solomons area altogether and so isolate Guadalcanal. Three battleships, 11 heavy and 3 light cruisers and 31 destroyers were to accompany the carriers *Shokaku*, *Zuikaku* and *Ryujo* with a total of 168 planes on board. The seaplane carrier *Chitose* brought along 12 seaplanes and there were others in the battleships and cruisers. The 25th Air Flotilla now had a hundred aeroplanes in the Rabaul area and a number of long-range flying-boats at the Shortlands. To oppose this fleet the Americans had the 3 carriers *Saratoga*, *Enterprise* and *Wasp* which between them carried 256 aircraft. They were escorted by the brand new battleship *North Carolina*, which had sufficient speed to keep up with them, and 5 heavy and 2 light cruisers and 18 destroyers. There was a sizeable striking force of fighters and dive-bombers on Henderson Field and the Catalina flying-boats based on seaplane tenders were established at Ndeni, Malaita and Nandi as well as at Nouméa and Esperitu Santo (See fig. 18(a), p. 254). As usual the Japanese plan was complicated: there were three major forces subdivided

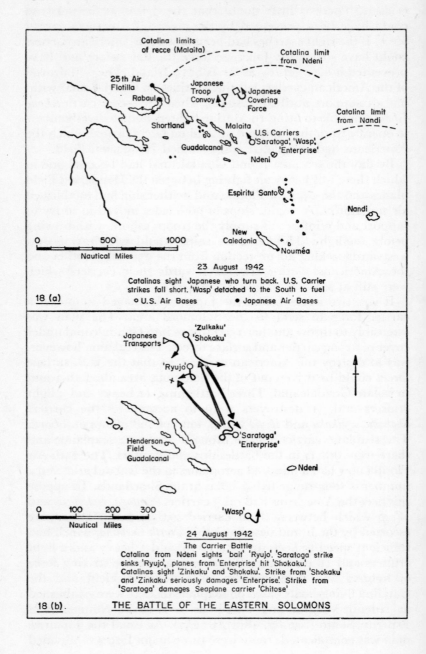

18 (a)

Catalina limits
of recce (Malaita)

Catalina limit
from Ndeni

25th Air
Flotilla

Rabaul

Japanese
Troop
Convoy

Japanese
Covering
Force

Shortland

Malaita

Catalina limit
from Nandi

Guadalcanal

U.S. Carriers
'Saratoga', 'Wasp',
'Enterprise'

Ndeni

Espiritu Santo

Nandi

New
Caledonia

Nouméa

0   500   1000
Nautical Miles

23 August 1942
Catalinas sight Japanese who turn back. U.S. Carrier
strikes fall short. 'Wasp' detached to the South to fuel
○ U.S. Air Bases   ● Japanese Air Bases

Japanese
Transports

'Zuikaku'
'Shokaku'

'Ryujo'

Henderson
Field
Guadalcanal

'Saratoga'
'Enterprise'

Ndeni

'Wasp'

0   100   200   300
Nautical Miles

24 August 1942
The Carrier Battle
Catalina from Ndeni sights 'bait' 'Ryujo'. 'Saratoga' strike
sinks 'Ryujo', planes from 'Enterprise' hit 'Shokaku'.
Catalinas sight 'Zinkaku' and 'Shokaku'. Strike from 'Shokaku'
and 'Zinkaku' seriously damages 'Enterprise'! Strike from
'Saratoga' damages Seaplane carrier 'Chitose'

18 (b).   THE BATTLE OF THE EASTERN SOLOMONS

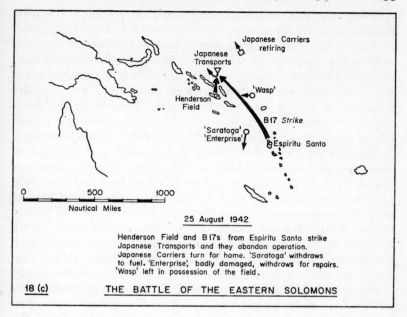

Japanese Carriers
retiring

Japanese
Transports

'Wasp'

Henderson
Field

B17 *Strike*

'Saratoga'
'Enterprise'

⊗ Espiritu Santo

0   500   1000

Nautical Miles

25 August 1942

Henderson Field and B17s from Espiritu Santo strike
Japanese Transports and they abandon operation.
Japanese Carriers turn for home. 'Saratoga' withdraws
to fuel. 'Enterprise', badly damaged, withdraws for repairs.
'Wasp' left in possession of the field.

18 (c)   THE BATTLE OF THE EASTERN SOLOMONS

into seven smaller groups, all with particular functions. This
fleet left Truk and entered the area to the north of the Solomons
on 23rd August. The light carrier *Ryujo* was to be detached
ahead as a 'bait' to draw the American carrier strikes upon
herself, while the two fleet carriers, 60 miles astern, would make
a devastating counter-attack.

The three American carriers had been given the duty of
protecting the sea communications into the Solomons and they
were positioned well to the east of the islands, behind the area
searched by the Catalina flying-boats from Malaita and Ndeni.
It was one of these Catalinas which sighted the enemy transport
force during the morning of 23rd August. Admiral Fletcher
ordered *Saratoga* to attack and she flew off 31 dive-bombers and
6 torpedo planes. At the same time 23 planes left Henderson
Field for the same purpose. Neither of these strikes found anything because the Japanese, realizing that they had been sighted,
reversed course and the *Saratoga* planes, short of fuel, had to land
at Henderson Field. At this point, Admiral Fletcher made the
serious miscalculation that no action was likely in the near
future and detached the aircraft-carrier *Wasp* and her group to
the south to fuel.

Dawn searches on the morning of the 24th at first revealed
nothing and the *Saratoga's* planes rejoined her from Henderson
Field (See fig. 18(b), p. 254). But at 09.05 a Catalina flying-boat
sighted the *Ryujo* group which had been sent forward as bait.
Admiral Fletcher, reluctant at first to swallow it, rightly suspected that there were other carriers about and ordered the
Enterprise to launch an extensive search. Meanwhile the *Ryujo*
flew off a strike at Henderson Field which, although it was
joined by planes of the 25th Air Flotilla, did very little damage.

Admiral Fletcher then changed his mind and decided to
swallow the bait and strike at the *Ryujo*. The *Saratoga* was
detailed for the job and flew off 30 dive-bombers and 8 torpedo
planes. At 13.45, shortly after the strike had left, both a Catalina
and an *Enterprise* search plane discovered the *Shokaku* and *Zuikaku*
to the north of *Ryujo*. Some of the *Enterprise* scouting planes
bombed the *Shokaku* without doing much damage and missed
the *Ryujo* with two torpedoes. The *Saratoga's* strike then found
Ryujo and made a devastating attack. Over four 1000-lb bombs
and a torpedo hit and she was disabled and came to a stop badly
on fire but she managed to keep afloat for several hours. Just
before the attack on *Ryujo* a Japanese float-plane from the
cruiser *Chikuma* located the *Saratoga* and *Enterprise*. This was just
what Admiral Nagumo had been waiting for and a heavy strike
was at once flown off both the large Japanese carriers, followed
by a second one an hour later.

The American carriers were in two separate groups ten miles
apart and the Japanese strike concentrated on the *Enterprise*.
Fifty-three defensive fighters were airborne and a very heavy air
battle resulted in which *Enterprise* was hit by three heavy bombs
damaging the flight deck, killing 74 men and setting her badly
on fire. Fortunately for her some of the enemy strike planes
could not resist attacking the *North Carolina* but without scoring
a hit. In an hour the *Enterprise* had got the fires under control
and had started to recover her aircraft but, as in *Lexington* at the
Coral Sea, there was a subsequent petrol explosion, her rudder
jammed and the fires broke out again. Fortunately the second
strike from the Japanese carriers failed to find her and, short of
fuel, had to return. The *Enterprise* however again recovered and
landed-on the rest of her planes except for a dozen or so which
landed at Henderson Field.

A few planes from *Saratoga* which had been ordered into the
air so that they would not be caught on deck during the air

attack, found one of the Japanese groups and considerably damaged the seaplane-carrier *Chitose* but she managed to get home. At sunset, Admiral Fletcher decided to withdraw to fuel and it is just as well that he did. The Japanese believed that they had sunk one carrier and damaged another as well as a battle-ship. Admiral Kondo, commanding the Japanese Fleet, concentrated his surface forces and advanced to force a night action and finish off any cripples, but when he had sighted nothing by midnight he abandoned the attempt and withdrew to the north.

The Japanese transport group continued to advance but next day was heavily attacked by Henderson Field planes and by Flying fortresses from Espiritu Santo (See fig. 18(c), p. 255). The light cruiser *Jintsu* was damaged and had to retire to Truk and the destroyer *Mutsuki* was sunk by the B17's in a high-level bombing attack. The attempt to reinforce Guadalcanal was then abandoned. The battle area was left in possession of the *Wasp* and her group, which, having completed fuelling, had arrived. She flew searches but could find no enemy ships. The Japanese had, in fact, retired to Truk.

So ended the third carrier action, known as the Battle of the Eastern Solomons, in which surface ships again made no contact. Undoubtedly the Americans got the best of it for, although the *Enterprise* was badly damaged, they had only lost seventeen aircraft and this was not such a loss as the sinking of the *Ryujo*. They also prevented the landing of the Japanese troops on Guadalcanal although this was only a temporary advantage and they were put ashore in destroyers a few days later. Nevertheless the result was disappointing for the Americans. They had had the same size of carrier force as at Midway whereas the Japanese had about half. They had full coverage of the whole battle area by flying-boat reconnaissance and felt that they should have done very much better. Admiral Fletcher was understandably nervous after losing *Lexington* and *Yorktown* and knowing that they could ill afford to lose another carrier. The detaching of the *Wasp* to fuel when her group had plenty with which to fight the action was the main reason for his lack of success.

The Japanese were even more understandably nervous after Midway. Their tactics of using a bait certainly succeeded with the result that they struck the American carriers while the *Shokaku* and *Zuikaku* were practically unscathed. But it was poor comfort for them to have scored a tactical success at the cost of the *Ryujo* and probably as many as a hundred planes from their air groups. As an attempt to destroy the American carriers and then drive the surface ships right out of the area and so isolate Guadalcanal, this battle clearly failed.

After the action the *Enterprise* left her air group ashore and made her way to Pearl Harbor for repairs. Shortly afterwards the *Hornet* arrived and with the *Wasp* and *Saratoga* reverted to the task of safeguarding supply and reinforcement convoys to Guadalcanal. This involved a great deal of counter-marching in the same general area and on 27th August, three days after the Battle of the Eastern Solomons, *Saratoga* was torpedoed by the Japanese submarine *I.26*, flooding a boiler room. She came to a stop and had to be towed by a cruiser but she managed to get her air group away to the shore where they were subsequently used at Henderson Field. *Saratoga* limped her way home for the second time since the outbreak of war and was out of action for three months.

The absence of these two carriers, however, did not have as much effect upon the campaign as might have been expected. In this situation carriers were not really exercising command of the sea in the Solomons. By day this was now being done by the aircraft at Henderson Field in the vicinity of Guadalcanal on the one side and by the 25th Air Flotilla in the vicinity of New Britain on the other. In this the air groups of the *Saratoga* and *Enterprise* working from Henderson Field played a considerable part. The *Hornet* and the *Wasp*, kept back to ensure the safety of convoys into the Solomons, were also available should Japanese carriers enter the area, but they were no counter to the 25th Air Flotilla. The distances were short enough for the Japanese cruisers and destroyers by night to leave the northern Solomons in which they had air cover and make forays into the disputed waters around Guadalcanal. During darkness the ships of both sides fought for control of the sea and aircraft had little part in this struggle. The Japanese, with superior night-fighting techniques, at first had the advantage and were able to reinforce Guadalcanal by bringing in troops in destroyers. The aircraft at Henderson Field and still more the American carriers were powerless to stop these excursions; when dawn broke the shore-based aircraft reigned supreme and the Japanese ships took care to be out of the area by then. The American convoys approached by day and were safe under the air protection from Henderson Field but they took care to be clear of the area by sunset. This cycle of the alternate command of the sea by each side every twelve hours could not be broken by air power alone either shore-based or carrier-borne. The Americans had to defeat the Japanese surface forces in a night action or the Japanese army had to capture Henderson Field.

During September and October the struggle ashore on Guadalcanal continued. Both sides brought in reinforcements, the Americans by day and the Japanese by night. The American carriers were still employed guarding the sea communications to the Solomons and on 15th September the aircraft-carrier *Wasp* was sunk by the Japanese submarine *I.19*. Another submarine, the *I.15*, narrowly missed the *Hornet* and succeeded in hitting the battleship *North Carolina* and a destroyer. Both *Enterprise* and *Saratoga* were still under repair and the *Hornet* was now the only American carrier left in the south-west Pacific. With so many Japanese submarines about the defensive role on which she was employed, involving much counter-marching, was clearly a dangerous one and on 5th October she was used to strike at the Japanese base in the Shortlands where transports were reported to be gathering, but failed to achieve anything due to bad weather. On 11th and 12th October the Americans won their first night surface action against the Japanese at the Battle of Cape Esperance. Nevertheless the Japanese continued to reinforce their troops by destroyers and two nights later the battleships *Kongo* and *Haruna*, in an attempt to cut down the American air superiority by day, bombarded Henderson Field. Shortly afterwards the Japanese started to use an airstrip which they had built on the southern end of Bougainville Island from which they could give their bombers fighter support over Henderson Field. On 15th October the Japanese carrier *Zuikaku* joined the fight and struck at the American supply line, sinking the destroyer *Meredith* and the next day the *Hornet* replied by destroying twelve seaplanes at a new base on Santa Isobel Island.

By mid-October, the Japanese Army had been reinforced to the extent that they expected to be able to take Henderson Field in the near future. The Japanese Navy put to sea in force partly to prevent American reinforcements being sent in but mainly to make another attempt to defeat the American carriers in battle. The Japanese Fleet, still under Admiral Kondo, was stronger than at the Battle of the Eastern Solomons; this time they committed 4 battleships and 5 aircraft-carriers with 267 aircraft, which were escorted by 8 heavy and 3 light cruisers and 37 destroyers. The 25th Air Flotilla in the Rabaul area had now been brought up to 220 planes and it was taken over by the headquarters of the 11th Air Fleet. The Japanese Army in fact never succeeded in taking Henderson Field and the Japanese fleet was at sea for some time waiting to the northwards outside the range of the Catalinas. They were split as usual into a number of groups, this time into five. The carriers *Zuikaku*, *Shokaku* and *Zuiho* were in the striking force under Admiral Nagumo and the *Junyo* and *Hiyo* were kept back with Admiral Kondo.

The American forces, now under Admiral Kinkaid, consisted of two carriers, the *Enterprise* having rejoined the *Hornet*, and between them they had 171 aircraft. Their surface warships included the two new battleships *South Dakota* and *Washington* with 4 heavy and 5 light cruisers and 20 destroyers. At Henderson Field there were now 60 aircraft, two-thirds of which were fighters, and some 32 Catalinas at Ndeni and Malaita provided the usual coverage of the area. Admiral Halsey had taken over command in the South Pacific area and he ordered the American carriers to make a sweep north of the Santa Cruz Islands on 25th October while the Japanese Fleet was still marking time some 300 miles north of Guadalcanal. The Japanese had meanwhile been reduced to four carriers when the new *Hiyo* developed engine trouble and had to return to Truk; her air group, however, joined the 11th Air Fleet at Rabaul.

At midday on 25th October one of the Catalinas sighted two Japanese carriers and Admiral Kinkaid ordered *Enterprise* to fly off a search. The Japanese, however, reversed course and they were not sighted. That night a Catalina found them again and narrowly missed the *Zuikaku* with a torpedo and Admiral Kinkaid was ordered by Admiral Halsey to attack. At dawn the carriers on both sides launched searches (see fig. 19(a), p. 262). The American planes sighted Admiral Nagumo's *Zuikaku*, *Shokaku* and *Zuiho* and the Japanese found the *Enterprise* and *Hornet*. The *Enterprises*'s search planes were carrying bombs and one of them hit the *Zuiho* and put her flight-deck out of action but not before all three Japanese carriers had flown off a strike

above: H.M.S. *Centaur*
A Light Fleet Carrier
with an 'interim'
angled deck and Sea
Vixen, Scimitar and
Gannet A.E.W.
aircraft

left: H.M.S. *Albion*
A Commando
Carrier. Royal
Marine Commandos
manning their
helicopters

ATE XI — THE FLEET AIR ARM IN THE SIXTIES
(Ministry of Defence Official Photographs)

PLATE XII –
U.S.S. *Enterprise*
The nuclear propelled 'super' carrier

totalling 65 planes (see Fig. 19(b), p. 262). The American carriers launched their strike some twenty minutes after the Japanese but it was slightly stronger with 73 aircraft in two waves. The opposing strikes actually sighted each other as they passed and some of the Japanese fighters succeeded in shooting down a number of the *Enterprise*'s strike aircraft.

The Japanese strike arrived first and was opposed by 38 fighters. As the attack was coming in the *Enterprise* entered a rain squall and the enemy therefore concentrated on the *Hornet*. In spite of a toll taken by the defending fighters and very heavy anti-aircraft fire, she was hit by three heavy bombs and two torpedoes and was brought to a stop badly on fire and completely out of action. In this vulnerable state she was again attacked and hit by three more bombs. While this was going on, the American air groups had arrived over the Japanese carriers. The dive-bombers scored at least three and probably six hits on the *Shokaku* and put her out of action as a carrier but she managed to limp away. A number of the American strike aircraft were unable to find the Japanese carriers and attacked other Japanese groups seriously damaging the cruiser *Chikuma*. The *Zuikaku* and *Shokaku* had, however, got away a second strike of 44 planes and these attacked the *Enterprise*. The *Enterprise* was hit by three bombs killing 44 men and starting fires but all the Japanese torpedoes were evaded. She was just getting the fires under control, making a remarkable recovery, when a 29 plane strike from the *Junyo*, which was operating beyond the main Japanese carrier force, arrived. Fortunately they only scored a near miss on *Enterprise* but hit the battleship *South Dakota* and the light cruiser *San Juan*. The *Enterprise* was then able to recover her planes and fly off some defensive fighters but she had had enough and turned to withdraw. The *Hornet* was still completely out of action but was being towed by the cruiser *Northampton*.

On the Japanese side neither the *Zuiho* nor the *Shokaku* were fit to operate aircraft and were also retiring from the battle but the *Junyo* and *Zuikaku* were undamaged. The *Junyo* managed to get away another strike of 15 planes and these hit the *Hornet* yet again with a torpedo and a bomb and she had to be abandoned (see fig. 19(c), p. 263). She refused to sink even after two American destroyers had torpedoed her and was left burning from end to end. As night fell, Admiral Kondo, as at the battle of the Eastern Solomons, decided to follow up with surface ships. Just after midnight they came up with the *Hornet*

s

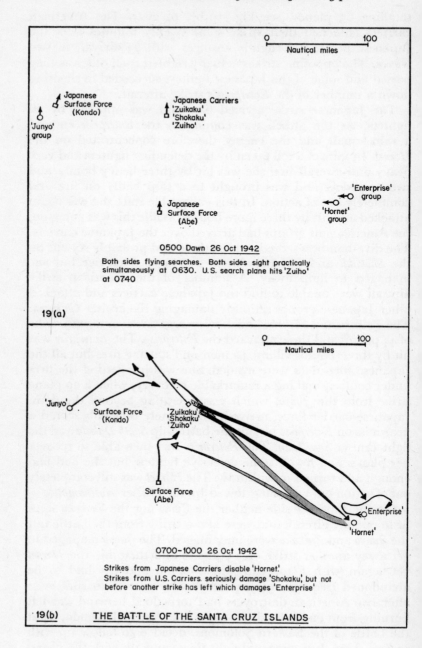

0   100
Nautical miles

Japanese
Surface Force
(Kondo)

Japanese Carriers
'Zuikaku'
'Shokaku'
'Zuiho'

'Junyo'
group

'Enterprise'
group

Japanese
Surface Force
(Abe)

'Hornet'
group

0500 Dawn 26 Oct 1942

Both sides flying searches. Both sides sight practically
simultaneously at 0630. U.S. search plane hits 'Zuiho'
at 0740

19(a)

0   100
Nautical miles

'Junyo'

Surface Force
(Kondo)

'Zuikaku'
'Shokaku'
'Zuiho'

Surface Force
(Abe)

'Enterprise'

'Hornet'

0700-1000 26 Oct 1942

Strikes from Japanese Carriers disable 'Hornet.'
Strikes from U.S. Carriers seriously damage 'Shokaku', but not
before another strike has left which damages 'Enterprise'

19(b)   THE BATTLE OF THE SANTA CRUZ ISLANDS

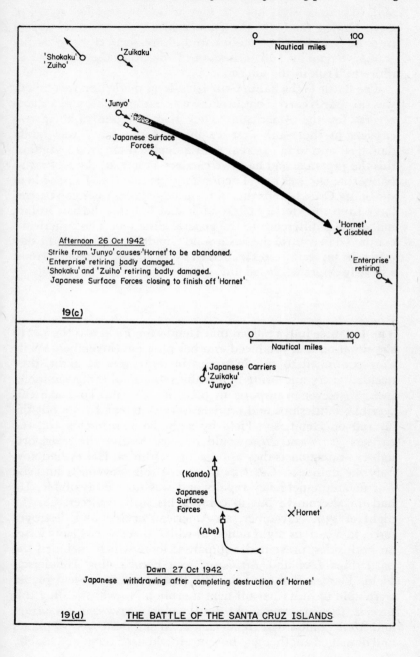

'Shokaku'
'Zuiho'

'Zuikaku'

0   100
Nautical miles

'Junyo'

Japanese Surface
Forces

'Hornet'
X disabled

'Enterprise'
retiring

Afternoon 26 Oct 1942
Strike from 'Junyo' causes 'Hornet' to be abandoned.
'Enterprise' retiring badly damaged.
'Shokaku' and 'Zuiho' retiring badly damaged.
Japanese Surface Forces closing to finish off 'Hornet'

19(c)

0   100
Nautical miles

Japanese Carriers
'Zuikaku'
'Junyo'

(Kondo)

Japanese
Surface
Forces

X 'Hornet'

(Abe)

Dawn 27 Oct 1942
Japanese withdrawing after completing destruction of 'Hornet'

19(d)   THE BATTLE OF THE SANTA CRUZ ISLANDS

and, having finished her off, they withdrew (see fig. 19(d), p. 263). During the night the *Zuikaku* was again narrowly missed by a torpedo from a Catalina and although next morning the Japanese remained in possession of the area, they decided to retire to Truk in the afternoon.

The Battle of the Santa Cruz Islands, as this action was called, was the fourth carrier battle of the war in the Pacific and a sharp reverse for the Americans. They had lost one of their two carriers in the south-west Pacific and the other was badly damaged and in full retreat with a depleted air group. Against this the Japanese had had two carriers damaged; the *Shokaku* so badly that she was out of action for nine months. The Battle of the Santa Cruz Islands therefore dislodged the American carrier force from the vicinity of Guadalcanal but this did not in fact make much difference to the general situation. The Americans continued to control the sea around Guadalcanal during the day using the aircraft from Henderson Field and the Japanese could still only approach by night.

The Japanese fully realized that Henderson Field was the key to the situation and produced another plan for November. While they continued to run in troops in destroyers at night, they decided to try and bring in a really substantial army reinforcement in eleven transports to take the airfield. To make this possible, battleships and cruisers were to try and wipe out the aircraft on Henderson Field by night bombardments and the carriers *Junyo* and *Hiyo* would be used to give the transports fighter protection as they approached. Admiral Halsey had now only the damaged *Enterprise* to oppose this movement and she was making emergency repairs and was not yet available. He had no alternative but to commit his surface forces. On the night of 13th November, an American cruiser and destroyer force, in a furious night action in which there were heavy losses on both sides, prevented a Japanese force which included the battleships *Hiei* and *Kirishima* from bombarding Henderson Field. They so damaged the *Hiei* that Henderson Field planes were able to finish her off next morning. Notwithstanding this success, the very next night the Japanese heavy cruisers *Suzuya* and *Maya* bombarded Henderson Field and destroyed 18 planes and damaged another 32, but the airfield itself remained usable.

Help however, was on the way and the *Enterprise* and the two new American battleships *South Dakota* and *Washington* were approaching. In the morning both the *Enterprise* and the surviving Henderson Field planes were able to fly strikes at the retiring Japanese bombardment force and they were joined by B17's from Espiritu Santo. They sank the cruiser *Kinugasa* and damaged the *Chokai* and *Maya* as well as a light cruiser and a destroyer. At the same time they struck at the eleven Japanese transports which were now approaching with fighter protection from the Japanese carriers. They succeeded in sinking seven transports during the day and that night the *South Dakota* and the *Washington* were in action with another Japanese bombarding force consisting of the battleship *Kirishima* and the heavy cruisers *Atago* and *Takao*. They sank the *Kirishima* and again prevented a bombardment.

The Japanese never succeeded in reinforcing Guadalcanal sufficiently to take Henderson Field and this in spite of a sharp reverse inflicted on the Americans in the night surface Battle of Tassaferonga at the end of November. Fighting continued for two more months and in February the Japanese evacuated Guadalcanal and so ended a bitter campaign started for the sole reason that the United States wished to prevent the Japanese using Henderson Field from which they could sever the sea communications between the U.S.A. and Australia and New Zealand.

There is unfortunately no space in this book to follow the whole Pacific campaign but two actions following the fall of Guadalcanal and illustrating the use of air power over the sea are of special interest. In March 1943 the Japanese decided to move some 7000 men from Rabaul into Lae in New Guinea in eight transports escorted by eight destroyers. The voyage of some 500 miles could not be completed during the hours of darkness and they knew that they would be heavily attacked by shore-based aircraft. The 11th Air Fleet was, however, able to produce strong fighter cover over the whole route and in spite of their experience when trying to reinforce Guadalcanal in November they believed it would get through. The American and Australian Air Forces in New Guinea were strong and now consisted of 207 bombers and 129 fighters. The Japanese convoy was at first shielded by bad weather but later, when it came into the open by day, in spite of a heavy fighter escort sometimes of as many as 40 Zeroes, it lost all eight transports and four of the eight destroyers and some 30–40 planes into the bargain. The Battle of the Bismarck Sea, as it was called, was a naval action but as at Midway and the other carrier actions, no ships made contact. The difference was that there were no carriers present and all the aircraft were shore-based. The Allied attack was made partly by B17 bombers flying at about 5000 feet and partly by medium bombers using mast-head skip bombing. All the attacks had ample fighter escort. This remarkable result was achieved with the loss of two bombers and three fighters. Here were aircraft commanding the sea in no uncertain terms even though the enemy had fighter protection. Furthermore the victory was not won by specialist maritime aircraft but by U.S. Army and Royal Australian Air Force planes.

In April Admiral Yamamoto came to the conclusion that the situation in this area could only be restored by the employment of massive shore-based air forces. He therefore landed 170 aircraft from his four operational carriers to join the 11th Air Fleet. They made very heavy air attacks in the Solomons and New Guinea areas but were heavily opposed by the defending fighters. By the middle of April, Admiral Yamamoto, believing he had destroyed about 175 planes, called off the offensive and re-embarked the carrier-borne planes; in fact he had only shot down about 25 American aircraft and had sunk a destroyer, a tanker and some minor war vessels. The offensive was not sufficiently strong or sustained to gain air superiority. A few days later Admiral Yamamoto himself, while on an inspection tour in the Solomons, was shot down and killed.

The Battles of the Coral Sea and Midway confirmed that carrier-borne aircraft had replaced battleships as the units of sea power. There could be no question that the strategic effect of both of those battles was far-reaching. In the Battles of the Eastern Solomons and the Santa Cruz Islands, although they were actions of the new type in which ships never sighted each other, the strategic effect was comparatively unimportant. In the Guadalcanal campaign, carrier-borne aircraft were not able to gain command of the sea on either side and the curious situation, whereby the American shore-based aircraft were supreme by day and the Japanese ships at night, held good for the whole campaign. Indeed it seemed as if surface warships, even battleships, had made a spectacular 'come-back', at any rate by night. The second naval Battle of Guadalcanal was one of the few actions between capital ships of the whole of the Second World War and the battleship momentarily reassumed its traditional role of being the only counter to the battleship. During this battle both sides held their carriers back in a supporting role and although their aircraft met over the Japanese transports, they made no attempt to strike at each other or indeed at the opposing battleships as they approached the area in daylight.

The reason for this policy seems to have been the shortage of aircraft-carriers on both sides and the fear of losing them. This is very understandable: the aircraft-carrier was proving very vulnerable not only to air but also to submarine attack and in five months the Americans had lost four of them and the Japanese six. After Santa Cruz the Americans were down to the damaged *Enterprise* and the escort carrier *Long Island*, in the whole of the Pacific. The *Saratoga* was still under repair and there was only the *Ranger* left in the Atlantic. The British were not very helpful in responding to American appeals to lend them a carrier and it was not until March 1943 that the *Victorious* arrived in the Pacific and by then the crisis was over: *Saratoga* and *Enterprise* were by then repaired and the escort carriers *Chenango* and *Suwannee* had arrived. Furthermore the *Essex*, first of the new fleet carriers, had commissioned in December 1942 and seven more escort carriers had by then been converted.

The Japanese were a little better off. After Santa Cruz the *Zuiho* was soon repaired and they also had the *Zuikaku*, *Junyo* and *Hiyo*. Nevertheless Admiral Yamomoto did not feel strong enough to challenge the U.S. shore-based air power at Espiritu Santo and Nouméa which were beyond the range of the 25th Air Flotilla. If the Japanese carriers could have gained air superiority over these places they could have isolated Guadalcanal and so have proved decisive. The same applied to American carrier-borne air power. If it was to have been decisive in the Guadalcanal campaign it would have to have been powerful enough, not only to defeat the Japanese carriers in action, but to challenge the 11th Air Fleet at Rabaul. In this way the supremacy of ships at night could have been countered because they would have been driven so far away that they could have been sunk in daylight as they approached.

In spite of the number of night surface battles, air power was the dominating feature of the fighting in the south-west Pacific. In the whole of the campaign aircraft were involved either directly or indirectly and the operations of ships were subsidiary or connected with attempts to obtain air superiority over the sea. For instance the attempt by the Japanese to use battleships at night to bombard Henderson Field had as its aim the securing of air superiority next day. The success of the Battle of the Bismarck Sea and the failure of Yamamoto's air offensive were incidents in the gradual establishment of air superiority in the area by the Americans. The fact that this was done more by shore-based aircraft than carrier-borne was partly because of the shortage of carriers and partly because geography made it possible. Air superiority thus gained led to command of the sea and was what enabled the Allies to win both at Guadalcanal and in New Guinea.

XI

Aircraft at Sea in European Waters

August, 1942 – 45

B Y AUGUST 1942 it had become essential to supply Malta
again. Upon the island's recovery and power to strike at
Rommel's supply lines across the Mediterranean much
depended. This operation was therefore given priority over all
others. With the withdrawal of most of Fliegerkorps II and the
regaining of air superiority over the island by the Spitfires in
July, the British submarines had been able to return, but more
supplies were required for the sustained operation of air and
surface striking forces. With the Eighth Army still right back at
El Alamein there seemed little chance of getting a convoy
through from the east and so it was decided to make a supreme
effort to send no fewer than fourteen ships through from the
west. A powerful escort was assembled, mostly from the Home
Fleet but also by withdrawing the aircraft-carrier *Indomitable*
round the Cape from the Indian Ocean, where the situation
had become much easier as a result of the Japanese defeat at the
Battle of Midway. This time the convoy was to be escorted by a
carrier task force consisting of the *Indomitable*, *Victorious* and
Eagle with 72 Sea Hurricane, Fulmar and Martlet fighters and
28 Albacores. The escort was also to include the battleships
Nelson and *Rodney* with 7 cruisers and 24 destroyers. The 80
Spitfires at Malta, upon which the convoy would depend in the
Sicilian narrows, were to be reinforced by 38 more which were
to be flown off the *Furious*. Malta had also 40 strike aircraft,
mostly Beaufort torpedo-bombers, and 16 reconnaissance aircraft
for this operation.

The Germans and Italians were determined to prevent the

convoy getting through but were handicapped by a severe shortage of oil fuel. They decided, therefore, that they could not use their superior fleet of six battleships and that they would have to rely on aircraft and light naval forces. They reinforced their air forces in Sardinia and Sicily to a strength of some 600 planes, one third of which were German, and allocated 21 German and Italian U-boats and 23 motor-torpedo-boats to the attacking force. A surface group of 6 cruisers and 11 destroyers, for which fuel could be found, was also to stand by in the western basin.

The 'Pedestal' convoy, as it was called, passed through the Straits of Gibraltar on 10th August and next day was located by Axis reconnaissance aircraft. Just as the *Furious* was flying off the Spitfires for Malta, the German submarine *U.73* torpedoed and sank the *Eagle* and she took nearly a quarter of the fighter force down with her. Nevertheless the 56 remaining fighters from the *Victorious* and *Indomitable* were able to get the convoy through the western basin with the loss of only one merchant ship and a destroyer. The air attacks from Sardinia and Sicily were very heavy and totalled some 246 sorties. On 12th August the *Victorious* was hit by a heavy bomb but it failed to penetrate her armoured flight deck. Later the same day, as the convoy reached the Sicilian narrows, the *Indomitable* was attacked by German dive-bombers and suffered three heavy bomb hits which put her flight deck completely out of action.

The main escort of battleships and aircraft-carriers turned back according to plan at the narrows and almost at once the convoy ran into a submarine trap. The cruiser *Cairo* was sunk and the *Nigeria* and a merchant ship were damaged: the *Nigeria* had to turn back for Gibraltar. An air attack by 20 German dive- and torpedo-bombers then came in, but the long range fighters from Malta which were overhead, now had no fighter direction as the *Nigeria* and *Cairo* were the only ships fitted to do this: as a result two more merchant ships were sunk and another was damaged. During the night of the 12th–13th August the convoy was heavily attacked by U-boats and motor-torpedo-boats and, being in some disarray, lost the cruiser *Manchester* and four merchant ships, another merchant ship being damaged. At dawn, aircraft renewed their attacks and sank two more merchant ships and damaged others. In the end five merchant ships, three of which were damaged, the only survivors of the fourteen which had started, reached Malta. It might have been worse, for the powerful Italian cruiser-destroyer force had in fact put to sea. The convoy was saved from this menace because the Italians would not face the threat of the powerful air striking force at Malta without fighter protection and this was not forthcoming as it was required as escort for the Axis bombers.

In getting this convoy through to Malta, the British lost an aircraft-carrier and had a second one badly damaged: they lost two cruisers and had two more damaged; a destroyer was sunk and 19 of their aircraft were shot down. These were grievous losses but on the other hand, with the fuel and stores landed, Malta could now hold out until December. The Axis lost 35 aircraft and two U-boats and two of their cruisers were badly damaged by a British submarine as they returned to base. The Axis U-boats and motor-torpedo-boats could claim a substantial success but in the air the operation was undoubtedly a victory for the British: they had a total of 256 aircraft against some 600. They would have done better still if U-boats had not sunk the *Eagle* and the *Cairo* and damaged the *Nigeria* thus adversely affecting their fighter defence. Nevertheless the employment of a proper carrier task force carrying 100 planes got the convoy as far as the narrows. Thereafter Malta-based planes were able to protect it from the Italian cruiser-destroyer force and against air attack when close to the island and subsequently after it had arrived. However they were not able to do much about the U-boats and motor-torpedo-boats.

In the autumn of 1942, it was of great political importance to start the Russian convoys again but this could not be considered until after the 'Pedestal' convoy operation to Malta. It was finally decided to sail 39 merchant ships in a single convoy early in September. The escort was to be very much stronger than before and was to consist of the anti-aircraft cruiser *Scylla*, 20 destroyers and 11 smaller escorts with 2 other anti-aircraft ships and 2 submarines. For the first time the convoy was to be accompanied by an escort carrier, H.M.S. *Avenger*, with 12 Sea Hurricanes and 3 Swordfish and also by the CAM ship *Empire Morn* with another Hurricane. It was hoped that this strong escort would be able to compete not only with air and submarine attacks but also with a surface attack even by the German heavy ships. To give additional protection 24

Hampden torpedo-bombers of Coastal Command were temporarily based in North Russia with some photo-reconnaissance Spitfires and some Catalina flying-boats. Three cruisers were as usual in support of the convoy but this time the battleship covering force remained behind in Iceland. Extra destroyers were then freed for the convoy escort and this strategy is of great interest. Protection of a convoy threatened by attack by heavy ships was left to shore-based aircraft and a light surface escort with some help from submarines as for the 'Vigorous' convoy in the Mediterranean. In this case a battle-fleet was available and this course was taken as the better solution rather than as an unfortunate necessity.

The convoy sailed from Loch Ewe on 2nd September and followed a route to the west of Iceland through the Denmark Strait and then as far north as the ice permitted and so to Archangel. The Luftwaffe found PQ18 on 8th September north of Iceland and twelve U-boats closed in. The convoy was helped by fog, snow and rain for several days but on 13th, the air attacks began at a distance of 450 miles from the Norwegian bases. Over the next six days, the 225 German strike aircraft based in North Norway flew 337 sorties against the convoy and sank ten ships, and U-boats accounted for three more. Most of the casualties were caused on 13th when forty torpedo planes attacked *en masse* and practically annihilated the two starboard columns of the convoy. At the time the *Avenger*'s fighters were busy with some high bombers but the escort's guns shot down five of the torpedo-bombers. Two similar attacks next day were opposed by 6 and 10 Hurricanes respectively: 22 German aircraft were destroyed and only one merchant ship was sunk. The Catalinas from Russia were able to escort the convoy against U-boats for the last four days. The U-boats were made to pay dearly for their successes and lost three of their number, all sunk by ships, but one with the help of a Swordfish from the *Avenger*. The air escort was also able to keep the U-boats down and can claim much of the credit for the light casualties from them.

The casualties in PQ18 were heavier even than in PQ16, amounting to a third of the convoy. Nevertheless the Luftwaffe lost 41 aircraft altogether which represented 18 per cent of the force engaged. The operation was an undoubted success for the British, who had re-opened the route to North Russia. It was largely due to the strong escort, to the fact that the convoy stayed together and to the presence of the *Avenger* and her aircraft. The great difference in air strength, 225 planes to 16 was to a large extent offset by the distance of the German air bases. After the passage of PQ18, North Russian convoys had to be suspended until the end of December because of the pre-occupation of the Home Fleet with the landings in North Africa. The North African operation, however, soon required most of the German strike aircraft in North Norway to move south as well and subsequent Russian convoys were harried more by surface ships and to a certain extent by U-boats than from the air.

We must now return to events in the Mediterranean where, after the relief of Malta in August, the disruption of Axis supplies to North Africa became of paramount importance. The retreat of the Eighth Army to El Alamein had made it difficult to attack this traffic from the east and only the few Liberators available could even reach Benghazi. Nevertheless attacks by Beauforts, Beaufighters and Wellingtons continued as best they could. With the arrival of the 'Pedestal' convoy, with the departure of Fliegerkorps II and the regaining of air superiority over Malta, the island rapidly recovered as a base for offensive operations. In consequence, between July and November, the assault on Rommel's supplies gradually yielded better results. In July 7 ships were sunk, 4 by aircraft, and in October 29 ships were disposed of, 11 by aircraft. Nearly every Axis convoy now had to fight its way through to North Africa and most of them suffered more than one air attack. Rommel's failure at Alam Halfa and to push through to Cairo was in large measure due to these attacks, which were by submarines as well as aircraft, on his communications. Nevertheless the air forces were never in sufficient strength to cut the supply lines altogether.

On 23rd October, General Montgomery attacked at El Alamein and on 8th November the Allies landed in French North Africa and the whole strategic situation rapidly changed. The Allied landings were supported by very large carrier forces: the British with the *Victorious, Formidable, Furious, Argus, Avenger, Biter* and *Dasher* and the Americans with the *Ranger, Sangamon, Chenango, Suwannee* and *Santee*. This was the first large-scale use of escort carriers, seven of which were present. These carrier forces provided some 400 aircraft to support the assault and protect the landings from air and submarine attack. This carrier-borne air support was to become a feature of most of the subsequent amphibious operations in the Mediterranean.

During the Eighth Army's advance to Tripoli, which it captured on 23rd January, attacks on Rommel's sea communications continued, sinking nearly half of the traffic. This was a substantial contribution to General Montgomery's success, but meanwhile Axis forces landed in Tunisia by sea and air. In December, 28 ships, 13 by air attack, were sunk on the Tunisian route but this was not enough to prevent General von Arnim from establishing himself. The Axis invasion of Tunisia emphasized that the Allies did not command the sea in the central Mediterranean as yet. Nevertheless the failure to prevent the landing was compensated for to a great extent by the final relief of Malta. On the same day that the Eighth Army recaptured Benghazi, a convoy arrived from the east without loss.

The Tunisian campaign lasted until May and the Axis sea communications were attacked heavily from both east and west during the whole period by both air and sea. Large offensive air forces were used which became more effective as the Allies advanced and established airfields nearer to the Sicilian narrows. In January and February a quarter of the supplies were sunk and in March, April and May the sinkings rose from under a half to over three quarters of the traffic. Of the hundred convoys which had sailed for Tunisia, half were attacked by submarines which sank 33 ships but nearly all were attacked by aircraft, often more than once, which sank 71 ships. In this period 48 per cent of the casualties were caused by aircraft, 29 per cent by submarine, 21 per cent by mines or capture in port or other causes and 2 per cent by surface ships. Well over half the ships sunk by aircraft were destroyed in harbour by bombing. There can be no doubt that this throttling of the supply lines of the Axis armies in Tunisia, in which aircraft played the major part, was a great contribution to their defeat. Tunis fell on 13th May and the Allies were able to prevent an evacuation so that the surrender of the Axis armies was complete.

On 17th May, the Allies were able to start convoys right through the Mediterranean. This traffic no longer needed full fleet operations to ensure its safety and took the form of normal merchant convoys with air and surface anti-submarine escort covered by fighters from the shore. At this point Great Britain and her Allies could claim that they had regained command of the sea in the Mediterranean. It had been done not by battle-fleet supremacy as in the past, or even by carrier task forces as used by the Americans in the Pacific, but by capturing the whole north coast of Africa. Here air bases could be established from which to gain air superiority over the fleets and convoys and so make their operations possible. It has to be admitted that the enemy's traffic across the Mediterranean was not finally stopped until they had lost North Africa and so did not need it. Hereafter the command of the sea in the Mediterranean ceased to be in dispute. Before the surrender of the Italian Fleet and indeed to a certain extent afterwards, precautions had to be taken but the Mediterranean was effectively in Allied hands for the rest of the war.

Throughout the three years of the struggle for command of the Mediterranean, the most important element was air power. This is not to say that submarines, battle-fleets and other forces did not exert any influence. It could for instance be asserted that it was the British battle-fleet which enabled the Malta convoys to be run and that when there was no battle-fleet in the eastern basin the convoys had to be stopped. Certainly the Italian battle-fleet can claim the credit for turning back the 'Vigorous' convoy which, in their turn, the British can claim would have got through if it had had battleship escort. Yet one cannot but be struck by the fact that the disablement of the Italian battle-fleet at Taranto and subsequently of the British battle-fleet in Alexandria at the end of 1941 made very little difference to the general strategic position. On the other hand the Luftwaffe, after the capture of Crete, prevented the British battle-fleet from escorting convoys to Malta and virtually blockaded it in the eastern basin.

If the British battle-fleet was to have truly commanded the sea it would have had to be able to stop the Axis traffic to Libya and to cover mercantile traffic through the Mediterranean. But to do this it would have needed air cover able to compete with first line shore-based air strength not only over itself but also over the merchant traffic. There is little doubt that in the event the battle-fleets were of secondary importance: for both sides the dominating factors in the war at sea in the Mediterranean were the strength of the air threat and the strength of the air cover available to counter this threat.

The aircraft-carrier did, of course, in theory provide an alternative way to gain air superiority over the sea even in the Mediterranean. Before the North African landings took place, the British committed eight aircraft-carriers at one time or another to the Mediterranean theatre. Two were sunk by U-boats and three were so badly damaged by air attack that they had to leave the station for extensive repairs. They achieved a great deal, notably by disabling half the Italian battle-fleet at Taranto and by making it possible for the British battle-fleet to come up with elements of the Italian fleet at Matapan. Their fighters contributed substantially to the protection of a number of Malta convoys and they flew a total of 670 R.A.F. fighters to Malta. They were never able, however, to compete with the Luftwaffe when it was in strength, and even when a task force of three of them was available for the 'Pedestal' convoy, only one returned undamaged to Gibraltar. If there had been more of them with larger complements of aircraft with higher perform-ance they could, of course, have done better; but something far superior in numbers and quality than could be provided by the British Fleet Air Arm was needed to gain full command of the sea. It is practically certain that this could be done more economically in this theatre by capturing territory on which to establish shore-based air forces, a system invented by the Ger-mans in 1941 and copied by the Allies with success in 1943.

Before finally leaving the Mediterranean, mention must be made of the German use of two forms of guided weapon from aircraft which, among other ships, sank the brand new Italian battleship *Roma*. The three battleships *Roma*, *Vittorio Veneto* and *Italia* (ex *Littorio*) with six cruisers and eight destroyers left La Spezia early on the morning of 9th September 1943 to surrender to the British. They followed a route down the west coast of Corsica and altered course to pass through the Straits of Bonifacio. The fleet was then attacked by eleven German Do217 aircraft of Kampfgeschwader 100 which were based near Marseilles. This was a special unit armed with the new FX1400 armour-piercing guided weapon. The FX1400 was a heavy free-falling bomb of 3300 lbs: it was designed to be dropped from high altitude but the aim could be corrected on the way down by radio from the bombing aircraft. The Italian fleet had no fighter cover and two of these bombs hit the *Roma* and one the

Italia. The *Italia* was struck forward and although 800 tons of water flooded in, she was able to reach Malta. The *Roma* was hit near the mainmast and also on the port side abreast the bridge. Both bombs penetrated the armoured deck, they damaged her machinery and she had to stop, and they set alight uncontrollable fires which soon reached the magazines with the result that she blew up and sank.

The FX1400 was used by the same German unit against the Salerno landings a few days later. The American cruiser *Philadelphia* was damaged by a near miss and then the *Savannah* was hit on a turret, the bomb passing right through the ship and blowing a large hole in the bottom. She lost all power and was very nearly destroyed but with the help of salvage tugs she was able to reach Malta. The same fate befell the British cruiser *Uganda.* Then five days later the battleship *Warspite* was also hit by three FX1400 bombs: the first penetrated the armour deck and blew out her bottom under a boiler room, letting in 5000 tons of water, the other two were near misses which ruptured the side plating. She was completely disabled and had to be towed back to Malta.

The Germans also produced a glider guided weapon, the HS293, which had a warhead of 1100 lbs but was not capable of piercing armour. It was used with effect against unarmoured ships and also sank the cruiser *Spartan* off Anzio. In the FX1400 and the HS293 the Germans had clearly developed extremely effective air-launched anti-ship weapons. It was fortunate that they did not produce them earlier. As it was they came at a time of great Allied air and technical superiority and they were countered mainly by shooting down the launching aircraft with fighters but partly by jamming or confusing the radio control system. The FX1400 and the HS293 never therefore became a serious menace. Nevertheless, their introduction is of great interest as it was the dawn of the guided missile era in naval warfare.

By midsummer 1942, the organization of proper anti-submarine measures on the American coast, especially the institution of convoy, had made U-boat operations there no more profitable than in the central Atlantic. It was clearly sense for the Germans to resume operations much closer to their own bases. Early in

August, therefore, they returned to attack the Atlantic convoys and now had some 150 U-boats for the task. With the first attack on 5th August on SC94, the pattern became clear (see Fig. 20, p. 279). The U-boats intended to exploit the 'Greenland Air Gap' to the full and they sank eleven ships in it from this convoy alone. It is of interest that after Liberators arrived as escorts from Northern Ireland some 800 miles away and Catalinas from Iceland, there were no more sinkings. The U-boats also exploited the 'Azores Air Gap' further south and in mid August SL118 lost three ships before a Liberator arrived from Cornwall 780 miles away, after which there was only one more casualty. Coastal Command, however, still only had No. 120 Squadron with sixteen very-long-range Liberators. The majority of maritime aircraft could only escort convoys out to 450 miles if they were to stay with them for any length of time. The Germans decided to use about a third of their U-boats to attack in the remoter areas such as the Caribbean, off Brazil and in the South Atlantic where they were often able to sink independent ships out of range of shore-based aircraft.

In November the Battle of the Atlantic was interrupted by the North African landings but in the period from August to December 1942, the U-boats were able to sink a monthly average of a hundred ships totalling half a million tons for the loss of ten U-boats. Of the 55 U-boats sunk, aircraft had disposed of 29, ships 18 and miscellaneous causes, including mines and submarines, 8. For the first time aircraft had sunk more than ships and had scored well over double the sinkings they had achieved in the first half of the year. The reason was that the combination of radar and the torpex depth charge, using the Leigh light at night, was proving a substantial U-boat killer. Eighteen of the U-boats sunk by aircraft had been destroyed in the vicinity of convoys and eleven by independent patrols such as those in the Bay of Biscay and north of Scotland. In October the effectiveness of aircraft was much decreased by the fitting in U-boats of a search receiver which gave warning of the approach of radar-fitted aircraft and allowed them time to dive. The number of U-boats being sunk was considerably less than the number being built and during 1942 they had destroyed the huge total of $7\frac{3}{4}$ million tons of shipping; Admiral Dönitz could therefore still view the situation with some optimism.

In January the menace of the U-boats was considered to be so great that Bomber Command was ordered to make them a

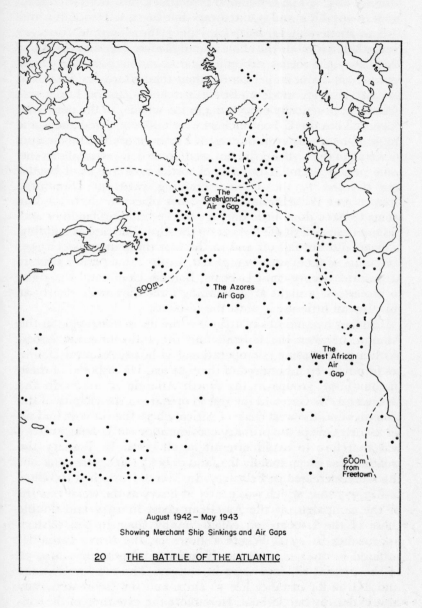

The
Greenland
Air Gap

600m

The Azores
Air Gap

The
West African
Air
Gap

600m
from
Freetown

August 1942 – May 1943

Showing Merchant Ship Sinkings and Air Gaps

20   THE BATTLE OF THE ATLANTIC

primary target. The Command protested that it was unlikely to achieve anything and pointed out that there were bomb-proof U-boat shelters at Lorient, La Pallice, Brest and St Nazaire. Nevertheless it made the effort and in the first four months of the year flew 3,568 sorties, dropping over 9000 tons of bombs which represented about 20 per cent of their total effort. Raids of over 400 aircraft were made on St Nazaire and again on Lorient in February; practically everything in the vicinity of the ports was destroyed but the U-boat shelters were not penetrated and not a single U-boat was even damaged. Furthermore the repair and servicing of the U-boats did not seem to be delayed at all. In the same period a similar number of sorties and weight of bombs was devoted to the U-boat building yards at Hamburg, Bremen and Wilhelmshaven and other places in Germany but seems to have done little damage to the building facilities and did not destroy a single U-boat on the slips. The rate of building U-boats did not fall off and in the first three months of 1943, sixty-nine of them were completed. In the same period aircraft at sea sank twenty-eight U-boats; Bomber Command was right to protest: as a night area-bombing force they were clearly at this time of little use against the U-boats.

The first five months of 1943 saw the decisive struggle in the Atlantic between the U-boats and the anti-submarine forces. With a fleet of some 200 operational U-boats, Admiral Dönitz could now keep a hundred of them at sea. He disposed of these in four large groups in the North Atlantic to attack in the Greenland Air Gap and the rest to operate in the vicinity of the Canaries and the west coast of Africa where the convoys had no air escort. One of the primary considerations of U-boat strategy was therefore to avoid aircraft at all costs. In January the weather was exceptionally bad and only 37 ships were sunk but the score increased in February. In March it reached 108 ships of 627,377 tons which was nearly as heavy as the worst months of the campaign off the American coast in 1942 and double those of the U-boats' so called 'happy time' in the western approaches in 1941. Whereas, however, the heavy casualties suffered in 1941–2 were mostly among independent ships, 72 of the present sinkings were in convoy. The convoys HX229 and SC122 for instance lost 21 ships and the escort were only able to destroy one U-boat. Heretofore the extension of the convoy system had always proved the answer to the U-boat but now it seemed to be failing as a strategy. There was, however, a ray of hope: only two of the ships sunk in convoy had been destroyed while air escort was present. Aircraft continued to be the cause of half the U-boat casualties and in the first quarter of the year sank 21 U-boats out of the total of 40 destroyed whereas ships were only responsible for 14. Clearly it was of exceptional importance to close the 'air gaps' and provide air escort for all convoys over the whole of their routes.

In March the means to do this began to become available. At the Casablanca Conference, the Chiefs of Staff had recommended that 80 very-long-range aircraft be provided for the 'Greenland Gap' and reinforcements began to arrive. Yet there were still no more than 30 V.L.R. Liberators available and a similar number of aircraft fitted with the Leigh light. In this same month, the escort carriers began to arrive: U.S.S. *Bogue* accompanied convoy SC123 and H.M. Ships *Biter*, *Dasher* and *Archer* joined the Western Approaches command. At the same time, the new 10 cm radar sets, which the search receivers of the U-boats could not pick up, began to be fitted in Coastal Command aircraft.

In April these measures, coupled with the use of support groups of ships which reinforced the escorts of threatened convoys, began to show results. Shipping casualties fell to half those of March and the rate of sinking U-boats was maintained. A V.L.R. Liberator sank *U.189* and an aircraft from the *Biter* assisted in the destruction of *U.203*. In May merchant ship casualties fell again and U-boat sinkings suddenly soared to 41, the highest of any month of the war. Of these, 20 were the work of aircraft, 12 of ships, and another 5 were shared between them. Twenty-seven of these U-boats were sunk in the vicinity of convoys by the air and surface escorts. In the middle of the month convoy SC130 when attacked lost no ships and two U-boats were sunk by V.L.R. Liberators, one by a Hudson and two by ships. The *Biter*, *Bogue* and *Archer* all escorted convoys: *U.569* and *U.752* were sunk by their aircraft and *U.89* was destroyed in co-operation with ships. These aircraft used rockets with success to attack U-boats caught on the surface. Ten more U-boats were sunk by air patrols, four in the Bay of Biscay and one off the Faroe Islands. The total number of U-boats sunk in May was far greater than the number being built and, with the fall in merchant ship sinkings, the U-boat campaign suddenly collapsed: on 22nd May, Admiral Dönitz conceded defeat and withdrew the U-boats from the North Atlantic. In fact, although the U-boats returned to the attack on a number of occasions, they were never again a serious menace and the 22nd May 1943 is now recognized as the date of victory in the Atlantic.

It would be wrong to claim that the defeat of the U-boats in the Battle of the Atlantic was entirely due to aircraft. There were many reasons: the huge American shipbuilding effort; the technical and tactical superiority of the Allies; radio warfare; intelligence; the endurance of the merchant ships' crews and the dogged determination of the surface escorts, to mention but a few. Nevertheless the development of aircraft from mere harassing agents in areas near the coast into lethal U-boat killers able to range right across the Atlantic was a major contribution. In the first five months of 1943, they sank 51 U-boats to 33 by ships and 12 by miscellaneous means. The contribution of aircraft was in fact greater even than these figures suggest. They denied the U-boats their surface mobility so that many never made contact with the convoys in consequence.

The air forces involved were very large: in February in all areas in the Atlantic, the Allies deployed 1120 machines. The greater part of these were still only of medium range and the U-boats were able to operate outside their radius of action. Twenty-eight U-boats were sunk by air escorts of convoys and 23 by air patrols on their passage routes or in areas not directly concerned with the passage of a convoy. It is difficult to draw firm conclusions from these figures but it is worth noticing that when on passage U-boats could give their whole attention to the problem of evading attack, whereas in the vicinity of convoys they were busy and had to make every effort to stay on the surface to use their mobility and were therefore more vulnerable. On the other hand, since most of the aircraft were medium-range the bulk of them could not be employed to escort convoys in the areas chosen by the U-boats for attack, whereas nearly all aircraft even of medium range could be used to patrol some part of their passage routes.

The need for very-long-range aircraft to close the air gaps was clearly of the first importance. There were ample aircraft of the right type being manufactured but nearly all of them were destined to be used as strategic bombers. At this time, strategic bombing was the only way to hit back at Germany at all and one cannot but have sympathy for the 'bomber' school. Furthermore if it had not been for bomber requirements, there would not have been any V.L.R. aircraft at all, for neither

Coastal Command nor the Admiralty had foreseen the need for such aircraft in the maritime role before the war. Nevertheless, the need to transfer V.L.R. aircraft from bombing to anti-submarine work was clearly vital if the U-boat was to be defeated. There was no doubt that they could contribute far more to the war at sea in this way than by attacking U-boat bases and building yards.

Although in May Admiral Dönitz had withdrawn from the North Atlantic, he still intended to attack in remote areas where the opposition was not so strong. He was, however, already having considerable difficulty in getting the U-boats in and out of their bases in the Bay of Biscay. The main reason for this was the new 10 cm radar which was being fitted in Allied aircraft and which could not be detected by the U-boat search receivers. By the end of April a number of U-boats had been attacked without picking up anything on their search receivers and Admiral Dönitz again ordered them to proceed submerged by night and only to surface by day when they would at least have a chance to sight approaching aircraft and dive. Coastal Command contacts at once fell off by night but naturally increased by day and during the month six U-boats were sunk. It was now taking the U-boats ten days to traverse the Bay and at the end of May they were ordered to press on by day on the surface and fight it out with aircraft with their anti-aircraft guns. In June, Coastal Command were able to reinforce the Bay patrols because of the lack of attacks on convoys in the Atlantic. The Germans, however, began to use Ju.88 fighters against them and Beaufighters and Mosquitos had to be brought to the area. At the same time, British surface escorts worked in the outer Bay to co-operate with the air patrols.

In July the Bay Offensive reached its climax. Thirty-one U-boats crossed the Bay undetected but 55 were sighted, 20 of which were sunk and 7 others badly damaged. All except one were victims of air attack either in the Bay itself or off the coast of Portugal. They succeeded in bringing down fourteen aircraft with their guns but at the end of the month a whole group of three U-boats was annihilated by a combined attack by aircraft and a surface support group. Early in August, Admiral Dönitz cancelled all group sailings and ordered the U-boats to creep along the Spanish coast and to surface at night only long enough to charge their batteries. As a result only six U-boats were sunk in the Bay in August. The attempt by the U-boats to stay on the surface and fight was foredoomed to failure. In spite of heavy anti-aircraft armaments which now generally consisted of two quadruple 20 mm mountings and one 37 mm single mounting, U-boats could never be expected to compete with air attacks of the strength that Coastal Command was able to mount.

But the Bay of Biscay was not the only place where the U-boats suffered defeat during the summer of 1943. With the move of the Atlantic U-boats west of the Azores, the American convoys from New York to Gibraltar were threatened. The escort carriers *Bogue*, *Card*, *Core*, *Santee* and *Croatan* were sent to provide air cover. They began by giving direct air escort to convoys but before long began to range further afield in search of U-boats. Their general area of operations was in fact where the U-boats working in distant areas fuelled from the U-tankers. Over a period of three months they sank no fewer than sixteen U-boats, half of which were U-tankers, for the loss of only three aircraft. The effect was that the U-boat operations in distant areas collapsed for lack of fuel: many U-boats had to return prematurely and others in their way out had to be used as emergency U-tankers.

The defeat of the U-boats in the Atlantic in May had been a victory for the air and surface escorts of the convoys. This was followed by a secondary victory in the Bay and off the Azores, gained almost entirely by air patrols. Seventy-nine U-boats were destroyed in June, July and August, of which 59 were sunk by aircraft to 15 by ships, and all of these except 9 were by air patrols rather than air escorts of convoys. The U-boats were as near to being blockaded in their bases as at any time in either world war and even their distant operations were inhibited. Twenty-one of the U-boat casualties were in the Bay of Biscay and were caused by rocket and depth charge attacks by Liberators, Halifaxes, Sunderlands, Catalinas and Wellingtons by day and night. An equal success was obtained by the American escort carriers which operated offensively as 'Hunter-killer' groups, oddly enough in the way the British had found so disastrous in the early part of the war. None of the American escort carriers was, however, even attacked by a U-boat and in any case they were cheap and mass-produced and it would not have been a disaster comparable to that of the *Courageous* if some of them had been sunk.

In September 1943, the U-boats had a new search receiver able to pick up the transmissions of the 10 cm radar. With this and a new acoustic homing torpedo they decided to return to the attack in the Atlantic. The plan was to evade approaching aircraft by diving as soon as their radar transmissions were detected, but, if it was essential to stay on the surface, to use their heavy anti-aircraft armament. The new search receiver certainly reduced their casualties from aircraft when on passage but it could not restore their mobility in the vicinity of convoys. If they stayed on the surface to fight it out they were likely to be sunk or at least damaged and if they stayed submerged the convoy would pass out of range in safety. In the seven months from September 1943 to March 1944 they lost another 120 U-boats, 60 of which were sunk by aircraft, 44 by ships and the rest by accident or miscellaneous causes. Sixteen of these U-boats were sunk by aircraft from the escort carriers and the greatest number of casualties was again in the vicinity of convoys. The U-boats completely failed to make a come-back and in the whole of the seven months sank only a little more than in the single month of March 1943.

Admiral Dönitz had, however, been experimenting with a more fundamental solution to his problems in the form of the schnorkel tube which allowed the U-boats to use their diesel engines submerged. With this invention it would be possible to make a passage without surfacing at all. The schnorkel was too small to be picked up by the wartime radar sets but it could still be seen as it left a wash like a periscope and was apt to emit smoke. The schnorkel began to appear operationally in the Spring of 1944, and in June the Germans decided not to use U-boats for operations unless they were fitted with it. In the last nine months of the war in Europe the U-boat campaign took an entirely different form. The U-boats soon found that, although the schnorkel gave them considerable protection from aircraft, they had not regained their surface mobility. There was therefore no chance of intercepting convoys in the open ocean but they found that they were able to return to coastal waters where they had a greater chance to find targets than in mid-Atlantic. Aircraft still made occasional kills, but in general were reduced to the passive role of preventing the U-boats surfacing to use their mobility. This is illustrated by the fact that from August 1944 to May 1945, aircraft at sea sank only 30 U-boats whereas ships destroyed 69. In the same period the U-boats were not able to sink more than 81 ships of under half a million tons.

The total U-boat losses in this period were very much higher than the above totals, 248 being in fact destroyed of which another 83 fell to aircraft. Twenty-seven were sunk by air strikes on U-boats in the Skagerrak and Kattegat right at the end of the war and 56 were sunk in harbour by strategic bombing. A third of these were destroyed by Bomber Command and two-thirds by the U.S. Air Force in daylight raids. Air minelaying was also responsible for 12 of the U-boats sunk and these successes were by no means the only contribution of strategic air forces to the undersea war.

In 1943 the Germans had decided to mass produce two new types of U-boat with high underwater speed which might well have restored their potency. The original plan was to commission prototypes in the Spring of 1944 and to have the types operational by the autumn. By producing forty of them a month they hoped to have over 500 by the Spring of 1945. In fact, by the end of the war, only seven of the new type were available for operations although another 181 had been completed. The German failure to meet their original programme was partly because it was too ambitious but also because of Allied bombing and mining. It is estimated that bombing prevented the completion of some eighty of the new U-boats in addition to the seventeen actually destroyed after completion. These results were obtained by widespread attacks on factories, communications and sources of power, as well as by direct attacks on the building yards which practically stopped production by April 1945. The mining campaign forced the abandonment of the trials area off Dantzig in January 1945 and made it very difficult to move the U-boats from port to port or for them to do any training. The strategic air commands can therefore claim that they did much to prevent the new U-boats coming into service as well as being the largest single killer of U-boats during the last nine months of the war.

At the end of the Second World War, aircraft could certainly claim that they had been the greatest destroyer of U-boats; of the grand total of 785 German submarines sunk, they had been responsible for the destruction of 368 against 255 by ships and another 48 shared between them. Allied submarines sank 21, whilst accidents, scuttling and unknown causes accounted for the remaining 93. It must be remembered that aircraft had virtually no success until the introduction of radar, the Leigh light and torpex depth charges in the summer of 1942 and until then asdic-fitted ships had caused nearly all the losses. The period of the great success of aircraft was during 1943 when they drove the U-boats from the surface of the sea and consequently were the main cause of their defeat. The massacre of U-boats in this period was, however, largely due to the German persistence with surface tactics long after air attack had rendered them practically suicidal. Strategic bombing of the U-boat bases and building facilities proved useless until the middle of 1944 and this form of attack can claim no credit for the winning of the Battle of the Atlantic; the bombers can claim, nevertheless, that they did much to prevent the new type U-boats from taking part in the war, a type against which aircraft at sea would have been of limited value. With the introduction of schnorkel in the spring of 1944 aircraft found their potency against submarines was greatly decreased, yet they were still able to prevent them returning to the surface tactics which were the secret of the U-boat's success in the Second World War.

The U-boats had proved by far the greatest destroyers of Allied commerce during the war and had accounted for 68 per cent of the tonnage sunk, whereas the figure for sinking by aircraft was 13 per cent, by surface warships 7 per cent and by mines $6\frac{1}{2}$ per cent. Three surface merchant raiders had put to sea from the Bay of Biscay along the Spanish coast during 1942 but they achieved little. With the extension of air patrols across the Atlantic in the middle of 1943 and of air cover in all the distant areas, any further forays were out of the question and U-boats with increased endurance now replaced surface raiders in the distant areas. Allied shipping losses from aircraft were never serious after 1942 mainly because the Luftwaffe was busy in Russia and in defending the homeland. All the same Allied defending aircraft must be given due credit too.

The British air campaign against German shipping continued up to the end of the war both in direct attack by strike wings of Coastal Command and in minelaying by Bomber Command. In conjunction with the operations of Allied submarines and coastal forces, they had by the end of the war virtually destroyed the German merchant marine. British aircraft laid over 48,000 mines in German waters during the war and sank over 600 merchant ships and 270 warships, mostly of small size, and they forced the enemy to maintain a huge mine-sweeping organization. Direct attacks made by Mosquitos and Beau-fighters in 1944 proved much more successful than before and right at the end actually did better than minelaying; yet they sank fewer than 300 ships and although these were of a greater average tonnage than those sunk by mines, their losses in aircraft were much heavier. Allied submarines, which were never in very great strength in Home waters, sank less than half the tonnage destroyed by direct air attacks, and coastal forces less still. Aircraft were therefore mainly responsible for the destruction of German seaborne merchant traffic and it is of interest that this was a task which ships could not do by themselves in either world war.

In April 1944 the British Fleet Air Arm had expanded to an operational strength of 993 aircraft. They now had five fleet aircraft carriers with two more nearing completion and fourteen of a new light fleet type under construction. They had some fifteen escort carriers in commission and many more building in the United States. There was a new generation of carrier-borne aircraft in service. The Swordfish and Albacore had been replaced during 1943 by the Fairey Barracuda, a monoplane strike aircraft with an engine of 1640 H.P. This aircraft was a great improvement and was a dive- as well as a torpedo-bomber which could also be used for reconnaissance, spotting and for anti-submarine patrols. Nevertheless it had suffered interminable delays during development and when it emerged was inferior in every way to its American counterpart, the Grumman Avenger. The Seafire, a naval edition of the Spitfire with folding wings, had replaced the Sea Hurricane during 1943 as the single-seat fighter. It had, except for its endurance, an excellent performance in the air but was not a great success for operation from carriers. Finally the Fairey Firefly, a two-seater fighter-reconnaissance plane had replaced the Fulmar with an all-round improvement in performance. During 1943, however, the Fleet Air Arm had taken delivery of a large number of American aircraft. They already had a number of Grumman Wildcats or Martlets as they called them, and to these were added Chance Vought Corsairs, Grumman Hellcats and Grumman Avengers. These types were superior to their British opposite numbers not only in performance but for

operation from carriers. In April 1944, 44 per cent of the Fleet
Air Arm was equipped with them.

The German heavy ships based in North Norway constituted
a constant threat to the convoys to North Russia. They were a
'fleet in being' which tied down battleships and aircraft-carriers
in the Home Fleet which were urgently needed in the Far East.
The situation was improved when *Tirpitz* was put out of action
by a British midget submarine attack in September 1943 and
when the *Scharnhorst* was sunk by the ships of the Home Fleet in
December. By April, however, the repairs to the *Tirpitz* were
nearing completion and she again became a threat. It was
therefore decided to make a heavy strike by carrier-borne
aircraft in the hope of putting her out of action for good. The
Tirpitz was closely surrounded by anti-torpedo nets so a torpedo-
bomber attack was not possible. The actual attack was therefore
to be made by 42 Barracudas by dive-bombing, 10 of them
carrying 1600-lb armour-piercing bombs. To compete with the
expected heavy fighter opposition, an escort of eighty Corsair,
Hellcat, Seafire and Wildcat fighters was provided. This strong
force of modern aircraft, six times the size of that used at
Taranto, was carried to the scene of action in the fleet carriers
Victorious and *Furious* and the escort-carriers *Searcher, Pursuer,
Emperor* and *Fencer*. They were flown off in two waves at dawn on
3rd April, 120 miles from the target. Their attack was a
complete surprise and the *Tirpitz*, which was in fact just
getting under way to do trials, was struck by fourteen bombs,
four of the 1600-lb armour-piercing type and the rest of 500-lbs.
No fighter opposition was encountered and only four aircraft
were lost.

The *Tirpitz* was severely damaged; she had 438 casualties
and was put out of action for three months. That she was still
afloat was very disappointing and the great skill with which the
attack had been made deserved a better result. The 500-lb
bombs were quite unable to do lethal damage and the 1600-lb
bombs failed to penetrate the armour deck probably because, in
a desire to hit at all costs, they had been dropped from too low
an altitude. The Admiralty were determined to try again but
bad weather on the next attempt on 24th April made an attack
impossible. Throughout the summer another seven attempts
were made but four were abortive due to weather and two
because of the shrouding of the target in dense smoke screens.
Two more bomb hits were obtained on 24th August by a strike

of 33 Barracudas escorted by 44 Corsairs, Hellcats and Fireflies from the *Indefatigable, Furious,* and *Formidable*. This time a 1600-lb bomb hit and penetrated the armoured deck but failed to explode.

It was now clear that, as it was unable to use torpedoes, the Fleet Air Arm had no weapon capable of destroying the *Tirpitz*. More damage had in fact been done by two midget submarines in a single attack. This series of carrier-borne air strikes are, however, of great interest. The *Tirpitz* was, in fact, defended more by the weather than anything else although the smoke screens and torpedo-nets played a large part; certainly she was not saved by the Luftwaffe and the British carrier task force operated close off the enemy coast with impunity. The loss of aircraft in all the attacks totalled no more than seventeen and the carrier force was never even counter-attacked. In a number of subsidiary attacks seven merchant ships were sunk off the Norwegian coast and five others were damaged. Although it must be admitted that the absence of the Luftwaffe was a major factor, the Home Fleet carriers had in spite of their failure to sink the *Tirpitz* regained freedom to operate wherever they liked in northern waters.

In August 1944, the R.A.F. had developed the 12,000-lb 'Tallboy' bomb which could be carried by the four-engined Lancaster bombers. Here was a weapon which should be able to finish off the *Tirpitz*. Bomber Command had already made four raids on the battleship with lighter bombs during the first half of 1942 when she was in Trondheim. These added up to 126 sorties by Halifaxes, Stirlings and Lancasters, but they were all delivered at night and they failed to hit her and lost twelve aircraft into the bargain. The plan now was to attack by day and, as Alten Fjord was out of range from the United Kingdom, to use a Russian base. Twenty-eight Lancasters took off from Yagodnik near Archangel on 15th September and surprised the *Tirpitz* before the usual smoke screen could be made. Only one aircraft was lost and they made one direct hit and two near misses causing such serious damage that her speed was reduced to 8 knots and a return to Germany made necessary to repair her. The result was that the Germans moved the *Tirpitz* to Tromsö to act as a floating battery and this put her within range of Lancasters from the United Kingdom. On 12th November, another raid of thirty-one Lancasters carrying Tallboys took off from Lossiemouth in Scotland and to make the long return flight they had to be 2 tons overweight. They approached the *Tirpitz* from over the land, succeeded in confusing the fighters which took off to oppose them and the smoke screen was again laid too late. Three of the 12,000-lb bombs, dropped from 14,000 feet, hit and another two were near misses: the after magazines exploded and the *Tirpitz* capsized. The final demise of this ship was of considerable strategic value and much of the Home Feet could now be moved to the Far East.

Aircraft had now sunk unaided the modern battleships *Prince of Wales*, *Roma*, *Tirpitz*, and also the Japanese *Mushashi* in the Pacific, as will be described in the next chapter. The *Prince of Wales* and the *Mushashi* were sunk by a large number of conventional bombs and torpedoes, probably more by torpedoes than bombs, thus restoring the reputation of the torpedo-bomber, which had suffered somewhat as a result of its poor performance against the *Scharnhorst* and *Gneisenau* in the Channel and later in the Battle of Midway. The *Tirpitz* and *Roma*, however, were destroyed by new and unconventional weapons. Attack had clearly overtaken defence and there could be little hope that ships could be designed to withstand air attack in the future.

XII

The Victory of Naval Air Power in the Pacific

1944 – 1945

TOWARDS THE END OF 1943 the great American amphibious advance across the Pacific began. Landings were made in the Gilberts in November, New Britain in December and in the Marshalls and Admiralty Islands in January and February 1944. In June it was the turn of the Marianas. All these operations were supported by massive air power, carrier-borne in the Ocean area and shore-based in the East Indies. The Japanese only opposed the advance with local resistance and did not make any serious attempt to dispute the general American command of the sea until June 1944 when the assault on the Marianas took place.

At the end of 1942, we left the U.S. carrier strength in the Pacific at its nadir. They were reduced to one fleet carrier and the British had had to lend them the *Victorious*. During 1943, however, as war production got into its stride, their strength grew rapidly. In just over a year eight of the new 'Essex' class fleet carriers, each with an air group of ninety planes, entered service. In the same period nine light fleet carriers of the 'Independence' class, converted from cruiser hulls and carrying thirty-five planes, joined the fleet. The great programme of escort carriers converted from merchant ship hulls got under way and in 1943 another twenty-five joined the ten already completed by the end of 1942. Aircraft were produced in ample numbers to fill these carriers and by the middle of 1943 there were nearly 17,000 naval planes in the Pacific theatre. The improvement was not only in numbers; during 1943 a new generation of carrier-borne aircraft had come into service. The Grumman Hellcat, of over twice the weight and horse-power of its predecessor, replaced the Wildcat as the standard carrier-borne fighter: it proved superior to the Japanese Zero and the equal of most shore-based high-performance fighters. The Grumman Avenger, also with an engine of twice the horse-power and 50 m.p.h. faster, replaced the Devastator as the torpedo-bomber. The Curtiss Helldiver, which replaced the Dauntless dive-bomber, the victor of Midway, was not such a success although it was also 50 m.p.h. faster and proved a powerful and robust aircraft.

After Midway, the Japanese set to work to rebuild their carrier fleet. They had one already building of pre-war design, the *Taiho*, which they completed in March 1944. She was an armoured aircraft-carrier after the British pattern and sacrificed capacity for protection, carrying only fifty-three planes. They laid down six new ships of the 'Unryu' class which were simplified 'Hiryu's', but it was clear that they needed carriers more quickly than those ships could be completed. They therefore converted the *Ryuho*[14] from the submarine depot ship *Taigei* and put a full deck on the seaplane-carriers *Chitose* and *Chiyoda*. At the same time they decided to convert the hull of the *Shinano*, a giant battleship of the 'Yamato' class and later, in November 1943, the *Ibuki*, a heavy cruiser hull on the slips. These were in addition to the five liners they converted into escort carriers during 1942 and 1943. After the Battle of Midway too they decided to increase the air component of the fleet still further by altering the battleships *Ise* and *Hyuga* into hybrids to carry twenty-two seaplanes each by taking out the two after gun turrets.

In the middle of 1944, the Mitsubishi Zero was still the standard fleet fighter but the Judy had replaced the Val as the dive-bomber and the Jill had replaced the Kate as the torpedo-bomber. Both these new types were 65 m.p.h. or so faster than their predecessors but they were both far slower than the Hellcat and had not got self-sealing petrol tanks. These aircraft had one great advantage over the new American types and that was their radius of action which was considerably greater. The Japanese took most of 1943 to recover from the decimation of their carrier air groups when they had been landed in the spring to take part in the New Guinea campaign. The production of planes and the training of pilots was far behind the Americans and by the middle of 1944 they were only just able to fill their carriers and then only with pilots of, on average, a quarter of the flying experience of the Americans.

With the relentless U.S. advance across the Pacific, it was obvious to the Japanese that they would have to stand and fight sooner or later. In the middle of 1944 they were ready to do so. Their First Mobile Fleet now consisted of 9 aircraft-carriers with 430 aircraft embarked and they were accompanied by 5 battleships including the huge *Yamato* and *Mushashi*, and 13 cruisers

Strategic Map

o U.S. Airfields
• Japanese Airfields

21(a)   THE BATTLE OF THE PHILIPPINE SEA

carrying another 43 float-planes. Finally the fleet was escorted by 28 destroyers. This fleet, formidable though it appears, was inferior to the Americans in practically every type of ship and aircraft, but the Japanese believed that they could redress the balance by the close co-operation of another 540 shore-based naval aircraft of what was now known as the First Air Fleet. These planes were disposed in a great semi-circle of bases from Chichijima in the Bonin Islands, through the Marianas and Palaus to the East Indies and the Philippines (see Fig. 21(a), p. 294). In emergency the Japanese hoped to be able to reinforce the First Air Fleet substantially from Japan and South-east Asia.

The First Mobile Fleet was now based at Tawi-tawi in the

southern Philippines where it was close to the oil supplies of Borneo. The plan was that at a suitable moment when the Americans were engaged in their next amphibious assault, they would put to sea and, in close co-operation with the shore-based aircraft, bring about a decisive fleet action. The Japanese intended to exploit the longer range of their aircraft to attack while American aircraft were still out of range. They also had an ingenious plan to extend their striking range still more by landing the carrier-borne aircraft at any of the ring of shore bases so that they could then re-arm and strike again on the way back. Using this technique the First Mobile Fleet could strike at an enemy anywhere in the Philippine Sea from a position roughly in the middle of it. It had the effect of giving them a strike range of 800 miles or so against the American 275 miles and so conferred upon them a substantial advantage which they hoped would more than make up for their inferiority in numbers.

The Americans landed on Saipan in the Marianas on 15th June supported by Task Force 58 which consisted of 7 fleet and 8 light fleet carriers with 891 aircraft embarked. The carriers were escorted by 7 modern battleships and 21 cruisers, carrying 65 float-planes, and 69 destroyers. The amphibious force itself was supported by another 7 escort carriers with 169 aircraft embarked. Long-range Liberator reconnaissance aircraft were based in the Admiralty Islands and Catalina flying-boats operated from the tender *Ballard* at Saipan. The Americans therefore had a substantial superiority in ships and carrier-borne aircraft but if the Japanese could bring all their shore-based aircraft to bear the disparity would not be so great.

Admiral Spruance, commanding the Fifth Fleet, had been using Task Force 58 since 11th June to gain air superiority in the Marianas and it had already destroyed a considerable proportion of the Japanese shore-based aircraft based there. The Japanese Commander-in-Chief ashore in Japan considered the moment had come for a decisive action and gave the executive order for his plan to be put into effect. The First Mobile Fleet sailed from Tawi-tawi on 13th June and passing north through the Philippines entered the Philippine Sea through the San Bernardino Strait. This movement did not escape the Americans. The submarine *Redfin* saw them leave Tawi-tawi, the *Flying-Fish* as they came out of the San Bernardino Strait and the *Seahorse* sighted a battleship component coming up from the East Indies to join them. Two American fast carrier groups had been detached to attack the Bonins and just had time to strike before the whole fleet was ordered to concentrate 180 miles west of Tinian to meet the Japanese attack.

The First Mobile Fleet was careful to keep out of range of the Liberators searching from the Admiralties but one of its supply groups was seen by the submarine *Cavalla* on 17th June. Admiral Ozawa refuelled his fleet from his tankers and pressed on to the north-east towards the Americans. On 18th June both sides flew air searches to try and locate the other's carriers. All the American searches fell short and saw nothing, but the Japanese midday search located the American carriers at a range of 360 miles. It was too late to strike that day and Admiral Ozawa decided to keep his distance during the night and make an overwhelming attack next morning. The Americans, however, got some information of the Japanese position as they obtained a 'fix' by radio direction finders when Admiral Ozawa made a signal to secure the co-operation of all the shore-based aircraft in his attack next day. Admiral Spruance decided not to advance towards this contact but to hold back in a defensive posture to ensure the safety of the invasion force on Saipan. The distance between the two fleets therefore remained much the same during the night. A night search by radar-fitted Avengers from the *Enterprise* also fell short of the Japanese; however, a Catalina from Saipan gained contact at 01.15 but his report failed to get through to Admiral Spruance until next morning.

At dawn on 19th June, both sides again flew off searches, the Japanese making full use of battleship and cruiser float-planes. As before the American searches, flying out to 325 miles, did not reach the enemy, but the Japanese searches flying out first to 350 miles and then to 560 miles sighted the Fifth Fleet and reported its position in spite of the fact that half of their scouting aircraft were shot down by American fighters. In the early morning the Americans, having no contacts and being bothered by a few Japanese planes from Guam and Rota which attacked Task Force 58, flew fighter patrols over the Japanese airfields. They intercepted air reinforcements arriving from Truk and shot down thirty-five Japanese planes.

With great skill, therefore, Admiral Ozawa had achieved a most advantageous position: he had located the American carriers without letting them know where he was. With Guam and Rota a hundred miles beyond the Americans he could launch his air strikes at their very maximum range and if necessary they could land at these shore bases. At 08.30 he began flying off the attacking planes and by 11.30 he had sent off 326 aircraft in four waves. They consisted of 169 dive-bombers and 48 torpedo-bombers escorted by 109 fighters. If these could be joined by powerful strikes from the 540 shore-based naval aircraft, the chance of an overwhelming victory was bright (see Fig. 21(b), p. 298).

Even before the last raid had been launched, however, Admiral Ozawa's troubles began. At 09.10 his flagship, the aircraft-carrier *Taiho*, was hit by a torpedo from the U.S. submarine *Albacore*. The damage did not at first seem serious but in fact the petrol system was badly damaged and she was soon in a very dangerous condition. At 12.20 the formation ran over the U.S. submarine *Cavalla* and she secured three torpedo hits on the aircraft-carrier *Shokaku*. The *Shokaku* caught fire, her speed was reduced and she had to be left behind. Nearly three hours later she suffered a number of heavy petrol explosions and sank. Very shortly afterwards the same fate overtook the *Taiho* and Admiral Ozawa had only just time to shift his flag to the cruiser *Haguro* before the carrier sank.

The second part of the plan to go awry was the failure of the Japanese shore-based air attacks to materialize. The 540 planes that were supposed to be available had begun to be drawn into the defence of Saipan as soon as the Americans landed and had suffered severely from the poundings they had received from Task Force 58 before the Japanese Fleet came on the scene. In the East Indies many naval aircraft had become involved in General McArthur's landing at Biak and had heavy losses. In the event only fifty planes or so were available on Guam and they were cut to pieces by the American fighter patrols over the islands which continued throughout the day. The shore-based air attack on which the Japanese plan relied to destroy a third of the American carriers therefore fizzled out and the few aircraft that approached the American Fleet were easily shot down by fighters.

With no knowledge of the position of the Japanese Fleet, Admiral Spruance could only continue to fly searches and prepare for an attack. The defensive posture assumed by Task Force 58 was one of great strength (see Fig. 21(c), p. 298). The Force was divided into five groups: four of these were made up

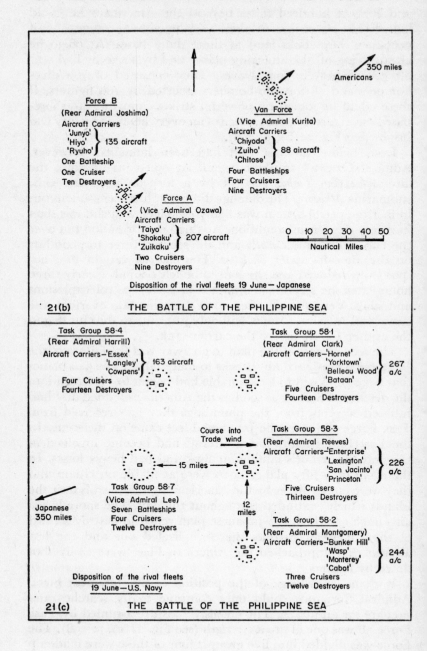

350 miles
Americans

Force B
(Rear Admiral Joshima)
Aircraft Carriers
'Junyo'
'Hiyo'   } 135 aircraft
'Ryuho'
One Battleship
One Cruiser
Ten Destroyers

Van Force
(Vice Admiral Kurita)
Aircraft Carriers
'Chiyoda'
'Zuiho'   } 88 aircraft
'Chitose'
Four Battleships
Four Cruisers
Nine Destroyers

Force A
(Vice Admiral Ozawa)
Aircraft Carriers
'Taiyo'
'Shokaku'   } 207 aircraft
'Zuikaku'
Two Cruisers
Nine Destroyers

0   10   20   30   40   50
Nautical Miles

Disposition of the rival fleets 19 June — Japanese

<u>21(b)</u>   THE BATTLE OF THE PHILIPPINE SEA

Task Group 58·4
(Rear Admiral Harrill)
Aircraft Carriers—'Essex'
'Langley'   } 163 aircraft
'Cowpens'
Four Cruisers
Fourteen Destroyers

Task Group 58·1
(Rear Admiral Clark)
Aircraft Carriers—'Hornet'
'Yorktown'
'Belleau Wood'   } 267 a/c
'Bataan'
Five Cruisers
Fourteen Destroyers

Course into
Trade wind

Task Group 58·3
(Rear Admiral Reeves)
Aircraft Carriers—'Enterprise'
'Lexington'
'San Jacinto'   } 226 a/c
'Princeton'
Five Cruisers
Thirteen Destroyers

← 15 miles →

Task Group 58·7
(Vice Admiral Lee)
Seven Battleships
Four Cruisers
Twelve Destroyers

Japanese
350 miles

12
miles

Task Group 58·2
(Rear Admiral Montgomery)
Aircraft Carriers—'Bunker Hill'
'Wasp'
'Monterey'   } 244 a/c
'Cabot'
Three Cruisers
Twelve Destroyers

Disposition of the rival fleets
19 June—U.S. Navy

<u>21 (c)</u>   THE BATTLE OF THE PHILIPPINE SEA

of the fifteen aircraft-carriers and their escorts and the fifth of the seven battleships with their accompanying cruisers and destroyers. The Battleship Group, under Admiral Lee, was thrown forward 15 miles in the direction of the enemy where it was hoped that it would absorb some of the attacks intended for the American carriers. Admiral Lee was given one of the carrier groups to supply fighter protection and with his very heavy anti-aircraft gun concentration was well able to accept this role. The other three carrier groups were in circular dispositions on a north-south line 12 miles apart where they could support each other. The whole force steadied on an easterly course into the trade wind and so could operate its 450 Hellcat fighters continuously. Radar was detecting aircraft 150 miles away and the fighter direction technique was by now highly developed in the fleet and extremely efficient. At the same time the sky was clear and the vapour trails of attacking aircraft could be seen clearly. Conditions were therefore ideal for air defence.

The Japanese strikes came in piecemeal over a period of four hours from 10.30 to 14.30 and the vast majority were intercepted by fighters well before they reached the ships. The Japanese lost 243 planes, mostly shot down by Hellcats and only 130 escaped. Altogether, including the planes lost on Guam and on board the *Taiho* and *Shokaku*, their casualties during the day were 315 aircraft. The Americans lost only 29 planes and not a single ship of the Task Force was sunk or seriously damaged. A number of ships were hit, notably the battleship *South Dakota* which had fifty casualties and the fleet carrier *Bunker Hill* in which there were 76 casualties. Both these ships were, however, soon operating again at full efficiency. Admiral Ozawa, with poor communications in his temporary flagship, did not at first realize that he had suffered a heavy defeat. He imagined that many of the missing aircraft were safe on Guam or Rota and believed his surviving pilots' exaggerated stories of the damage they had done. He therefore decided, after refuelling his fleet from his tankers, to renew the attack next day.

Admiral Spruance also intended to attack next day, but he found with an easterly wind, into which the carriers had to turn to operate aircraft, difficulty in closing with the enemy. He did his best to close during the night and as usual at dawn both sides flew off searches but the fleets were out of range of each other and no contact was made. The Japanese replenishment was badly delayed by lack of control from the temporary flagship and the Americans were able to close the range during the day. Soon after midday both sides again flew searches and this time both sides made contact. Admiral Ozawa, now in the carrier *Zuikaku*, found he had only a hundred planes left in his seven carriers and he decided to postpone a strike until next day. As soon as he realized he had been sighted he stopped fuelling and prepared to defend himself.

Admiral Mitscher, commanding Task Force 58, realized that this would probably be his last chance to strike at the Japanese Fleet. The distance was 275 miles which was the very maximum for his aircraft and the day was already so far advanced that they could not return before dark. In spite of the fact that his pilots were not trained in night landing, he decided he must take the risk and at 16.20 he launched 77 Helldivers and 54 Avengers escorted by 85 Hellcats. They found the Japanese just before dark in four groups, three of carriers and one of oilers. The Japanese flew off 75 planes to oppose the strike and 20 American aircraft were shot down by them or by anti-aircraft fire, but this did not save them from serious damage. The aircraft-carrier *Hiyo* was torpedoed, caught fire and sank. The *Zuikaku* was missed by torpedoes but hit by several heavy bombs and was very nearly lost. The light carrier *Chiyoda* was seriously damaged, the battleship *Haruna* and the cruiser *Maya* were also hit and two of the oilers were sunk. Furthermore 65 Japanese aircraft were lost. The American strike aircraft did not arrive back over their carriers until well after dark and they were all very short of fuel. Although the Task Force switched on all its lights, 80 planes were lost but by intense rescue operations on the following day all but 49 of the aircrew were saved.

The Americans pursued the enemy throughout the 21st June but gave over the chase at sunset. The Japanese First Mobile Fleet was out of range and in full retreat at 20 knots for Okinawa. It had lost three of its large aircraft-carriers and two of the survivors were seriously damaged. The total number of aircraft left fit for operations in the whole fleet had fallen to the pitiful figure of thirty-five. Against this the Americans had not lost a ship and their whole fleet was still fully operational. They had indeed lost 130 aircraft but they still had over 700 planes on board ready for action and they were left in full and undisputed command of the Philippine Sea.

The Battle of the Philippine Sea was one of the decisive naval victories of the war. Not only did the American Fleet protect the landing operations at Saipan but it drove the Japanese main fleet out of the area with losses from which it never recovered. The Battle has great strategic interest but there is no space here to enter into the controversy over whether Admiral Spruance was right to hold back in a defensive posture or should have advanced to the attack. As it turned out it led to one of the supreme examples of the ascendancy on occasion of defence over attack. The great victory of the Hellcat fighters was not only due to their high performance, superior training and superb direction but because they outnumbered outright the attackers who came in a series of raids spread over several hours and so exposed themselves to attack in detail.

All the same, the American naval aviators were disappointed: they had hoped to annihilate the Japanese carrier force. The Japanese had proved greatly superior to the Americans in air reconnaissance in spite of the fact that a proportion of the U.S. planes were fitted with radar and so were capable of searching at night. Admiral Ozawa exploited the greater range of his planes with consummate skill but luck was not on his side. All three of the Japanese carriers that were sunk were destroyed by torpedo attack followed by petrol explosions showing a design weakness as disastrous to them as a similar one to the British battle-cruisers at Jutland. The American deficiencies in air reconnaissance and strike were, however, more than made up by the contribution of their submarines, with whom the Hellcat fighters had to share the credit for the victory. It was the fore-sight with which Admiral Lockwood, who commanded the Pacific Fleet Submarine Force at Pearl Harbor, positioned his submarines which turned the Battle of the Philippine Sea into one of the decisive battles of the Pacific war. It was the fifth and last great carrier battle and in the size of the forces engaged, by far the largest.

In October 1944, four months after the Battle of the Philippine Sea, the Americans landed on Leyte. The great drives by General MacArthur through the East Indies and by Admiral Nimitz through the Pacific Islands had converged on this island in the Philippines. The Japanese could not afford to lose the Philippines as this would cut them off altogether from their oil supplies in the south. They realized that they would have to fight again to try to regain command of the sea if they were not to lose the war. They were, however, by no means ready and had not recovered from the Battle of the Philippine Sea. As before, their surface fleet had retired south to be near the oil supplies but this time to Lingga Roads in the East Indies. Their carrier force was still in Japan desperately trying to re-form and train new air groups. They had already made up most of their losses in aircraft-carriers: the new *Amagi* and *Katsuragi* of the modified 'Hiryu' type were in commission and the *Shinano* and *Unryu* were nearing completion. Nevertheless even if all the ships had had trained air groups, their carrier force would only have been half the size of the American. The battle-fleet in the south on which they must rely until the carrier air groups were ready, consisted of 7 battleships, 13 cruisers and 19 destroyers. This was a powerful force and stronger than at the Philippine Sea, but was also inferior to the U.S. Pacific Fleet's surface forces which were centred round twelve battleships. As before they hoped to make up for their inferiority by their large shore-based air forces in the area. The Second Air Fleet stationed in southern Japan, the Ryuku Islands and Formosa had 737 aircraft, nearly a third of which were fighters. The First Air Fleet in the Philippines was much smaller with 203 aircraft, over half of which were fighters, but there was also the Fourth Air Army with another 237 planes.

Early in October, before the landings on Leyte took place, Task Force 38 consisted of 9 fleet and 8 light fleet carriers escorted by 6 battleships, 13 cruisers and 58 destroyers. It was therefore stronger than Task Force 58 at the Battle of the Philippine Sea. There were 1081 aircraft in the carriers, over half of which were fighters and the rest dive- and torpedo-bombers. There were some sixty float-planes in the battleships and cruisers and another 300 or so planes in reserve carried in eleven escort carriers with the replenishment force. The total number of aircraft on each side at the start of the operation was therefore almost the same. Task Force 38 was, as usual, divided into four groups, the composition of which varied, but a group generally consisted of two fleet and two light fleet carriers with 250 planes and an escort of some battleships or cruisers and fifteen destroyers.

The attack by Task Force 38 began on 10th October on the airfields and shipping in the Ryuku Islands and then moved to Formosa and the Philippines and continued until the landings on 20th October. In the three-day attack on Formosa they destroyed 500 Japanese planes; they had already claimed over 100 destroyed in the Ryukus and by the 20th October there were only a hundred operational aircraft to oppose the landings. The Japanese brought up massive reinforcements from Japan itself, China and other areas to make up their losses and, what was most significant, they landed half their carrier air groups to join the fight ashore. The American strikes stimulated Japanese counter-attacks on the carrier groups, which were heavy but not very successful and the Japanese believed that they had done much more damage than they had. Between the 12th–16th October, Task Force 38 was subjected to some of the heaviest air attacks of the war; nearly half the planes however were intercepted before they reached the ships and the most serious casualties were the torpedoing of the cruisers *Canberra* and *Houston*. Both cruisers eventually got home and the only other damage was to the carrier *Franklin* which suffered casualties but was soon operational again. In all these operations Task Force 38 lost only just over a hundred aircraft and these were easily replaced from the replenishment carriers.

Some assistance in securing air superiority over the Philippines was received from the U.S. Air Forces in China and from the South-West Pacific area but most of the work was done by the American carrier force. The Pacific Fleet had faced shore-based aircraft many times before, but this was the first real confrontation with the main Japanese air strength. It showed that there was no fundamental reason why aircraft-carriers should not face shore-based air power provided their aircraft were of the right type and in sufficient strength. The American carrier-borne aircraft in 1944 were actually superior to the Japanese shore-based types and it is of interest that they brought as many fighters across thousands of miles of ocean in their carriers as Great Britain had in Fighter Command at the height of the Battle of Britain. This was a very different business from the twelve Fulmars with which Admiral Cunningham had to fight Fliegerkorps VIII in Crete.

The landings on Leyte on 20th October were supported by another 500 aircraft carried in another eighteen escort carriers. When the landing took place, the Japanese put into effect their plan (see Fig. 22(a), pp. 306–7) which was basically for the battle-fleet from Lingga to steam through the middle of the Philippines in two groups and, executing a pincer movement, to annihilate the amphibious forces off the beaches. They were to act in close co-operation with the shore-based First Air Fleet in the Philippines, who were to provide fighter protection and air reconnaissance for them. Such an attack had little chance of success against the thousand aircraft of Task Force 38 and so the Japanese aircraft-carriers, with what aircraft they could scrape together, were to make a diversionary attack from Japan to draw the American carriers away to the north and so prevent them from intervening.

As it came north the Japanese battle-fleet was as usual sighted by alert U.S. submarines which succeeded in sinking the cruisers *Atago* and *Maya* and severely damaging the *Takao* (see Fig. 22(a), pp. 306–7). These attacks were made in the Palawan Passage on the larger of the two forces, under Admiral Kurita, which intended to pass through the Sibuyan Sea and San Bernardino Strait and to attack from the north. The second force, under Admiral Nishimura, was to pass through the Sulu Sea and attack through the Surigao Strait from the south. It was to be followed by a smaller force from Formosa under Admiral Shima.

On 24th October, the American carrier force was, as usual, split into its four groups (see Fig. 22(b), p. 308). Admiral McCain's Group 38.1 had been sent away to the eastwards to refuel and replenish; Admiral Sherman's Group 38.3 was off Luzon; Admiral Bogan's Group 38.2 was off the San Bernardino Strait and Admiral Davison's Group 38.4 was off Leyte. All the groups were flying searches over a wide area and across on the other side of the Philippines. The Japanese were very unlucky and Task Force 38 planes sighted both Admiral Kurita's force in the Sibuyan Sea and Admiral Nishimura's in the Sulu Sea. At the same time shore-based U.S. planes detected the approach of Admiral Shima's squadron. The only force that was not sighted was the one the Japanese hoped would be. Admiral Ozawa in the Japanese carriers with 166 aircraft on board had left the Inland Sea on 20th October but there were no U.S. submarines to report him and so his main aim of diverting the attention of Task Force 38 from the other Japanese Forces initially failed.

As Admiral Kurita entered the Sibuyan Sea he was within striking range of three-quarters of Task Force 38. He made many requests for fighter cover from the First Air Fleet but Admiral Fukudome who commanded it concentrated most of his effort on attacking the American carriers. He got away three raids totalling some 150 aircraft which attacked Admiral Sherman's group off Luzon; they were all intercepted by fighters but one dive-bomber got through and hit the light fleet carrier *Princeton*. She caught fire and it spread so badly that she had to be abandoned and was in the end sunk by U.S. forces. Unknown to Admiral Sherman, Admiral Ozawa was now in range to the northwards; he launched a strike but it was broken up by the American fighters. Many of the inexperienced Japanese pilots were shot down and only twenty-nine aircraft got back to their carriers and 15 others managed to land in Luzon.

While these attacks were being made on Admiral Sherman, Admirals Bogan and Davison concentrated upon attacking Admiral Kurita in the Sibuyan Sea. They launched five waves totalling 118 dive- and torpedo-bombers escorted by 92 fighters. They concentrated upon the huge battleship *Mushashi* and finally sank her after scoring 19 hits with torpedoes and 17 with heavy bombs. They succeeded in damaging the battleships *Yamato*, *Nagato* and *Haruna* as well but their fighting efficiency was not appreciably diminished. The cruiser *Myoko*, however, was hit by a torpedo and had to return to base. Admiral Kurita, who had already been sunk once in the *Takao*, understandably lost his nerve and reversed course. After receiving encouragement from Admiral Toyoda, the Commander-in-Chief ashore in Japan, he turned again to the attack with four battleships, eight cruisers and a dozen or so destroyers which were what was left of his force. Meanwhile Admirals Nishimura and Shima with between them two battleships, four cruisers and eight destroyers pressed on through the Sulu Sea unmolested except for a few bombs from the aircraft which had originally sighted them.

In the late afternoon a search plane from Group 38.4 at last sighted Admiral Ozawa's force of four aircraft-carriers escorted by the two hybrid battleships *Ise* and *Hyuga*, three cruisers and eight destroyers. Admiral Halsey swallowed the bait whole. He concentrated the groups of Admirals Bogan, Davison and Sherman and steamed north at full speed to engage. He took with him Admiral Lee's six modern battleships and left the entrance to the San Bernardino Strait completely unmarked. Admiral Halsey believed from the reports of his pilots, which were somewhat optimistic, that Admiral Kurita was no longer a danger. He was not to know that Admiral Ozawa's carriers were 'paper tigers' and had only twenty-nine planes left between

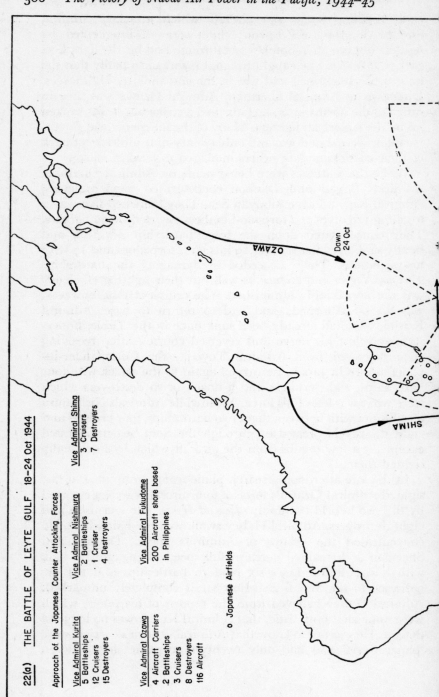

22(a)   THE BATTLE OF LEYTE GULF 18–24 Oct 1944

Approach of the Japanese Counter Attacking Forces

Vice Admiral Kurita
5 Battleships
12 Cruisers
15 Destroyers

Vice Admiral Nishimura
2 Battleships
1 Cruiser
4 Destroyers

Vice Admiral Shima
3 Cruisers
7 Destroyers

Vice Admiral Fukudome
200 Aircraft shore based
in Philippines

Vice Admiral Ozawa
4 Aircraft Carriers
2 Battleships
3 Cruisers
8 Destroyers
116 Aircraft

o Japanese Airfields

OZAWA

Down
24 Oct

SHIMA

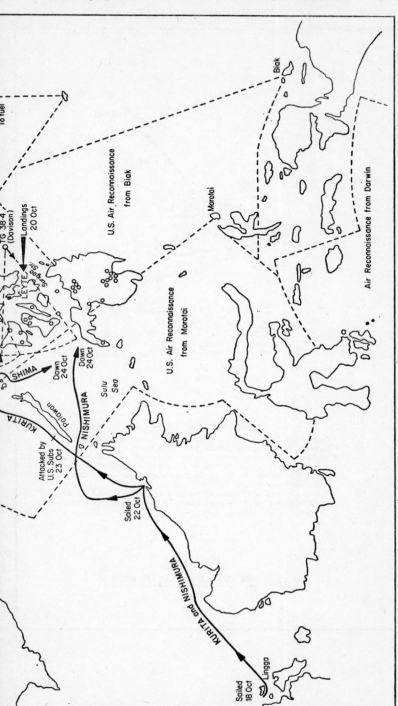

To fuel

Biak

U.S. Air Reconnaissance from Biak

Air Reconnaissance from Darwin

Morotai

TG 38·4 (Davison)

Landings 20 Oct

LEYTE

Surigao Str.

U.S. Air Reconnaissance from Morotai

SHIMA

Down 24 Oct

Down 24 Oct

NISHIMURA

Sulu Sea

KURITA

Palawan

Attacked by U.S. Subs 23 Oct

Sailed 22 Oct

KURITA and NISHIMURA

Lingga

Sailed 18 Oct

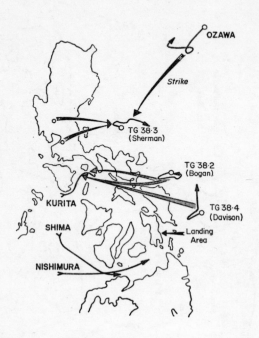

Daylight hours 24 Oct 1944

Task Groups 38·2 and 38·4 attack Kurita in the Sibuyan Sea, sink
'Mushashi' and damage other ships but fail to stop him.
Nishimura and Shima, although sighted by aircraft which bomb them,
are subsequently unopposed from the air, and advance towards Landing Area.
TG 38·3 is heavily attacked by Japanese shore based aircraft and
loses the 'Princeton'
Ozawa also gets within range of TG 38·3 and strikes at 1640, but does
no damage and has heavy plane losses, TG 38·3 at last sights Ozawa's
carriers.

22(b)   THE BATTLE OF LEYTE GULF

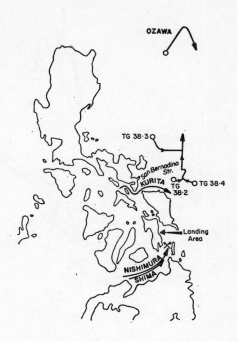

Night 24–25 Oct 1944

Halsey concentrates the U.S. Fast Carrier Task Force (38) and
hurries North to engage Ozawa.
Kurita emerges unopposed from the San Bernadino Strait and
turns South.
Nishimura is practically annihilated and Shima repulsed by U.S.
Surface forces in Surigao Strait.

22(c)   THE BATTLE OF LEYTE GULF

x

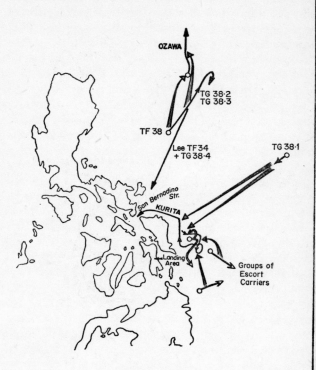

25 Oct 1944

Kurita surprises Escort Carriers and sinks 'Gambier Bay!' Escort Carrier
Air Groups attack and seriously maul him and he gives up any idea of
attacking the landing area and retires. He is also attacked by Admiral
McCain's TG 38·1 returning from fuelling.

Halsey (TF 38) sinks all four of Ozawa's carriers. He detaches Admiral
Lee with the Battleship Force to deal with Kurita, but Kurita escapes
through San Bernadino Strait before he can get there.

22(d)   THE BATTLE OF LEYTE GULF

them. He believed that they had sunk the *Princeton* and he was determined that this time he would annihilate them. In fact, although he had mauled Kurita, Admiral Halsey hadn't stopped him and all three Japanese surface forces, after being within range of Task Force 38 all day, were still converging on the landing area.

By midnight Admiral Halsey had concentrated five fleet and five light fleet carriers with about 600 aircraft escorted by 6 modern battleships, 8 cruisers and 41 destroyers. An hour later the *Independence* flew off a night search of radar-fitted aircraft which contacted Ozawa soon after 02.00. Admiral Mitscher, the commander of Task Force 38 under Halsey, now in tactical command, decided to wait and re-locate the enemy at dawn and then strike. During the night the forces (see Fig. 22(c), p. 309) under Admirals Nishimura and Shima attacked the Americans at Leyte through the Surigao Strait and were practically annihilated by the six old battleships and other surface forces which had been supporting the landings. While this battle was in progress Admiral Kurita emerged from the unguarded San Bernardino Strait and turned south.

At 07.10 on 25th October, Admiral Mitscher's search planes re-located Ozawa 145 miles away and a strike of about 150 aircraft which had taken off in anticipation of a contact, went straight in to the attack (see Fig. 22(d), p. 310). They were opposed by only a dozen fighters, they sank the light aircraft-carrier *Chitose* and the destroyer *Akitsuki*, damaged the carriers *Zuikaku* and *Zuiho* and shot down nine of the Japanese fighters. Just as Task Force 38 was launching 36 more planes to the attack, news came in that Admiral Kurita had broken out of the San Bernardino Strait and was attacking the escort carrier groups off Leyte. Admiral Halsey contented himself with ordering Admiral McCain's group, which was fuelling to the eastwards, to help and continued in hot pursuit of Ozawa with his whole force including Admiral Lee's battleships.

At daylight, Admiral Kurita had sighted the northernmost group of six escort carriers and had given chase. All sixteen escort carriers of all three groups at once launched all their available aircraft to defend themselves and recalled those which had already left to support operations ashore. In a running fight lasting two and a half hours in which the escort destroyers used smoke screens and torpedoes, they lost only the escort carrier *Gambier Bay* and the destroyers *Hoel*, *Johnston*, and *Roberts* and succeeded in torpedoing the Japanese cruiser *Kumano*. Their aircraft attacked and harassed the Japanese with bombs, torpedoes and machine-guns and succeeded in sinking the heavy cruisers *Chokai*, *Chikuma* and *Suzuya* and damaging the *Haguro*, *Nagato* and *Tone*. Admiral Kurita called off the pursuit soon after nine and by 13.00 was in full retreat for the San Bernardino Strait.

Meanwhile in the battle between Task Force 38 and Admiral Ozawa to the north, the 36 American planes we left in the air succeeded in damaging the light carrier *Chiyoda* and the light cruiser *Tama*, the first so heavily that she stopped. As they completed their attack, Admiral Halsey, after a query from Admiral Nimitz at Pearl Harbor, detached Admiral Lee's battleships to the south to deal with Admiral Kurita, with Admiral Bogan's carrier group to give him air cover. The rest of Task Force 38 continued the pursuit, launching four more strikes totalling some 450 planes, which sank the fleet carrier *Zuikaku* and the light carrier *Zuiho* but only succeeded in near-missing the two hybrid battleships. U.S. cruisers and destroyers then finished off the *Chiyoda* and sank the destroyer *Hatsuzuki*, and the U.S. submarine *Jallao* completed the destruction of the *Tama*. Admiral Ozawa, in the cruiser *Oyodo* with the *Ise*, *Hyuga* and five destroyers, escaped to Japan.

It was not only Admiral Lee's battle-fleet that was trying to intercept Admiral Kurita's retreat. Admiral McCain's Task Group 38.1 returning from fuelling was also coming up at full speed from the north-eastwards. At 13.16 his first strike of 147 planes was launched at maximum range and damaged the cruiser *Tone* again but that was all. Admiral Kurita retired through the San Bernardino Strait at 21.40, three hours ahead of Admiral Lee, who only arrived in time to sink the destroyer *Nowake* which was straggling. With such a start there was little to be done until dawn when Admiral Bogan's and Admiral McCain's groups attacked with 257 planes, hitting the *Kumano* again and sinking the light cruiser *Noshiro*, but the rest of Admiral Kurita's force escaped.

The Battle of Leyte Gulf, from the size of the forces engaged was one of the greatest contests at sea of all time. It confirmed the American command of the sea in the Far East and reduced the Japanese Navy to remnants which had little effect on the rest of the war. Leyte Gulf, however, does not rank with Midway and the Philippine Sea as one of the great carrier actions: it is of more interest as a battle between aircraft and surface ships. This is no place to discuss Admiral Halsey's strategy fully, in particular whether he should have left Admiral Lee to block San Bernardino Strait. It seems, however, that the pendulum had swung too far. Although the war in the Pacific had shown that aircraft over the sea had replaced the battleship as the 'unit of sea power' this did not mean that battleships were impotent. Indeed in this battle they re-established their reputation to quite an extent at the expense of the carriers.

Admiral Kurita's 'death ride' merely confirmed the folly of pitting a surface force without fighter protection against overwhelming air power and this had been demonstrated many times before; but it is only fair to say that he expected fighter protection from the First Air Fleet and got very little. It was remarkable that a fast and powerful surface force which got within gun range of slow and vulnerable escort carriers could do no better. The American destroyers with their smoke screens were very skilful but the main reason for the escape of the escort carriers was the attack by their 400 aircraft. The Americans found that even with overwhelming air power, it proved very difficult to stop a determined and heavily armed surface fleet. Half the Japanese attacking forces had to be left to surface ships to deal with and in the event these proved a more effective stopping agent than aircraft. The sinking of Ozawa's carriers was comparatively easy: without aircraft, aircraft-carriers are useless. With plenty of good aircraft, on the other hand, even simple and slow escort carriers can be very effective as the attack by Kurita on the American escort carriers shows. In other words it is aircraft that count rather than the aircraft-carriers and it is aircraft operating over the sea, and not, as is often wrongly believed, aircraft-carriers that have replaced the battleship.

After the Battle of Leyte Gulf, the Japanese Navy made one more bid to regain command of the sea and save Japan from invasion. The surface fleet was in no state to try again and their shore-based aircraft were no match for the American fighters. They therefore turned to the Kamikaze or suicide aircraft. Apart from the fact that the Japanese temperament was suited to this form of warfare, the Kamikaze attack had many advantages. They could use inexperienced pilots; they could expend their obsolete aircraft and, as they did not need to return, their strike range was automatically doubled. A hit was more certain than with conventional bombs or torpedoes and the damage, especially the incendiary effect on carriers, was expected to be much greater. The Kamikaze was in fact an early guided missile, the guidance being human.

The first attack, although it is not certain that it was premeditated, was on H.M.A.S. *Australia* on 21st October during the Leyte operation. Four days later, just after the Battle of Leyte Gulf, deliberate Kamikaze attacks were made on the escort carriers off the beaches. The *St Lo* was sunk and five others were damaged and the whole force had to withdraw next day to Manus, their duties being taken over by the Third Fleet carriers. On 29th-30th October, it was the Third Fleet's turn and the carriers *Intrepid, Franklin* and *Belleau Wood* were all hit. On 5th November the *Intrepid* was hit again as well as the *Lexington, Hancock, Essex,* and *Cabot.* Some planned operations had to be cancelled and Task Group 38.2 had to withdraw to Ulithi: subsequently five of the carriers needed extensive repairs. In the course of their attempt to defend the Philippines, the Japanese expended some 650 Kamikaze planes, sinking 16 ships including the escort carrier *Ommaney Bay* and damaging 150 others, mostly minor warships. Nevertheless the American operations went ahead as planned.

In the capture of Iwojima, the Japanese made one Kamikaze attack on 21st-22nd February using twenty-five planes. They sank the escort carrier *Bismarck Sea* and damaged the veteran *Saratoga* which was hit by four of them with over 300 casualties and forced to return to the United States for repairs. In March, twenty-five twin-engined Kamikaze bombers set off on the 1400 mile trip to Ulithi to attack the American carrier force in harbour. Only fifteen of them arrived but they seriously damaged the aircraft-carrier *Randolph.* So far the Kamikazes had done a lot of damage, they had sunk three escort carriers and damaged fourteen other carriers, half of which were of the fleet type. Although the majority of these ships were back in service for the Okinawa campaign, the Kamikazes can claim that they were the main cause of the premature withdrawal of the escort carriers and Task Group 38.2 at the end of the Leyte campaign. In spite of this they failed to delay, let alone stop, the main American amphibious advance.

In March the British Pacific Fleet, after attacking the oil refineries at Palembang on the way, arrived in the Far East. It consisted of the four armoured aircraft-carriers *Indomitable*, *Victorious*, *Indefatigable* and *Illustrious*, carrying 238 aircraft, with the battleships *King George V* and *Howe*, 5 cruisers and 14 destroyers. It was equivalent, therefore, to one of the four groups of the U.S. Fast Carrier Force. Their air groups were now mostly composed of American aircraft: Avengers, Hellcats and the Chance-Vought Corsair, which was a fighter of exceptional performance. British types were represented by the Seafire and Firefly. By adopting American methods of operating aircraft and accepting a deck park, the air groups were larger than early in the war; *Victorious* and *Illustrious*, for instance, now carried 52 instead of 33 aircraft. The *Indefatigable*, which was of an enlarged type completed in 1944, carried 69 aircraft. Later on the *Formidable* and the *Implacable*, sister ship of the *Indefatigable*, joined the fleet and a number of escort carriers became available to bring up replacement aircraft and for other auxiliary purposes.

In the operations against Okinawa planned for April, it was obvious that the Kamikazes were going to make a supreme effort. No special counter-measures to them had been devised except to step up every existing air defence measure. Kamikaze tactics were to approach either very high at 20,000 feet or so, to make interception difficult, or very low to get underneath the radar beams. It was fortunate that the proximity fuse for anti-aircraft ammunition came into service in 1944 as this very greatly increased the lethality of anti-aircraft guns. The U.S. Navy did, however, increase the proportion of fighters in its carriers; for instance the air group of an *Essex* class carrier now consisted of 72 Corsairs and Hellcats and only 30 dive- or torpedo-bombers. At the same time the number of 40 mm anti-aircraft weapons was increased wherever possible in every type of ship. Tactically it was found necessary to throw out picket destroyers to extend radar cover for low-flying aircraft and to step up attacks on the airfields from which the suicide planes operated.

The landings on Okinawa took place on 1st April after very heavy attacks by the U.S. Fast Carriers on airfields in Kyushu and other parts of Japan, in which they claim to have destroyed over 500 aircraft, and by the British Pacific Fleet on airfields to the south of the island. In these operations the carriers *Wasp*,

Franklin, Enterprise and *Yorktown* were all damaged by conventional bombers acting in a suicidal manner. The *Franklin* suffered nearly a thousand casualties and it was a miracle that she was saved; but her repairs were not completed by the end of the war. At this time a new type of Kamikaze appeared which became known as the Baka bomb. It was, in fact, a small single-seater suicide aircraft carried to the scene of action by a medium bomber. It had a range of twenty miles, a speed, with rocket propulsion, of 600 m.p.h. and its warhead consisted of 4000 lbs of high explosive. This can be said to be an early form of 'stand off' guided weapon with human guidance. Once released it was very difficult to shoot down but fortunately while still attached to its parent aircraft the combination was so unwieldy that it was very vulnerable to fighters.

The Japanese did not mount a serious Kamikaze attack on the Okinawa landing forces until 6th–7th April when 355 of them took off from airfields in Kyushu accompanied by a similar number of conventional aircraft. Some 250 of these Kamikazes were shot down by carrier-borne fighters and another 40 by anti-aircraft fire and it was only 28 that finally hit the ships. Three destroyers, an L.S.T. and two ammunition ships were sunk and a battleship, an aircraft-carrier, eleven destroyers and two minesweepers were damaged, four of the destroyers so badly that they were beyond repair.

The Japanese did not confine their suicide attacks to aircraft. On 6th April, to coincide with the heavy Kamikaze attack, they sent the battleship *Yamato* to sea escorted by the cruiser *Yahagi* and eight destroyers to attack the American forces off Okinawa. The intention was essentially suicidal and to fight to a finish doing as much damage as possible. If the ships were still afloat when they ran out of ammunition they were to beach themselves and land their crews to join the army in the defence of Okinawa. The force was sighted by U.S. submarines as it left the Bungo Strait and three of the four fast carrier groups of Task Force 58 under Admiral Mitscher concentrated to strike next morning. The *Yamato* group was found by a search plane from the *Enterprise* at 08.23 on 7th April and 380 planes were launched to the attack. Of these 75 were dive-bombers and 131 were torpedo-bombers escorted by 180 fighters. The *Yamato* had no fighter protection since it had been decided, as at the Battle of Leyte, to concentrate all planes to attack the American carriers. Ten torpedoes and five bombs hit the *Yamato* and she capsized with the loss of 2500 men. The *Yahagi* absorbed seven torpedoes and twelve bombs before she sank taking four of the destroyers with her. The other four destroyers, all damaged, escaped to Sasebo. This Battle of the East China Sea, as it was subsequently called, cost the Americans twelve aircraft. The result was no surprise to anyone, indeed it was expected, for the Fast Carrier Force of seven fleet and six light fleet carriers with over 700 aircraft was a heavier force than was brought to bear on Kurita at Leyte Gulf. But the reputation of aircraft for being able to stop such an attack dead in its tracks was restored.

During the rest of April and May there were seven more major Kamikaze raids, the largest of 185 planes and the smallest of 110, as well as a large number of minor attacks. In the whole Okinawa campaign some 1900 Kamikaze sorties were flown against the American and British forces and about 15 per cent of them got home. Twenty-five small ships were sunk, the largest being destroyers and the radar pickets being the worst sufferers. One hundred and fifty ships were damaged and these included four aircraft-carriers, two battleships and a cruiser. The aircraft-carrier *Bunker Hill*, Admiral Mitscher's flagship, was hit by two Kamikaze planes, caught fire and had 660 casualties and had to return to the U.S.A. for repair. The British carriers *Formidable*, *Indomitable* and *Victorious* were all hit by suicide planes but in every case their armoured flight decks saved them from serious damage. The British carrier design showed up well under these conditions; and although their ships carried fewer aircraft than the Americans, these were now of sufficiently high performance to compete with a shore-based air force. In the middle of April, one of the American carrier task groups had to be dissolved because of losses and Task Force 58 thereafter operated in three instead of four groups. At this juncture the threat of the Kamikazes was greater than at any other time. If casualties of this order had continued it might well have made operations off Japan too expensive.

Admiral Mitscher now returned to the offensive and renewed his attacks on the Kyushu airfields and the XXI Bomber Command of the U.S. Air Force was asked to help. As a result, in May there were far fewer Kamikaze attacks and in fact most of their fury was spent. They had run out of volunteers in April and were already very short of fuel and planes; so the crisis soon passed.

The Japanese Kamikaze offensive nevertheless proved a very

serious menace and caused a very great deal of damage and casualties. Whether manned by the Japanese Army or Navy the Kamikaze was used almost exclusively as an anti-ship weapon. The official British naval historian rates them as the most effective airborne anti-ship weapon of the war. Nevertheless in spite of their successes they never really came within sight of regaining command of the sea. While they were operating, the American Fast Carrier Task Force was able to finish off the last of the Japanese Fleet in harbour including the battleships *Haruna, Ise* and *Hyuga* and the carriers *Katsuragi* and *Ryuho*. They also made heavy raids on Japan itself as well as covering the Okinawa campaign. The defeat of the Kamikazes was mainly due to the immense superiority of the Americans in the air, especially their excellent fighters, the Corsair and the Hellcat, and also to the great number of ships they possessed and their ample reserves. The target was simply too big for the Japanese to destroy before they ran out of Kamikaze machines and pilots.

At the end of this final chapter dealing with the Second World War, it is time to try to sum up the influence of aircraft upon sea power during the struggle. It will be remembered that Douhet, in the twenties, held the view that future wars would be won by bombing the enemy's cities and that sea power would virtually be irrelevant. It is now generally accepted that, although strategic bombing was an important factor in winning the war, it certainly was not the only one. Germany was defeated by a combination of a maritime blockade and massive military operations from east and west as well as by strategic bombing. Sea power was of paramount importance in mounting the blockade and in supporting practically every campaign, especially the Allied landings on the continent which would not have been possible without it. Douhet of course could argue that he had always maintained that it was essential to gain what he called 'command of the air' at the outset and that then bombing would be decisive. Air superiority was not obtained over Germany until towards the end, when bombing certainly became very much more effective. Nevertheless Douhet was completely wrong in his belief that fighters could not compete with bombers and it is probably this error which misled him.

This is no place to discuss bomber strategy in detail but we are on firm ground and can say with confidence that bombing in the Second World War did not make sea power irrelevant; indeed sea power was if anything more important than in the past.

Douhet also maintained that strategic bombing could dominate a Navy by the destruction of its bases and infrastructure. In this he was also proved wrong. Bomber Command had a negligible effect on sea power for the first two years of the war and failed to have any effect on the major German warships or interfere in any way with the U-boat campaign. In 1941, however, strategic bombing did prevent the *Scharnhorst* and *Gneisenau* joining the *Bismarck* in the Atlantic and in 1944–5 strategic bombers did much to prevent the new type of U-boats with high underwater speed from taking part in the war, and this was probably their most important effect on sea power.

The most fundamental question that had been answered by the war was whether aircraft could sink ships. Beginning with the dive-bombing and sinking of the *Königsberg* by the British Fleet Air Arm in the Norwegian Campaign they had gone from strength to strength. Of the twenty-eight capital ships on both sides sunk during the war, sixteen were disposed of by aircraft. Their greatest rivals were other battleships which sank eight of their own kind and U-boats which sank three. Ten of the sixteen capital ships sunk by aircraft were attacked in harbour, a place where battle-fleets in the past had been relatively safe, and only six at sea. The first capital ship to be sunk by aircraft was the elderly though modernized *Cavour* at Taranto but by the end of the war the brand new battleships *Prince of Wales*, *Roma*, *Tirpitz*, *Mushashi* and *Yamato* had all been disposed of, unaided, by aircraft. Against these heavily armoured ships the air-launched torpedo proved the most successful weapon and was the primary cause of the loss of over half of the battleships sunk, the dive-bomber came next, and lastly the high-bomber which proved almost useless except against ships in harbour. New weapons such as the German guided bomb FX1400 and the British 12,000-lb Tallboy each had a success and might well have done better if they had been introduced earlier and used in greater quantity. Aircraft also sank more cruisers and destroyers than U-boats or other ships. 'Billy' Mitchell, however doubtful his assertions were in the early twenties, was therefore proved right by the Second World War.

Douhet was right when he said that maritime operations could not be continued without the 'command of the air' over the sea. Of the five modern battleships which were sunk by aircraft, only the *Mushashi* had any fighter protection at the time and she had very little. The war proved that when fighter protection was provided in sufficient strength, naval forces could operate even within range of first-line shore-based air strength as in the dash of the *Scharnhorst* and *Gneisenau* up the Channel. Whether air superiority over the sea was obtained by ship or shore-based aircraft was mainly a matter of geography. The Germans in Norway and Crete and the Japanese in their advance through South East Asia were very successful with shore-based aircraft, while the Americans in their advance across the Pacific were no less successful with carrier-borne aircraft. When carrier-borne aircraft contested air superiority with shore-based aircraft the winner was generally, as one would expect, the side that brought the largest number of the best aircraft to bear.

The United States Navy developed naval aviation from 1922 onwards with the express purpose of securing air superiority wherever they went. During the war we can discern three clear stages in the development of naval aviation. Firstly the British approach where, until the British Pacific Fleet was formed, a carrier accompanied a battle-fleet to give it air support; in this stage, the value of aircraft for reconnaissance and all the traditional purposes was fully appreciated but the battle-fleet remained the principal weapon. Fighter protection for the fleet was realized to be an absolute necessity but was assessed as equivalent to an anti-submarine screen or the use of destroyers to counter a surface torpedo attack. Aircraft used in this way were successful in the sinking of the *Bismarck* and at Matapan but were never able to compete with the Luftwaffe. Secondly comes the stage reached in the early part of the Pacific war in which the carrier air group had definitely replaced the battleship as the unit of sea power and the whole operation of the fleet was subordinated to its needs. This was the way in which Midway and the other great carrier battles of the Pacific war were won and by which command of the sea was gained in the Pacific. The third stage was reached at the end of the Pacific war when aircraft-carriers brought a whole air force across the ocean to challenge first-line shore-based air power. This was more an air force embarked in a fleet than a fleet with an air arm. This was the key to the U.S. amphibious advance across the Pacific that led to the defeat of Japan. The American strength in the Pacific, however, was exceptional and in general it can be said that it was easier to bring aircraft to bear in the right place at the right time from aircraft-carriers, but shore-bases were generally able to sustain operations for longer and were not so vulnerable.

Aircraft-carriers were indeed vulnerable. Forty-two of various types were sunk during the war, the cause being equally divided between U-boats and aircraft which disposed of nineteen each, while ships came a bad third with three. In this particular case of carriers the dive-bomber was more effective than the torpedo-bomber, although in many cases both weapons were used. The Kamikaze, although it sank three escort carriers and damaged a considerable number of larger aircraft-carriers, did not succeed in sinking any of the fleet type.

Douhet was again wrong when he maintained that aircraft should not be wasted in an 'auxiliary' role in support of naval operations. In such operations aircraft proved themselves in the Second World War to be as important as ships. The Battle of the Atlantic against the U-boats would certainly have been lost without them. The participation of R.A.F. Fighter Command in the defence of shipping in the North Sea and of Bomber Command minelaying in the destruction of the German merchant marine are examples of the potent use of aircraft in the 'auxiliary' maritime role.

The flexibility of air forces, a principle widely adumbrated by aviators before the war, proved a very variable quantity. It was probably at its best with the German Fliegerkorps VIII and X in Norway and the Mediterranean which were able to switch with conspicuous success from army support or strategic bombing to the attack of ships at sea and back again. This very flexibility, however, frequently meant that the aircraft that should have been commanding the sea were often doing something else as when Fliegerkorps X permitted the British Mediterranean Fleet to bombard Tripoli in 1941. The main Italian Bombing Force was certainly not flexible. Although it was able to exert a considerable effect on operations, when it operated over the sea it seldom hit anything. The British Bomber Command, trained for night work over the land, was almost a complete failure in the maritime role as was revealed by the escape of the *Scharnhorst* and *Gneisenau* up the Channel in

February 1942. They certainly sank the *Tirpitz*, put the *Gneisenau* out of the war in 1942 and sank several other German warships in harbour and many U-boats right at the end, but it is fair to say that most of these successes were in the nature of mopping-up operations and they cannot claim to have contributed more than a useful bonus to the war at sea.

In the Second World War, therefore, aircraft had certainly not rendered sea power irrelevant but they took what was undoubtedly a dominant part in it. They replaced the battleship as the strength element or the 'unit of sea power' and the role of ships became firstly one of carrying air power to sea, secondly of co-operating with aircraft in the exercise of sea power and thirdly of exploiting the use of the sea when command of it had been won.

XIII

Air Power Over the Sea since the Second World War
1945 – 1969

THE LESSONS OF the war had scarcely been discussed before the explosions of the atom bombs at Hiroshima and Nagasaki threw the whole business back into the melting pot. The eclipse of the battleship by aircraft was clearly a small matter compared with this revolution in the nature of war itself, whereby Douhet's theories might well come true. The United States decided, without delay, to carry out a large-scale test to determine the effects of the atom bomb on naval vessels. In the summer of 1946, they assembled some ninety target vessels at Bikini Atoll in the Marshall Islands. Sensing that inter-service rivalry might lead to such difficulties as occurred in the bomb tests after the First World War, they wisely put the whole trial under a single command. In July 1946 two atomic bombs, each equivalent to 20,000 tons of T.N.T., were exploded in separate tests, the first being an air burst and the second underwater. The battleship *Arkansas* virtually disappeared at 500 feet from the underwater burst. The light carrier *Independence* was gutted by fire after the air burst and the veteran *Saratoga* sank slowly after the second trial. Other ships within half a mile were either sunk or so badly damaged that they would have needed a major overhaul in a dockyard; and those within a mile, although they suffered less damage, would have had serious radiation casualties among their crews. The general opinion at the time was that ships at sea in dispersed formations would be relatively safe and that one atomic bomb would only sink one ship. The real danger was believed to be to ports and dockyards and to ships berthed close together in harbour.

Atomic bombs were still very scarce and at the time only the U.S.A. was capable of making them. Such bombs as were available would probably be used on more rewarding targets than ships and the study of naval warfare, modified in minor ways by the possibility of nuclear attack, went on much as before.

The British Fleet Air Arm ended the war with 959 operational aircraft, three-quarters of which were fighters. Coastal Command had 785 aircraft, 225 of which were long-range land-planes and 131 were flying-boats. The rest were of medium or short range or designed for attacks on enemy shipping. The British had seven fleet carriers, five light fleet carriers and thirty-eight escort carriers. They had twenty new aircraft-carriers building: four of the 'Ark Royal' class; three of the very large 'Gibraltar' class; and thirteen more light fleet carriers. Two of the 'Ark Royal' class, all the 'Gibraltar' class and four of the light fleets were cancelled and work was stopped on all the others except the *Theseus, Triumph* and *Warrior* which were nearing completion. The veterans *Furious* and *Argus* were scrapped and practically all of the escort carriers, having been supplied under Lend-Lease, were returned to the U.S.A. Half of the Fleet Air Arm and Coastal Command aircraft were also Lend-Lease and had to be returned. This left the Fleet Air Arm with British aircraft only, mostly Seafires, Fireflies and Barracudas of which the performance was subsequently improved by fitting more powerful engines. There were also some Blackburn Firebrands which were single-seater torpedo planes converted from a design originally intended for a fighter. The policy was, however, to give priority to long-term research and only two new types of aircraft, the design of which had been started during the war, came into service in the immediate post-war period: the first was a twin-engined fighter, the de Havilland Sea Hornet, and the second was a single-engined fighter, the Hawker Sea Fury. Few Sea Hornets were built but the Sea Fury, which proved an outstanding success, became the standard fighter and replaced the Seafire. Long-term research was concentrated on experiments to operate jet fighters from carriers and on 3rd December 1945 a Sea Vampire successfully landed on H.M.S. *Ocean.*

The United States ended the war with 20 fleet carriers, 8 light fleet carriers and 71 escort carriers. They had a grand total of forty thousand maritime aircraft in use for operations and training and sixty thousand pilots. Six fleet carriers of a new class, which were armoured, were building as well as thirteen of the 'Essex' class and two light carriers. Eleven of these were cancelled but two large fleet carriers of the 'Midway' class were completed in 1945 and a third in 1947. The excellent Avenger, Corsair and Hellcat aircraft remained in service for some years after the war and were joined by the Douglas AD-1 Skyraider, a single-seat strike aircraft and the Grumman F8F Bearcat, an improved single-seater fighter. Experiments were made with several types of jet fighter, a McDonnell FD Phantom[15] landing on the new carrier *Franklin D. Roosevelt* in July 1946 and the first operational squadron of North American FJ-1 Furies going to sea in 1948. Mention must also be made of the Lockheed P2V Neptune, an outstanding twin-engined shore-based patrol plane. One of this type called the 'Truculent Turtle' established a world distance record in September 1946 by flying 11,236 miles.

By 1950, on the eve of the Korean War, the aircraft-carrier was accepted in the Royal Navy as the backbone of the fleet. All the battleships were in reserve or had been scrapped except the *Vanguard* which was used as a sea-going training ship. The Commander-in-Chief of the Home Fleet now flew his flag in the aircraft-carrier *Implacable* and the light fleet carriers *Vengeance*, *Theseus*, *Glory* and *Triumph* were in full commission. Their air groups consisted mostly of Sea Furies and Fireflies with a few Seafires, Sea Hornets and Firebrands. Four other aircraft-carriers were in commission in the second line, the *Illustrious* for deck landing training and trials, the *Warrior* for experimental work, the *Ocean* as a troopship and the *Indefatigable* for the training of seamen. The *Victorious*, *Indomitable* and *Formidable* were in reserve. There were another fifteen squadrons of aircraft in the second line, mostly for training, and twelve reserve squadrons manned by the R.N.V.R. In emergency therefore, there were plenty of aircraft for air groups for the second-line carriers. By 1950, British naval air policy, with Soviet Russia as practically the only potential enemy, was to put anti-submarine warfare first followed by the air defence of the fleet and convoys, and lastly air strike against land or ship targets. The requirements for a new generation of fleet air arm aircraft were already laid down. The new anti-submarine aircraft, the first ever

Y

specially designed for this role in the fleet air arm, was to be a three-seater turbo-prop machine to carry radar and sono-buoys and to be armed with rockets, depth charges and, as soon as it could be developed, a homing torpedo. Until this aircraft could be produced, the Barracuda and Firefly were adapted for the purpose. There were to be two kinds of fighter, both jet-propelled, one a single-seater high-performance aircraft for use by day and the other a two-seater night or all-weather machine. These roles were at present carried out by the Sea Fury and night-fighter versions of the Sea Hornet and Firefly. The new strike aircraft was to be a single-seater turbo-prop machine capable of carrying a torpedo, bombs or rockets, the role presently carried out by the Firebrand.

In 1950, the United States had a larger force of aircraft-carriers than that with which they had finished the war. They had 27 fleet, 9 light fleet and 66 escort carriers. The vast majority of these ships, however, were now in reserve. They had the three 'Midway' class and four 'Essex' class attack carriers in commission with four light fleet and four escort carriers. American naval air policy had now diverged substantially from that of Great Britain. In 1948, Admiral Nimitz, on retiring from the post of Chief of Naval Operations, wrote to the Secretary of the Navy. In his memorandum he pointed out that until shore-based bombers were available with sufficient range to attack any place in the world from the United States, this task could only be done by carrier-borne aircraft. He drew attention to their success in this way in the Pacific in the Second World War and pointed out that carrier task forces supplied the means to provide fully equipped airfields at short notice any-where in the world. He also said that aircraft-carriers, by using their radar and fighters, could be used for the air defence of the United States. It was clear that this offensive role for the Navy meant that their aircraft must be able to carry nuclear weapons. The atom bomb still weighed some 10,000 lbs and could not be carried by any existing carrier-borne aircraft. Trials were made with the P2V Neptune, normally a shore-based aircraft and weighing 33 tons: one of these machines took off from the carrier *Coral Sea* early in 1949 off the coast of Florida and flew 5060 miles to land at San Francisco. In April 1949, the 75,000-ton aircraft-carrier *United States*, over a thousand feet long, was laid down; she was to carry twenty-four even larger aircraft but the project led to an inter-service controversy over the responsibility for strategic bombing, and was cancelled by the Government before much progress had been made. This led to the resignation of the Chief of Naval Operations. Nevertheless during 1949 the North American AJ-1 Savage aircraft was delivered; this was able to carry a nuclear weapon and, although it was much smaller than the Neptune, it still weighed 25 tons. One of these aircraft successfully landed on the aircraft-carrier *Coral Sea* in 1950 and plans were made to convert the other two ships of the 'Midway' class and three of the 'Essex' class to carry them.

When the fighting broke out in Korea in 1950, the aircraft-carriers U.S.S. *Valley Forge* and H.M.S. *Triumph* were in the Far East. They were used on both coasts of Korea to strike at shore targets and this was kept up by relays of aircraft-carriers for the three years until the armistice, the British sending a succession of light fleet carriers which operated on the west coast. The Americans commissioned more carriers and by 1953 eleven attack carriers had seen service in Korea and a total of eighteen were in commission. The British used piston-engined aircraft throughout the campaign, mostly Sea Furies and Fireflies with some Seafires early on. The Americans, in addition to the Skyraider and other piston-engined types, used the Grumman F9F Panther and the McDonnell F2H Banshee jet fighters. Aircraft-carriers in Korea were used almost entirely to support operations on shore and to help keep air superiority; they also maintained a blockade and prevented any enemy movement by sea. The enemy's few ships never seriously challenged the United Nations' complete command of the sea and this was largely due to the cover provided by the carriers.

One of the aircraft-carriers used in Korea was H.M.A.S. *Sydney*, an Australian ship. Four more countries had by now acquired aircraft-carriers. Australia had a second carrier, the *Melbourne*, completing in Great Britain; Canada had the *Magnificent*; France the *Arromanches* and *Dixmude* and the Netherlands the *Karel Doorman*. All of these except the *Dixmude*, which was an escort carrier, were ex-British light fleet carriers.

During the Korean War, research and development proceeded apace in both Great Britain and the U.S.A. The problem of operating jet aircraft from carriers was complicated by their high landing speed and very short endurance; they were also heavier than piston-engined types, and they needed a long take-off run. The use of jet aircraft, even from the larger Second

World War carriers, was a marginal operation. The American solution was to lay down in 1952 the *Forrestal*, a very large aircraft-carrier of 78,000 tons, 1100 feet long, thereby obtaining their very large aircraft-carrier on another pretext (see Fig. 23, p. 329). It was three British developments, however, which solved the problem of operating jet aircraft from carriers. These were the angled deck, the mirror landing aid and the steam catapult. The angled deck made it possible for a deck park of aircraft to be kept forward and at the same time have a clear deck for landing. An aircraft which missed the arrestor wires would not therefore crash into a barrier but could go round again. The mirror landing aid was an automatic device to guide in aircraft to land and took the place of a landing signals officer or 'batsman'. The steam catapult was a great improvement on former compressed air or cordite varieties and was sufficiently powerful to launch any new jet aircraft. All these inventions were embodied in the *Forrestal* before she was completed in 1955 and five more similar gigantic carriers were laid down at the rate of one each year. The new generation of aircraft which came into service in 1956–7 for these carriers were all jet-propelled. The Douglas A3D Skywarrior, designed to carry a nuclear weapon for 1000 miles at over 600 m.p.h. weighed 26 tons. The Douglas A4 Skyhawk was designed as a result of the Korean War as a much smaller attack aircraft; it weighed only 10 tons but could carry 8200 lbs of weapons. The Douglas F4D Skyray was a delta-wing interceptor fighter with a rate of climb of 18,000 feet a minute and a speed of 695 m.p.h. at 30,000 feet, but the L.T.V. F8 Crusader, which came into service only a year later, was supersonic with a speed of 1120 m.p.h. at 40,000 feet.

There is no doubt that the 'super carrier' with its 85 jet aircraft, some of which carried nuclear weapons, was fully capable of meeting Admiral Nimitz's strategic bombing plans of 1948. By 1955, however, the U.S.A.F. had the B52 strategic bomber which with flight refuelling could reach any target in the world. For some time they had also had bases in Europe and the Far East from which they could reach any point in Russia and so the concept of strategic bombing by carrier planes was out of date by the time it could be put into practice. By this time, however, the North Atlantic Treaty had been signed and the sea communications across the North Atlantic were of increased importance to the U.S.A. The strategy was therefore

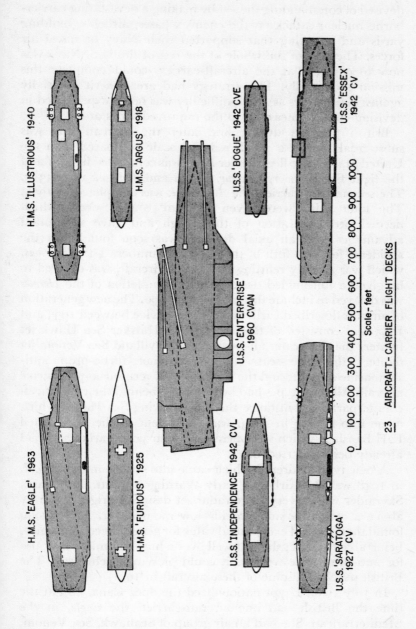

H.M.S. 'ILLUSTRIOUS' 1940

H.M.S. 'ARGUS' 1918

U.S.S. 'BOGUE' 1942 CVE

U.S.S. 'ESSEX' 1942 CV

H.M.S. 'EAGLE' 1963

H.M.S. 'FURIOUS' 1925

U.S.S. 'ENTERPRISE' 1960 CVAN

U.S.S. 'INDEPENDENCE' 1942 CVL

U.S.S. 'SARATOGA' 1927 CV

Scale—feet

0   100   200   300   400   500   600   700   800   900   1000

23   AIRCRAFT-CARRIER FLIGHT DECKS

devised of commanding the sea by making a devastating carrier-borne nuclear attack on the enemy's bases, airfields, building yards and anything that supported their Navy or naval air forces. The task of the whole of the rest of the U.S. Navy was now to ensure that the aircraft-carriers could complete this mission successfully. This strategy had great merit especially against submarines as much difficulty was being experienced in devising counter measures to the improved types at sea.

British progress during and after the Korean war was substantial but on a much reduced scale compared with the United States. The fleet carrier *Eagle* was completed in 1951 and the light fleet carriers *Centaur*, *Albion* and *Bulwark* in 1953–4. The second fleet carrier, the *Ark Royal*, was completed in 1955. The later ships were given 'interim' angled decks which needed little alteration of the design and gave substantial advantages over an axial deck. It was soon found that the accidents fell to a fifth of their former number. The *Victorious*, which was already refitting to take jet aircraft, was ordered to be given a full angled deck and the completion of the *Hermes* was delayed to include this modification too. The new generation of aircraft described earlier came into service between 1953 and 1955 and consisted of the single-seater Hawker Sea Hawk jet fighter; the two-seater all-weather de Havilland Sea Venom jet fighter; the three-seater Fairey Gannet turbo-prop anti-submarine aircraft; and the Westland Wyvern turbo-prop strike aircraft. All these types had begun to fall behind their American counterparts. For instance the straight wing Sea Hawk, which came into service in 1953, was equivalent to the McDonnell F2H Banshee, which was operational four years earlier and had already been superseded.

A new type of aircraft which came into service in the U.S.N. in 1948 was the Airborne Early Warning aircraft. This was a Skyraider with a very large radar set designed originally to fly above a task force and provide low radar cover. It was soon found that it was of exceptional value for general reconnaissance, being able to detect ships at well over a hundred miles, and also for anti-submarine work as it could pick up a schnorkel. The British purchased some of these aircraft in 1951.

In July 1956, Egypt nationalized the Suez Canal and at the time the British had one aircraft-carrier, the *Eagle*, in the Mediterranean. She had an air group of Seahawk, Sea Venom, Wyvern and Gannet aircraft. During the autumn her Gannets were replaced by additional fighters and the light fleet carriers *Bulwark* and *Albion* were sent out from the United Kingdom. The air groups of the three carriers totalled six squadrons of Seahawks, five squadrons of Sea Venoms and two squadrons of Wyverns. After Israel had attacked Egypt on 29th October, this carrier task force, escorted by six destroyers, was used in the Anglo-French intervention to gain air superiority over Egypt and to support the landings. Aircraft from the two French aircraft-carriers *Arromanches* and *Dixmude* and R.A.F. fighters based in Cyprus co-operated. At Suez this powerful and self-contained carrier force showed its great value, not only for its aircraft but for its communications and the fact that it brought everything with it. In this operation the light fleet carriers *Theseus* and *Ocean*, in commission for training seamen, embarked No. 845 naval helicopter squadron and some army and air force helicopters and were used to land Royal Marine Commandos.

It was in the nineteen-fifties that the helicopter became an important maritime aircraft. Obviously its ability to land in a very small space and to hover made it particularly useful. The U.S.N. had used them for rescue duties and general communications in aircraft-carriers from 1950 when the Sikorsky S.55 came into service. In 1951 a troop-carrying variety was used by the U.S. Marine Corps and in 1954 an anti-submarine type with the 'dunking sonar' was produced. Dunking sonar was an asdic set which could be lowered from a hovering helicopter into the sea to detect submerged and silent submarines. The British acquired some American helicopters and then the Sikorsky types were built under licence in England by Westlands. A training squadron was formed in 1950 and an operational squadron in 1952 which was used ashore in the Malayan emergency. The first British anti-submarine squadron was formed in 1954.

The British exploded their first nuclear weapon in 1952 and their thoughts soon turned to its use by carrier-borne aircraft. Although a full-scale offensive against enemy bases in the American style was never contemplated, a system was thought out in which the mobility of the aircraft carrier could be exploited had to close the enemy coast at unexpected times from unexpected directions to launch a low-flying jet aircraft to approach below the enemy radar beams. The actual attack would be made by a toss bombing technique which would allow the aircraft to get clear before the explosion. Such a system could be of use at sea against an enemy surface force as well.

Development of a special aircraft strong enough to fly fast at low altitudes and of a suitable nuclear weapon were put in hand.

Shore-based maritime aircraft were now all of the very long range type and mostly landplanes although a few flying-boats remained in service. The British used the Avro Shackleton and the Americans the P2V Neptune. Their main function was anti-submarine: they were equipped with radar for detecting schnorkelling submarines, with sono-buoys and sometimes with a magnetic anomaly detector to find them when submerged. They were able to detect a submarine on the surface at considerable distances and could often pick up the schnorkel of a submerged submarine, but once a submarine had submerged completely and was proceeding at slow and silent speed, they had little chance to find it. The U.S.N. formed special 'hunter-killer' groups in which an aircraft-carrier co-operated with a group of escort vessels. In 1954 the carrier-borne Grumman S2 Tracker, a conventional twin-engined aircraft, the equivalent of the British Gannet, entered service equipped in the same way as the Neptune.

With the advent of the Royal Air Force high-flying jet 'V' bombers, the laying of mines from the air became impracticable. Indeed it was doubtful whether minelaying had any relevance in the context of the short and devastating nuclear war envisaged at the time. Nuclear attack on ships by this type of aircraft was, however, a distinct possibility; they were hardly slower than jet fighters and they flew too high for anti-aircraft gunnery. It was to meet this kind of attack that the American Terrier and the British Seaslug shipborne guided missiles were designed.

In 1962 the American carrier striking force was at its zenith. It now consisted of the super-carriers *Forrestal, Saratoga, Ranger, Independence, Kitty Hawk, Constellation* and *Enterprise*. These were backed by the three 'Midway' class and eight of the modernized 'Essex' class. The air groups, of up to 85 planes in the largest carriers and 60 in the smallest, consisted of a new generation of aircraft just coming into service. The North American A5 Vigilante was a new all-weather supersonic nuclear strike aircraft; it weighed 27 tons, with a speed of 1385 m.p.h. at 40,000 feet and a range of 3000 miles. Another, the McDonnell F4 Phantom II, was an outstanding aircraft in every way; an 'air superiority' supersonic all-weather fighter carrying Sparrow III air-to-air missiles, it had a speed of 1,485 m.p.h. at 48,000 feet, a ceiling of 62,000 feet and a rate of climb of 28,000 feet a minute. It also had an exceptional strike capability and could carry 16,000 lbs of bombs, rockets or missiles. A third new type was the Grumman A6 Intruder which was a low-level, long-range strike aircraft designed to attack below the radar beams and so have almost complete immunity from ground-to-air missiles. It had a speed of 640 m.p.h. at sea level and could carry 18,000 lbs of bombs or missiles for 3000 miles or so. The Crusader supersonic fighter and the Skyhawk light attack aircraft remained in service as did a number of earlier types until they could be replaced.

By 1962 the U.S.N. had surface-to-air guided missiles in service to protect the carriers. The first was the Terrier, already mentioned, with a range of 17 miles, weighing 3000 lbs and able to compete with attack by aircraft using free-falling bombs. The second was Talos, a much larger weapon weighing 7000 lbs with a range of 50 miles, and so able to compete with stand-off-weapon launching aircraft or indeed with guided missiles of the cruise type. A twin Talos launcher with about 40 missiles was fitted in three existing light cruisers, the *Galveston*, *Little Rock* and *Oklahoma City* and the new guided missile cruiser *Long Beach*. Twin Terrier launchers were fitted in the carriers *Kitty Hawk* and *Constellation*, in the converted heavy cruisers *Boston* and *Canberra* and in those of the 'Providence' class and as secondary armament in the *Long Beach*. They were also fitted in the new guided-missile destroyer *Bainbridge* of 8580 tons and in the ten ships of the new 'Farragut' class of 5800 tons.

The super carrier *Enterprise*, the guided missile cruiser *Long Beach* and the guided missile destroyer *Bainbridge* were all nuclear powered. These ships cost 37 per cent or so more than if they had been conventionally powered and were no faster. Their principal advantage was that they had, for practical purposes, an unlimited endurance and at high speed at that. The *Enterprise* had more room for aircraft and their fuel and emitted no funnel smoke, but these advantages were very marginal and it was by no means certain that they were worth their extra cost.

The United States Navy in 1962, based on the super carrier, was therefore a force of immense power. It had a devastating nuclear strike capability by supersonic high-flying jet aircraft or subsonic low-flying 'under the radar' machines. With replenishment at sea, it had a world-wide endurance and could strike any place on earth which was within a thousand miles of the sea. Its capacity to defend itself with surface-to-air guided missiles and supersonic fighters using air-to-air missiles was the latest that could be devised. Moreover it did not have to use nuclear bombs but had a substantial striking power with conventional weapons as well. It was therefore a flexible force able to carry out many roles. It could contribute to the strategic nuclear deterrent in the way Admiral Nimitz had visualized in 1948 and it had the power to command the sea by wiping out the enemy's airfields, bases and building yards. As a fleet at sea it was all powerful: it could locate by high-flying radar-fitted aircraft anything on the surface of the sea over a vast area; it could sink any body of ships afloat several times over. It could provide air defence and support for amphibious operations and forces ashore and was a magnificent instrument to provide a U.S. military presence.

It was in 1962, however, that a number of doubts, which had in fact existed for some time, came to a head. The first of these was the carrier force's position as a contributor to the nuclear deterrent. A substantial number of inter-continental ballistic missiles had been added to the U.S.A.F. strategic bombing force and a gradually increasing force of Polaris submarines was at sea. The readiness of these systems and their relative invulnerability made them greatly superior to carrier-borne aircraft as a means to mount the deterrent. This role of nuclear deterrence was therefore deleted from the functions of the U.S. carrier force. This still left the function of commanding the sea by a nuclear strike on the enemy's bases, but this was also in trouble. Ever since the Russians first exploded a nuclear weapon in 1949, they had been building up their nuclear striking forces. A position of nuclear stalemate had now been reached in which the U.S.A. could not use nuclear weapons against Russia without the risk of a devastating nuclear counter-attack on the U.S.A. itself. In the circumstances a nuclear attack on naval and air bases in Russia was clearly liable to escalate into all-out nuclear war and was no longer possible as a strategy.

The threat of the American carrier striking force to the Soviet Union had, for some years, stimulated counter measures in the Russian Navy. By the late nineteen-fifties they already had 475 conventional submarines, mostly of ocean-going types. In

1958 they began to convert some of their larger conventional submarines to carry ballistic missiles. Although at first they were armed with an adaptation of a land-based weapon with a range of under 400 miles and they had to surface to launch them, they constituted a threat to the United States. The Russians developed nuclear-propelled submarines in the early nineteen-sixties, and soon designed a ballistic missile version. They also produced what was clearly a tactical cruise missile to be used from submarines against ships at sea in place of the torpedo. It seemed quite possible that these tactical missiles and probably also some of their conventional torpedoes, had nuclear warheads. At the same time a new type of large destroyer with a surface-to-surface missile armament came into service. It was known that the Russian naval air force had a number of Tu.16 Badger jet bombers with a range of over 3000 miles and that they were armed with stand-off missiles probably with nuclear warheads with a range of 50 miles. There were also a number of long-range four-engined turbo-prop Tu.95 Bear aircraft capable, with flight refuelling, of reaching the latitude of Gibraltar from bases in North Russia.

It was clear that the development of the Russian Navy posed two entirely new problems. The first was a threat to bombard the United States with missiles launched from submarines and the second was to give battle to the American carrier striking force if it came anywhere near the Russian coasts. The threat to the United States was, in fact, taken care of by the general deterrent policy but nevertheless it stimulated thought on anti-submarine warfare in the United States. The Russian missile destroyers, especially if their weapons had nuclear warheads, were clearly a serious menace to the American carrier strike forces. With A.E.W. reconnaissance machines, the U.S.N. should certainly be able to find them and their many strike aircraft should be able to attack them. The Russians, however, were beginning to mount surface-to-air guided missiles in their ships and the Bullpup, the American air-to-surface missile, had a range of under seven miles. The Russian Badger aircraft with their stand-off weapons were a more serious menace. The American fighters, with their supersonic performance and air-to-air missiles, would undoubtedly be able to shoot many of them down before they reached the launching point but the Talos missile would scarcely have the chance to engage them in time. Nevertheless if the Russian aircraft used nuclear warheads on their missiles the threat was extremely serious. Air defence would have to be virtually one hundred per cent effective if disaster was to be averted. Even the most modern of the American carriers could be completely destroyed by a nuclear explosion within half a mile and could be damaged so that it would be unable to operate aircraft from a greater distance. The submarine threat was probably the greatest of all. They were likely, in spite of the best anti-submarine measures, to be able to launch their missiles and torpedoes without being detected and if they had nuclear heads the effect could be devastating. The nuclear submarines had considerable mobility and, working with reconnaissance aircraft, had a good chance to make contact. To sum up, the U.S. carrier force might win if the Russians used conventional warheads on their missiles but the threat was appalling if the Russians used nuclear warheads.

The need for the American carrier striking forces to try conclusions with the Russian Northern Fleet in its own waters was by 1962 much less. The U.S. carriers were no longer required to contribute to the general nuclear deterrent and they could only bombard the Russian naval and air bases in the unlikely 'broken-backed' phase of a war. It was still planned that they should give air support to NATO forces ashore in Europe but it seemed doubtful if this was worth the risk. In 1962 they became engaged in two more limited actions of a type that were to become their principal role; the first was the blockade of Cuba in which the super carriers *Enterprise* and *Independence* took part and the second was in the war in Vietnam in which relays of aircraft-carriers were used in the same way as in Korea.

The British development of their aircraft-carrier force was on a much more modest scale. In the 1957 Statement on Defence, the British put their faith in a nuclear deterrent strategy based on the 'V' bombers and the inter-continental ballistic missile Blue Streak. The Statement included the somewhat damping remark that the role of naval forces in total war was uncertain. They had already decided, however, to have three aircraft-carriers with a fully angled deck, steam catapults, mirror landing aids and a new three-dimensional radar. In 1958 the *Victorious*, already seventeen years old, emerged from a seven-year refit with all these alterations; she was followed in 1959 by the completion of the much smaller light fleet carrier *Hermes* which had been building ever since the war. At the same time the *Eagle* was taken in hand for modernization; meanwhile the *Ark Royal* and *Centaur* with interim angled decks continued in service. The *Bulwark* and *Albion* were converted to Commando helicopter carriers but all earlier aircraft-carriers were by now either sold, scrapped or on the disposal list.

At this time a new generation of aircraft came into service. They were the Supermarine Scimitar, a single-seat swept-wing fighter which was also a strike aircraft and could carry a nuclear weapon; and the de Havilland Sea Vixen, a two-seat swept-wing all-weather fighter. Both of these aircraft carried air-to-air guided missiles and had a greatly improved performance over the Seahawk and the Sea Venom. Nevertheless they were subsonic, except in a dive, and not in the same class as the American Crusader and Phantom II. Two years later they were joined by the Blackburn Buccaneer low-level high-speed strike aircraft, also able to carry a nuclear weapon but also subsonic. The anti-submarine Gannet was now replaced entirely by the Westland Whirlwind helicopter, which, in its turn was superseded by the larger turbine-engined Westland Wessex. These helicopters were used both for landing Commandos and for anti-submarine work. Finally a version of the Gannet replaced the Skyraider as the Airborne Early Warning aircraft.

The British carrier force was therefore very small compared with the American. Nevertheless with its nuclear strike capability, its swept-wing fighters with air-to-air missiles directed by its three dimensional radar and its A.E.W. aircraft for reconnaissance, it was still a force of immense power. It was further strengthened by the advent of the Seaslug surface-to-air missile, equivalent to the American Terrier, carried in guided missile destroyers of the 'Hampshire' class which provided a significant improvement in the air defence capability of the carrier force. One carrier, however, had always to be kept east of Suez and another was generally refitting and so the three modern carriers were seldom available to act as a single force. The role of nuclear strike in general war, although practised by the Buccaneers, was never taken very seriously and the case for the carriers became more and more their use in limited wars such as the Kuwait crisis and the East African mutinies.

France completed the new light fleet carriers *Foch* and *Clemenceau* of 32,800 tons in 1961 and 1963 carrying Etendard IV fighter-bombers, Breguet Alise turbo-prop anti-submarine and reconnaissance machines and some American Crusaders.

They also completed the *Jeanne D'Arc,* a special helicopter carrier, in 1964. Argentina, Brazil and India acquired the ex-British light fleet carriers *Warrior, Vengeance* and *Hercules* between 1956 and 1961. The Indian carrier renamed *Vikrant* had an air group which included Seahawks but the South American carriers were little more than status symbols. The Netherlands used the *Karel Doorman* as an anti-submarine carrier with Grumman Tracker aircraft and Seabat helicopters and the Canadians followed the same policy with the *Bonaventure.* The Australians, on grounds of expense, decided to discontinue their naval air arm in 1960 and the *Sydney* became a fast military transport and training ship.

The cost of maintaining aircraft-carriers also began to worry the British. The *Victorious* had originally been completed in 1940 and would have to be replaced before long. The cost of yet another generation of aircraft at £2 M. each also promised to be formidable. The 1962 Statement on Defence had emphasized the tri-service aspect above all else. With the Kuwait crisis fresh in their minds, the Government stated a requirement for amphibious concentrations based on the Commando carrier and a new type of assault ship to carry tanks and heavy equipment, and laid down that the function of the Fleet Air Arm was to carry out reconnaissance, tactical strike, close support and air defence for such operations. This policy not only had the full support of the British Fleet Air Arm but was the Admiralty's choice of what they called a maritime strategy. The main strength of the Royal Navy was thereby tied up in a peace-keeping' or 'fire brigade' role east of Suez.

In 1966 the axe fell: the British Government put a ceiling of two thousand million pounds on the Defence budget and it was the aircraft-carriers which had to go. Their reasoning was that carriers would only be required to support an opposed landing against sophisticated opposition outside the range of shore-based aircraft. They decided not to build a new aircraft-carrier to replace the *Victorious* but to modernize the *Ark Royal* to take the new F4 Phantom aircraft recently ordered in America. They intended gradually to phase aircraft-carriers out of the Royal Navy but did not expect all of them to go before 1975. This decision caused somewhat of a political upheaval. Both Mr Mayhew, the Minister for the Navy, and Admiral Luce, the First Sea Lord, resigned. The British forces were still to have a peacemaking role east of Suez and Mr Mayhew felt that if this was to be so, carriers were essential. The contention of some naval commentators that F.111 aircraft, working from shore bases, in spite of their great endurance, were no substitute for carriers, had much merit. The argument that the 'fleet' must have carrier air support was, however, weakened by the fact that if the carriers went it was difficult to see what constituted the 'fleet'. In the end, the F.111 aircraft were never put to the test as Great Britain decided to withdraw from east of Suez altogether and this decision sealed the fate of the British aircraft-carriers. The decision to phase out the carriers was made mainly on financial grounds: costs had risen sharply since the war, for instance the *Hermes* completed in 1960 cost £18M against the £3·8M for the larger *Illustrious* in 1940. Her air group cost over £10M against a mere £600,000 for that of the *Illustrious*. The policy for the Fleet Air Arm had always been to have nothing but the best and this proved very expensive. The Admiralty rejected a policy of cheaper and simpler aircraft of the American Skyhawk type on the score that our aircraft were out of date anyway by the time they were in service and that the situation would be worse still if they designed them to be inferior at the outset. They could also show that the likely opposition already had sophisticated equipment supplied either by Russia or the west. Nevertheless if the role was really to be peacekeeping east of Suez, an air group of sophisticated nuclear attack aircraft can only be seen as an expensive luxury. There was also much debate about the size of the proposed new carrier *CVA.01*. Obviously the smaller the carrier the cheaper it would be but the number of aircraft it could carry decreased dramatically as the size of carrier got smaller. For instance the *Hermes* of 27,800 tons carried only eighteen Scimitars and Sea Vixens and a few helicopters, whereas the *Eagle* of 54,100 tons carried 41 Buccaneers, Sea Vixens and Wessex Helicopters. The size finally decided upon for *CVA.01* was 53,000 tons to carry 36 Phantoms and Buccaneers and 11 Helicopters. This was much smaller than the American super carriers and slightly smaller than the *Eagle*, nevertheless the initial cost of such a ship and her air group was substantial and rather than face it, the Government adjusted their strategy east of Suez to do without it.

The Admiralty had based their whole case for the retention of aircraft-carriers on this peacekeeping role east of Suez and were unable to justify the retention of carriers on any other grounds. The Fleet Air Arm had never relished the idea of a nuclear attack on the enemy bases in general war and they were undoubtedly right. They tacitly agreed with the 1957 Statement on Defence that the role of naval forces in general nuclear war was uncertain. They came more and more to rely on a case for the retention of carriers based on their use in limited war in which they would not have to face the threat of nuclear weapons. They thereby made the case for carriers on a requirement which was secondary and by no means vital. Since the Second World War, the need to command the sea had become obscured by limited war operations in which it had not been challenged. The weakness of the Russian Fleet in 1945 and the overwhelming superiority of the Americans meant that it had come to be taken for granted. The evolution of a nuclear-missile-armed Russian Fleet in the nineteen-sixties made little impact: it was almost ignored as irrelevant while a so-called maritime strategy was devised to land and support a small military force in the Indian Ocean. One wonders whether there was not some truth in the suggestion that it was not the survival of an amphibious task force which depended upon aircraft-carriers so much as the survival of aircraft-carriers which depended upon the concept of an amphibious task force. To others it was clear that if aircraft-carriers, which constituted the main strength of the British Navy, could not face the main threat to the command of the sea, then they had better make way for some system which could. The growth of this subsidiary role until it involved the main strength of the Royal Navy will probably be seen by history as a major strategic blunder. It is likely to be quoted as an example of over-zealous inter-service co-operation to the neglect of the basic principle that a fleet must command the sea before it embarks upon an amphibious operation.

The Government clearly had doubts whether an amphibious attack outside shore-based air range against sophisticated opposition could in any case be effectively supported by the British aircraft-carriers. One carrier east of Suez could only bring forty aircraft to bear at the most. If the opposition had surface-to-air missiles, casualties would mount and the strength of the air group would rapidly dwindle. At the same time the end product of the so-called maritime strategy was to land no more than a brigade group and they found it difficult to visualize many occasions where such a small force could be effective. They had, therefore, apart from a political antipathy to

'gunboat' diplomacy, serious doubts about its military effectiveness. If this strategy was the only remaining justification for
carriers then they felt that it was better to save the money and
do without them. Nevertheless Great Britain had undoubtedly
lost an effective way to intervene to a limited extent east of Suez
and the protagonists of the carrier were able subsequently to
point to the success of H.M.S. *Eagle* in the blockade of Beira.
The blow was, however, more to the prestige of the Royal Navy
than anything else. The Fleet Air Arm, though small, was
extremely efficient and it was a bitter pill for the inventors of the
aircraft-carrier to have to swallow to be among the first to have
to give them up.

The decision not to replace the British carriers did not mean
the end of aircraft operating over the sea. Three carriers, the
Eagle, *Ark Royal* and *Hermes* were still in existence in 1969 and
only the *Victorious* had been scrapped. The two large carriers
had the splendid F4 Phantom II and the Buccaneer in their air
groups as well as helicopters. The two Commando carriers were
to remain in service with their helicopters and the guided missile
destroyers and the frigates, which already carried some sixty
helicopters, were to continue to do so. The helicopter force was,
in fact, to be increased by the conversion of the three 'Tiger'
class cruisers to helicopter carriers, and by the time the aircraft-
carriers are phased out in 1975 or thereabouts will constitute the
New Fleet Air Arm. The Shackletons of Coastal Command are
to be replaced by a new jet aircraft, the Nimrod, and their role
is laid down as the surveillance of vital sea areas and the anti-
submarine protection of surface forces and merchant shipping.
The air defence of the fleet is also to be taken over by the Royal
Air Force using fighters from the shore. As this book goes to press
the Royal Navy is beginning to show interest in VTOL aircraft[16],
which, although they cannot compete with the very sophisticated supersonic fighters such as the F4 Phantom II, are very
much better than nothing and can be used from most of the
ships now carrying helicopters. The capacity of the rest of the
fleet to deal with aircraft is also greatly improved. The British
now have eight guided missile destroyers of the 'Hampshire'
class armed with the Seaslug medium-range and the Seacat
short-range surface-to-air guided missiles. Under construction is
the *Bristol*, a new 6750-ton frigate with the new Seadart missile.
The Seadart is expected to have a better performance than the
Seaslug at very high and very low levels and to have an anti-ship

z

capability. Nevertheless it has not the range to compete with aircraft using stand-off weapons.

The United States Navy has been unmoved so far by the British decision to build no more aircraft-carriers. A conventionally propelled super carrier, the *America*, had been completed in 1965 and another the *John F. Kennedy* was commissioned in 1968. On the stocks is another nuclear-propelled aircraft-carrier, the *Nimitz*, and two more are proposed. At the time of writing, the U.S. Navy has a total of 24 aircraft-carriers, 15 of which are attack carriers and 8 are anti-submarine support carriers. They also have a training carrier and some Commando carriers with helicopters. Eight of the attack carriers are super carriers of the 'Forrestal' and succeeding types, three are 'Midways' and five still of the 'Essex' class. Their air groups are composed of supersonic fighters, either F4 Phantom II or F8 Crusaders, but the heavy attack planes of the Vigilante and Skywarrior types have all been replaced by subsonic Skyhawks or the Skyhawk's replacement, the A7 Corsair; and by A6 Intruders which still have a nuclear capability. Their air groups also include radar reconnaissance planes and aerial tankers as well as helicopters. The U.S. Navy now has some seven ships armed with the long-range Talos missile and thirty-three with Terrier, all available to escort the carriers. This carrier force is still of immense strength and sufficiently flexible to be of use over a wide range of operations from a nuclear striking force to counter insurgency. The attack carriers are divided into four fleets: the First in the Pacific; the Second in the Atlantic; the Sixth in the Mediterranean and the Seventh in the Far East. The Seventh Fleet provides carriers for the Vietnam War and would, in emergency, be able to mount a heavy nuclear attack on Communist China, and so still acts as a local deterrent force. The Second Fleet is available to support the Centre and the Northern flank of NATO and the Sixth Fleet the southern flank: it also provides a force well able to compete with the Russian ships which have recently taken up their station in the Mediterranean. The eight anti-submarine carriers include both fixed-wing anti-submarine aircraft and helicopters in their air groups and they are supported by over 400 Neptune and Orion patrol planes based ashore.

In the other Navies of the world, the Netherlands have given up their only carrier but the Australians have changed their minds and have kept one of their carriers for anti-submarine work. There was a general increase in helicopters and helicopter carriers and many ships especially in NATO are now armed with American Terrier and Tartar surface-to-air guided missiles.

The Soviet Navy is now the second largest in the world. It includes seven ships of the 'Kresta' and 'Kynda' classes with a new Shaddock surface-to-surface cruise missile and also the Goa surface-to-air missiles. There are sixteen ships of the 'Krupny' and 'Kildin' classes with the older Strela surface-to-surface cruise missiles and ten of the 'Kashin' class with surface-to-air missiles only. Of great interest are two new helicopter carriers of 18,000 tons, the *Leningrad* and *Moskva* carrying 30 helicopters each. Its main strength is, however, still in its 380 submarines, 50 of which are nuclear-propelled: 43 submarines carry ballistic missiles and 45 carry anti-ship cruise missiles with a range of 300 miles. The naval air force includes 300 Tu-16 Badger jet aircraft with air-to-surface missiles and fifty Tu-95 Bears for reconnaissance. There are another hundred flying boats and torpedo planes and a hundred anti-submarine helicopters.

XIV

The Future of Aircraft at Sea

IT NOW ONLY REMAINS to speculate on the part that aircraft are likely to play in war at sea in the future. Certainly the number of maritime aircraft now based ashore or carried in ships by the navies of the world seems to indicate that there is confidence at present that they are still very necessary. There are, however, a number of questions about the future to be answered.

Air forces all over the world are agreed that the day of the manned strategic bomber is past and that it has been replaced by the ballistic missile. Have aircraft in fact any real part to play at sea in the nuclear missile age or are the days of aircraft over the sea numbered too? If still required, are they best operated from aircraft-carriers, from other ships or from the shore? It is difficult to obtain a complete answer to these questions solely from a study of the past, and it is a study of the past which has formed the bulk of this book. The advance of science is so rapid that it makes assessment of the future harder still, nevertheless a great deal can be foretold from trends and from the history of the development of aircraft; these may give us a line which may be obscured to the professionals by their proximity to the subject and by the formidable technicalities.

Before making an attempt to predict the future of aircraft and sea power it will, of course, be essential to discuss the nature of war itself in the missile nuclear age. There are many possible types of conflict and opinions vary widely on which kind is most likely. It is, however, often the unexpected that happens in war and in any case recent history shows that ideas about the future change rapidly. It is therefore most necessary to approach the subject with a flexible mind and some possibilities need to be considered, whether they seem likely at this moment or not.

The range of possible conflicts is very wide, it extends from total nuclear war to small police actions like that in Anguilla in 1969. Unlimited general nuclear war is now accepted by both the super powers as mutual suicide and a calamity to be avoided at all costs. Little consideration is therefore now given to fighting this type of war; faith is put into a policy of deterrence. This has the curious effect that if the forces allocated for the purpose have to be used they will have failed. The main weapons of deterrence today are ballistic missiles but the sea has become involved ever since the American aircraft-carriers embarked aircraft which could carry nuclear weapons. Seaborne strategic missiles are now carried by both sides mainly in submarines and this has stimulated anti-submarine warfare on both sides as a protection against them. In the same way it was the American carrier striking forces' nuclear capability which stimulated anti-ship counter measures which the Russian Navy have made their main concern for the last two decades.

If such a disaster as a total nuclear war ever overtook mankind it seems unlikely that sea-power would have much purpose. The continuation of the struggle in a 'broken backed' phase does not seem likely and could not continue for long with sophisticated weapons which had no industrial backing. Nevertheless war under such conditions has been studied and thought about a great deal in the last twenty years. There is little doubt that the American carriers, when they ceased to contribute to the deterrent forces, believed that they had a good chance to survive a nuclear exchange and so be available with their nuclear capability to throw in their weight and decide the issue. It is conceivable that, if the American nuclear attack was concentrated on cities or missile launching sites, the Russian fleet with its naval and air bases could also have escaped destruction. A sea fight might therefore take place, if for no other reason than that there would be no one left ashore to tell them to stop! It is difficult, however, to believe that, after a nuclear exchange, such a use of the American carrier force would make much difference. Certainly there is no case to retain them for this role alone which could be better done by a small expansion of the Polaris force.

Limited wars short of general war are now considered to be much more likely to break out. Limited wars, such as Korea, Suez and Vietnam have been limited not only in a geographical sense but in the scale of weapons that have been used. In all of them sea power has been of extreme importance and without it the West would not have been able to intervene. This sea power has, however, never been challenged except in a very minor way. There seems no reason in the future why this truce should continue. What is more significant, there seems no reason why tactical nuclear weapons should not be used to dispute the command of the sea.

If general nuclear war is accepted as the ultimate catastrophe to be avoided at all costs then both sides will be very reluctant to escalate; the danger of escalation is therefore much reduced and the overall situation is inherently stable. The belief that if one side has its 'vital' interests threatened, it will inevitably resort to an all-out nuclear attack is no longer valid. Such an attack would be suicidal and the effects far worse than the loss of any 'vital' interest. The only occasion when either side is likely to unleash a nuclear holocaust is if it believes that its opponent has already set its nuclear attack in motion. Escalation may therefore occur if even a small nuclear attack is made on enemy territory. Thus, a nuclear attack on enemy air or naval bases as a way to command the sea cannot be contemplated. On the other hand the use of tactical nuclear weapons actually at sea seems unlikely to escalate into all-out nuclear war. Is it conceivable that one country would say to the other, 'You have used a nuclear depth charge on one of my submarines and so I will reply with unlimited nuclear war on your country' and in parenthesis 'on mine as well'? The nuclear powers are doing their best to prevent the proliferation of nuclear weapons and they may well succeed. If they do not, then of course the danger of the use of nuclear weapons in limited war is very greatly increased. Whatever the outcome of these deliberations may be, it would seem that the use of such weapons at sea in limited war of the future must be considered as a distinct possibility.

In the last few years the likelihood of a new kind of limited war which would be limited in the sense that it would only be maritime, has been put forward. It is suggested that this could vary from a guerrilla-type campaign against shipping confined to a certain area to a full-scale attempt to cut off Europe from America. It has been pointed out that the use of Polaris could not be contemplated to stave off defeat in such a campaign. This would be like saying: 'If you continue to starve me, I will commit suicide'. The only valid defence would be to regain the command of the sea in the time honoured manner. There seems no reason why the same arguments on the use of nuclear weapons put forward in the last paragraph should not apply to this kind of struggle and that tactical nuclear weapons could be used without fear of escalation. If this is accepted then it is clear that maritime forces in the future must be prepared to fight both with and without nuclear weapons and, as it will never be known whether they will be used, must always be in an 'anti-nuclear posture'. By an 'anti-nuclear posture' is meant that they must be composed of units of the right type disposed in such a way that they cannot be knocked out completely by a surprise nuclear attack.

It seems probable that the aim of actual operations in a war at sea in the future will be much as in the past. They will be concerned with attacking and defending trade, the mounting and protection of amphibious operations and defence against invasion. They will continue to support forces on shore, to enforce blockades and carry out many other conventional operations. To these will be added the all-important mounting of the seaborne deterrent and attempts to find and destroy the enemy's nuclear deterrent forces. In all of these, ships, submarines and aircraft may well be involved with or without missiles which may or may not have nuclear warheads. The nature of such a struggle can probably best be illustrated by following an imaginary contest between the American Second Fleet and the Russian Northern Fleet.

We will take it that the American Second Fleet, consisting of four 'super carriers' with 350 aircraft embarked and escorted by three Talos-equipped cruisers and ten Terrier-equipped destroyers, advances into the Norwegian Sea to support the northern flank of NATO. Nuclear weapons have not so far been used ashore and the Americans have instructions not to use them first. The Second Fleet is supported by shore-based maritime aircraft in Iceland and the United Kingdom, by nuclear submarines sent on ahead and by two anti-submarine warfare carrier support groups. The Russian Fleet consists of 170 submarines, 30 of which are nuclear-propelled and the same number fitted with anti-ship cruise missiles with nuclear warheads. It has 150 Tu-16 Badgers with fifty Tu-95 Bears with 50 mile stand-off weapons with nuclear heads. Finally there is a surface force built round 15 cruisers and destroyers armed with anti-ship missiles with nuclear heads with a range of 300 miles and also with medium range surface to-air-missiles.

The outcome of such a battle cannot be predicted with

certainty and would depend much on training and weapon efficiency and probably more still on luck. It would start with an approach phase in which the supersonic American fighters with their air-to-air missiles would probably shoot down many of the Russian Bear reconnaissance machines. In dispersed formation, the American Second Fleet would cover an area as large as Iceland and some of the units, if they avoided detection by aircraft, would be likely to be seen or heard by the many Russian submarines on patrol. In addition there would be electronic listening devices and in the future probably satellite reconnaissance as well. It would therefore be expecting too much to hope that no reports would get through to the Russian command at all.

Once detected the Americans could expect a heavy three-pronged attack and much would depend on whether the Russians decided to use tactical nuclear weapons. The air attack by the Badgers with their fifty-mile stand-off missiles would be a serious menace although the fighters and the Talos missile ships should destroy a number of them. The missiles themselves could also be engaged by Terrier but it seems unlikely that, with the weight of attack available, these defence measures could prevent all the missiles getting through. The submarine attack would, in some ways, be more dangerous as there would be little warning before the explosion of the missiles. Intensive anti-submarine measures in the area could cut down the threat especially from the conventionally propelled submarines but it would probably be ineffective against the nuclears. Certainly the Americans could do much to prevent the submarines using aircraft for reconnaissance or to direct their missiles. In general, it is probable that some of the submarine-launched missiles would get through as well. The attack by the Russian missile-armed surface vessels would be the easiest to compete with. The American carrier air groups should have little difficulty in finding and striking at the ships. If the Russians had already used tactical nuclear weapons, it would be easy to dispatch them in the same way, but if they had not then there would be a contest between the Russian surface-to-air guided missiles and the American aircraft using the short range Bullpup stand-off missile. The American air striking groups would still probably be numerous enough to win but casualties would be heavy. Again the Russian co-operating aircraft could be shot down and the ships attacked by American nuclear submarines. As the

American carriers should be able to keep outside the range of the Russian surface-to-surface missiles, it is possible that defence against this threat could be made one hundred per cent.

The outcome of such a battle using conventional warheads could favour either side. The splendid American equipment and training would stand them in good stead but the sinking of the Israeli destroyer *Eilat* in 1967 shows that the Russian missiles must not be underestimated. Even if the American carrier force could survive, however, it is doubtful that it could maintain its position for long periods to support NATO without unacceptable losses. On the other hand if *nuclear* warheads were used the Russians would almost certainly win. Their total 'broadside' would be over three hundred tactical nuclear missiles to which would probably be added some conventional torpedoes with nuclear warheads fired by submarines. Even the smallest known nuclear warhead of about half a kiloton can completely destroy a super carrier with a near miss, whilst a 50 megaton explosion a dozen miles away could render them unfit to operate or control aircraft. Moreover only four bombs out of the three hundred available need to be well placed to cause complete disaster. The use of tactical nuclear weapons in retaliation would not redress the balance. The most effective use would be against the enemy air bases but this is ruled out by the danger of escalation. True the use of nuclear depth charges and nuclear warheads for Talos would improve anti-submarine and air defence but not to an extent that would make them a hundred per cent effective and nothing less than this could make such operations viable. The effectiveness of the Russian Navy would therefore be enormously increased by the use of tactical nuclear weapons and the temptation for the Russians to use them is very great. If they did so, it is doubtful if even the American super carriers could survive. It was no doubt this kind of battle to which Khruschev was referring when he described the American super carriers as incinerators.

It seems that the point has now been reached when even the American super carriers are unlikely to be able to compete in a war in which tactical nuclear weapons are used at sea. They still provide self-contained airfields with world-wide strategic mobility and fulfil the basic requirement for the future that their aircraft can fight with or without nuclear weapons. It is doubtful, however, by their very nature, whether they can ever be said to be in an 'anti-nuclear posture'. They will, no doubt, as in the past in Korea, Vietnam, Cuba, Lebanon and many other places, show their value in limited wars of intervention; it may well be some time before they meet in limited war the aircraft and submarines with nuclear missiles that are a match for them.

There is, however, the danger that the Russian forces described above will sortie to attack trade and, if there are no forces to prevent them, there is no reason why they should not gain complete command of the sea. Fortunately shore-based aircraft can provide a great deal of the answer. Shore air bases, unlike aircraft carriers, are secured from nuclear attack by the cover provided by the Polaris deterrent submarines and by the fear of escalation. Modern jet aircraft, refuelled in the air if necessary by tankers, can cover the whole Norwegian Sea. Russian aircraft of the Bear or Badger types, or indeed their successor the Blinder, could be opposed by a shore-based fighter system from NATO bases in Norway, Scotland, Iceland and Greenland. Using supersonic fighters with air-to-air missiles, directed if necessary by A.E.W. aircraft, they could inflict an unacceptable casualty rate on Russian maritime aircraft. Against the Russian missile-firing ships, shore-based radar-fitted reconnaissance aircraft of the Nimrod type followed up by jet bombers with stand-off weapons as well as by the attack of our own nuclear submarines should be more than enough. To compete with enemy submarines is a more difficult problem. Fixed wing aircraft have been of limited value since the advent of the true submarine which leaves nothing above the surface to be detected. Helicopters, however, with their 'dunking sonar' can detect them; they have the tactical mobility to follow them and can be carried to the scene in ships. Our own nuclear submarines are probably the most effective weapons against other nuclear submarines and it should be possible for a team of all these anti-submarine agents to work behind the shore-based fighter system with a chance of success. There seems therefore to be a way to deal with the Russian northern fleet and it is predominantly by aircraft based ashore. They would fulfil the requirement to be able to fight with or without nuclear weapons and would be in an 'anti-nuclear posture'. History warns us, however, that the aircraft must be permanently assigned to such a force, as otherwise they would not be properly trained and probably would not be there when wanted.

The demise of the British aircraft-carriers, therefore, may

well be a step in the right direction in spite of the fact that it came about for different reasons. It has already had the beneficial effect of forcing the Royal Navy and Royal Air Force to concentrate on meeting the main maritime threat and already there are grounds for optimism. There will undoubtedly continue to be the possibility of small limited wars of intervention and police actions of the type with which we are familiar. The new Navy will not be quite so well equipped to deal with them. There is little doubt, however, that the former navy, designed only to fight small limited non-nuclear wars could have had little chance to meet the major threat and to command the sea.

The dominating factors in war at sea in the future are likely to be missiles and nuclear weapons. The missiles will be of many kinds: surface-to-air; surface-to-surface; air-to-surface and air-to-air; and the definition includes such weapons as guided or homing torpedoes. Nuclear weapons can vary from the small tactical half-a-kiloton to the monster fifty megaton. Many nuclear weapons can be made relatively clean and so sea warfare, where civilians do not suffer so much and where misses only hurt the fish, is a likely place for their power to be exploited. The new British Navy with its shore based R.A.F. support will be better able to compete in such conditions than its predecessor based on the aircraft carrier.

It is possible to imagine a time when the surveillance of the surface of the sea will be done by satellite and the regions below the surface by some as yet unknown method. It is also possible to imagine a system of missiles mounted ashore which could be directed at any object detected by the surveillance systems. When this time comes it may be possible to do away with aircraft in maritime war. Such speculation is, however, for the present rather more in the realms of science fiction than serious strategic study. It seems probable that manned aircraft will continue for some time to be an important factor in war over the sea. They will be less likely to be shot down by surface-to-air guided weapons than aircraft working over the land: they should normally be able to keep their distance from ships which mount these weapons, with the possible exception of submarines which may be developed to carry them. Even if satellites can provide a constant picture of what is going on at sea, manned aircraft will almost certainly be necessary to probe and clarify the situation. They will also be required to carry and monitor anti-submarine detecting devices as far as can be seen in the future. Provided they are progressively armed with up-to-date missiles of sufficient range they will continue to be an important factor in war over the sea especially for air defence and attack on ships. The aircraft-carrier will only continue to be the base for the shorter range types for a few years longer, and when, due to the spread of missiles and nuclear weapons, it becomes too vulnerable, aircraft will do better to rely on flight re-fuelling from shore bases where they will be secure against nuclear attack.

Bibliography

Bacon, Admiral Sir Reginald, *From 1900 Onwards*. London: Hutchinson 1940.

Bell Davies, Vice-Admiral Richard, *Sailor in the Air*. London: Peter Davies 1967.

Bragadin, Marc Antonio, *The Italian Navy in World War II*. Trans by Gale Hoffman. Annapolis: United States Naval Institute 1957.

Brassey's Naval Annual. 1911–14. London: William Clowes.

Chatfield, Admiral of the Fleet Lord, *It Might Happen Again*. London: Heinemann 1947.

Churchill, Winston S., *The Great War* vols 1–6. London: Cassell 1948–54.

Ciano's Diary. Ed. by Malcolm Muggeridge. London: Heinemann 1947.

Corbett, Sir Julian S., *Naval Operations*, vols 1–3. London: Longmans Green 1920–3.

Derry, T. K., *The Campaign in Norway*. London: H.M. Stationery Office 1952.

Dönitz, Admiral, *Memoirs*. Trans. by R. H. Stevens. London: Weidenfeld and Nicolson 1959.

Douhet, General Guilio, *The Command of the Air*. Trans. by Dino Ferrari. London: Faber and Faber 1943.

Encyclopaedia Britannica 23 vols. London: William Benton 1963.

Fisher, Admiral of the Fleet Lord, *Memories*. London: Hodder and Stoughton 1919.

Gamble, C. F. Snowden, *The Story of a North Sea Air Station*. London: Oxford University Press, Humphrey Milford 1928.

Grahame-White, Claude and Harper, Harry, *The Aeroplane*. London: T. C. and E. C. Jack 1914.

Higham, Robin, *The British Rigid Airship 1908–1931*. London: G. T. Foulis 1961.

Jane's Fighting Ships 1914–1968. London: Sampson Low, Marston.

Jellicoe, Admiral Viscount, *The Grand Fleet, 1914-1916*. London: Cassell 1919.

Jellicoe, Admiral of the Fleet Viscount, *The Crisis of the Naval War*. London: Cassell 1920.

Jones, H. A., *The War in the Air* vols. 2–6. Oxford: The Clarendon Press 1928–37.

Joubert de la Ferté, Air Chief Marshal Sir Philip, *Birds and Fishes*. London: Hutchinson 1960.

Kemp, Lieutenant Commander P. K., *Fleet Air Arm*. London: Herbert Jenkins 1954.

Kerr, Admiral Mark, *Land, Sea and Air*. London: Longmans, Green 1927.

Killen, John, *The Luftwaffe*. London: Frederick Muller 1967.

Lanstrom, Bjorn, *The Ship*. London: Allen and Unwin 1961.

Marder, Arthur J., *Fear God and Dread Nought* vols. 2 and 3. London: Jonathan Cape 1956 and 1959.

Martin, L. W., *The Sea in Modern Strategy*. London: Chatto and Windus 1967.

Mitchell, William, *Winged Defence*. G. P. Putnam's Sons, New York and London 1925.

Morison, Samuel E., *History of the United States Naval Operations in World War II* 15 vols. London: Oxford University Press 1948–62.

Navy Records Society, *The Papers of Admiral Sir John Fisher* vol. 2. Ed. by Lieut. Commander P. K. Kemp. London 1964.

Navy Records Society, *The Jellicoe Papers* vol. 1, 1893–1916, Ed. by A. Temple Patterson. London: 1966.

Neon – The Great Delusion. London: Ernest Benn 1927.

Neumann, Major Georg Paul, *The German Air Force in the Great War*. Trans. by J. E. Gordon. London: Hodder and Stoughton 1920.

Newbolt, Henry, *Naval Operations* vols. 4 and 5. London: Longmans Green 1928–31.

Parsons, Nels A. Jr., *Missiles and the Revolution in Warfare*. Cambridge, Mass.: Harvard University Press 1962.

Playfair, Major General I.S.O., *The Mediterranean and Middle East* vols. 1–3. London: H.M. Stationery Office 1954–60.

Raeder, Grand Admiral, *Struggle for the Sea*. Trans. by Edward Fitzgerald. London: William Kimber 1959.

Raleigh, Walter, *The War in the Air* vol. 1. Oxford: The Clarendon Press 1922.

Robinson, Douglas H., *The Zeppelin in Combat*. London: G. T. Foulis and Co. 1962.

Roskill, Captain S. W., *The War at Sea 1939–45* 3 vols. London: H.M. Stationary Office 1954–61.

Roskill, Stephen, *Naval Policy Between the Wars* vol. 1 1919–1929. London: Collins 1968.

Scheer, Admiral, *Germany's High Sea Fleet in the World War*. London: Cassell and Co. Ltd. 1920.

Sinclair, Captain J. A., *Airships in Peace and War*. London: Rich and Cowan 1934.

Sueter, Rear Admiral Murray F., *Airmen or Noahs*. London: Sir Isaac Pitman and Sons 1928.

Swanborough, Gordon and Bowers, Peter M., *United States Navy Aircraft since 1911*. London: Putnam 1968.

Thetford, Owen, *British Naval Aircraft since 1912*. London: Putnam 1962.

Tirpitz, Grand Admiral von., *My Memoirs* vols. 1 and 2. London: Hurst and Blackett.

Turnbull, Archibald D., and Lord, Clifford L., *History of United States Naval Aviation*. New Haven: Yale University Press 1949.

Webster, Sir Charles and Frankland, Noble, *The Strategic Air Offensive against Germany 1939–45* vols. 1–4. London; H.M. Stationery Office 1961.

Woodburn Kirby, Major General S., *The War against Japan* vols. 1–3. London: H.M. Stationary Office 1957–61.

Notes

1. Captain Sueter was an officer of an inventive and progressive turn of mind, who had had much to do with submarines, mines and torpedoes. He remained, with one short break, the Director of the Air Department of the Admiralty until 1917. He was the leading personality in naval aviation, but was a difficult subordinate and fell foul of his superiors. Nevertheless he was promoted to Rear Admiral but was then retired. He did not, like so many of his colleagues of the R.N.A.S., transfer to the Royal Air Force and later became a Member of Parliament and was knighted.

2. The introduction of these small airships is described in Chapter IV page 86.

3. Ripping a non-rigid airship was a method of releasing all the gas quickly to save the blimp from being carried off by the wind.

4. Up until now the airships had simply been numbered 'No. 23', 'No. 24' etc. By 1917, however, the various classes of blimps, such as 'S.S.' and 'C', had had prefixes for some time. Rigids, therefore, were now given the prefix 'R' and this was made retrospective – 'No. 9' becoming 'R.9' etc.

5. The *Cavendish* was renamed *Vindictive* before completion.

6. There are a number of explanations for the origin of the word 'blimp'. The most probable seems to be that a staff officer wrote a paper in the early days of aviation saying 'There are two kinds of airship: – (A) Rigid. (B) Limp' and (B) limp it has been ever since. What is certain is that elderly Colonels took their name from the airship and not vice-versa.

7. A recent study in America denies that the R.N.A.S. sank any U-boats in World War I. These sinkings are taken from the British Official History *The War in the Air*. If the American study is accepted it makes no difference to the general conclusions which I have arrived at.

8. The Osprey was also introduced to navigate for the single-seater Nimrods, and there was one in each flight of fighters. The scheme failed because the Ospreys were far too slow to keep up with the Nimrods.

9. Japanese aircraft are more commonly known by their American intelligence nicknames. For instance the Japanese Navy Bomber type 96 was known as 'Nell' and all other types had girl's or boy's names too and will normally be referred to by them in this book.

10. H.M.S. *Albatross* was built in Sydney in 1926–8 for the Royal Australian Navy. She was a seaplane carrier of 4800 tons with catapults and cranes to carry six Seagull flying-boats. She was transferred to the Royal Navy in 1938 and in World War II carried six Supermarine Walrus amphibians.

11. It has not been easy to find the reason for what seems to us now to have been an extraordinary business. There were three squadrons of Blackburn Skua fighter dive-bombers allocated to the Home Fleet, two for the *Ark Royal* and one for the *Furious*. When the carriers were in harbour, the Skuas were landed at the nearest naval air station and it seems that the naval bases became dependent upon them for air defence and this became an important part of their duties. When the *Ark Royal* left for the Mediterranean in March 1940, she left her two Skua squadrons at Hatson in the Orkneys. When the Norwegian campaign began, the *Furious* was in the Clyde with her two Swordfish squadrons on board but with her Skua squadron landed at Evanton near Invergordon. The Commander-in-Chief, when he sailed on 7th April with the Home Fleet to try and intercept the German invading forces, gave the *Furious* no orders. Her Captain raised steam on his own initiative and he was eventually ordered to sea by the Admiralty. He sent urgent orders for the Skua squadron to embark but for some reason they did not do so. The suggestion that they were too far away is obviously untrue as the *Furious* must have passed within a hundred miles of Evanton on her way up the west coast. It seems, therefore, that they were retained ashore for air defence.

12. The Fairey Albacore was an improved version of the Swordfish with a more powerful engine and an enclosed cockpit. It was a biplane and was, however, of much the same performance.

13. A Napoleonic maxim.

AA

14. Not to be confused with the earlier *Ryujo*, sunk at the Battle of the Eastern Solomons.

15. Not to be confused with the later supersonic F4 Phantom II.

16. Vertical take off and landing. The R.A.F. Hawker Harrier is an example of this type of aircraft.